Prentice Hall Advanced Reference Series

Physical and Life Sciences

ADSORPTION EQUILIBRIUM DATA HANDBOOK

Diego P. Valenzuela

Alan L. Myers

Department of Chemical Engineering
University of Pennsylvania
Philadelphia

PRENTICE HALL, Englewood Cliffs, New Jersey 07632

LIBRARY OF CONGRESS CATALOGING-IN-PUBLICATION DATA

VALENZUELA, DIEGO P.
    Adsorption equilibrium data handbook.

    (Prentice Hall advanced reference series)
    Bibliography: p.
    Includes index.
    1. Adsorption.   2. Surface chemistry.   I. Myers,
Alan L.   II. Title.   III. Series.
TP156.A35V35 1989   660.2'9453   88-25381
ISBN   0-13-008815-3

Editorial/production supervision: BARBARA MARTTINE

Cover design: LUNDGREN GRAPHICS, LTD.

Manufacturing buyer: MARY ANN GLORIANDE

Prentice Hall Advanced Reference Series

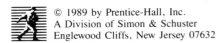
© 1989 by Prentice-Hall, Inc.
A Division of Simon & Schuster
Englewood Cliffs, New Jersey 07632

The publisher offers discounts on this book when ordered
in bulk quantities. For more information, write:

> Special Sales/College Marketing
> College Technical and Reference Division
> Prentice Hall
> Englewood Cliffs, New Jersey 07632

Printed in the United States of America

10   9   8   7   6   5   4   3   2   1

ISBN   0-13-008815-3

Prentice-Hall International (UK) Limited, *London*
Prentice-Hall of Australia Pty. Limited, *Sydney*
Prentice-Hall Canada Inc., *Toronto*
Prentice-Hall Hispanoamericana, S.A., *Mexico*
Prentice-Hall of India Private Limited, *New Delhi*
Prentice-Hall of Japan, Inc., *Tokyo*
Simon & Schuster Asia Pte. Ltd., *Singapore*
Editora Prentice-Hall do Brasil, Ltda., *Rio de Janeiro*

To

Pat
and
Irmgard

CONTENTS

Appendices

PREFACE

This book is a collection of data on adsorption equilibrium. The intent is to provide a reference for engineers and scientists interested in the application of adsorption to the separation and purification of gas and liquid mixtures.

The justification for publishing this book is that single-gas adsorption isotherms cannot be calculated from first principles and therefore must be measured experimentally. The accumulation of a data bank is a first step toward the development of theories for calculating adsorption isotherms from gas-gas and gas-solid potential energies. In the meantime, single-gas adsorption isotherms provide the basis for correlating and predicting adsorption from gas and liquid mixtures.

Chemisorption is excluded because it is irreversible and therefore unsuitable for adsorptive separation processes.

Characterization of the adsorbent is a problem. One of the most useful adsorbents is activated carbon, but there are many different brands to choose from and their properties depend upon the starting material and the manufacturing process. In addition, the equilibrium adsorption of a particular gas depends to a degree upon the degassing procedure employed to clean the adsorbent before it is used. Fortunately, the ability of an adsorbent to separate a particular mixture is not as sensitive to its history.

Fundamental constants including virial coefficients and heats of adsorption, which are related by the principles of statistical mechanics to the structure and composition of the solid and the molecular properties of the adsorbate, are tabulated in Chapter 2. The adsorption second virial coefficient (B_{1S}) characterizes the interaction of the surface with a single adsorbate molecule. The temperature derivative of B_{1S} is related to the heat of adsorption at the limit of zero surface coverage. For vapors, free energies and enthalpies of immersion of the adsorbent in the saturated liquid provide information about the strength of adsorbate-adsorbent interactions over a range of temperature. Surface area cannot be measured precisely for high-capacity adsorbents like zeolites and activated carbon and therefore is not a fundamental constant.

An important application of an adsorption data bank is estimating adsorption from mixtures using single-gas adsorption isotherms. The necessity of predicting adsorption of mixtures becomes apparent when one considers the alternative of experimental measurements. For example, a set of 10 gases and 3 adsorbents can be characterized by 30 adsorption isotherms of single gases: a difficult, but not impossible experiment. For the same set, there are 135 isotherms for adsorption of binary mixtures, and 360 isotherms for adsorption of ternary mixtures. Each of these mixture isotherms is a function of pressure and gas-phase composition. Thus experiments on mixed-gas adsorption are usually reserved for systems of special importance.

Chapter 1 is a summary of thermodynamic equations for adsorption of gases and liquids and their mixtures. These equations are applied to experimental data in Chapters 2 through 4.

Data for adsorption of single gases is in Chapter 2. Adsorption isotherms were fitted by analytical equations whose constants were derived by minimizing the root-mean-square deviations. Tabular and graphical comparisons of the data with the equations are provided for each adsorption isotherm.

Mixed-gas adsorption data are tabulated in Chapter 3. Only those mixtures accompanied by measurements of the single-gas isotherms are included, so that the mixture data can be checked for thermodynamic consistency. Numerical methods are provided for predicting mixed-gas adsorption from single-gas data.

Data for adsorption from liquid mixtures is given in Chapter 4 in terms of the surface excess. Experimental points are not fitted with equations because they are either too complicated, contain too many unknown parameters, or do no apply to both U- and S-shaped isotherms.

References are abbreviated by the first two letters of the last name of the first author, the page number of the article, and its year of publication. For example, [Ta1263(86)] refers to the article by O. Talu and I. Zwiebel, *AIChE J.* **32**, 1263 (1986). For books, the page number refers to the particular page of interest. Full references are listed alphabetically in Appendix G.

Isotherms contained in this book are indexed in Appendices A, B, and C. Appendices D, E, and F contain a more complete compilation of references to experimental data, some of which are not tabulated in this book.

Graphical data are excluded because of the uncertainty associated with conversion to tabular form. Representative sets of data for a wide variety of systems of practical importance are included so that this compilation may serve as a guide toward further experimentation and development of theory. There are wide gaps in published experimental data on systems of commercial importance such as adsorption of ethanol, water, sulfur compounds, oxygen, nitrogen, and xylene isomers.

Since some journals discourage the tabulation of experimental data in articles, authors are invited to send their data to us for inclusion in the next edition of this book.

Diego P. Valenzuela
Alan L. Myers

Department of Chemical Engineering
University of Pennsylvania
Philadelphia, PA, 19104

LIST OF SYMBOLS

Computer printout symbols are listed in Tables 2-1 and 3-1.

$\tilde{a}$ molar area of adsorbate, m^2/mol

a activity in liquid phase

A specific surface area, m^2/kg

b constant, Eq. (2-1), Eq. (3-17)

B_{1S} adsorption 2nd virial coefficient of adsorbate no. 1, m^3/kg

B_{ij} 2nd virial coefficient in gas phase for molecules i and j, m^3/mol

c constant, Eq. (2-2)

C constant, Eq. (3-19)

$\tilde{E}_G$ molar internal energy in gas phase, J/mol

$\tilde{E}_S$ molar internal energy in adsorbed phase, J/mol

$\bar{E}_S$ differential internal energy in adsorbed phase, J/mol

ΔE integral heat of adsorption, J/mol

f fugacity, Pa; function, Eq. (2-13)

$\mathcal{F}$ function of ψ

g^e excess Gibbs free energy in adsorbed phase, J/mol

G function of P_i^o, Fig. 3-2, mol/kg

ΔG_{im} $[= \phi]$, specific free energy of immersion of adsorbent, J/kg

$\mathcal{G}$ excess Gibbs free energy function, Eq. (3-19), J/kg

H Henry constant, $mol/(kg\text{-}Pa)$

$\tilde{H}_G$ molar enthalpy in gas phase, J/mol

$\tilde{H}_S$ molar enthalpy in adsorbed phase, J/mol

$\bar{H}_S$ differential enthalpy in adsorbed phase, J/mol

ΔH equilibrium heat of adsorption, J/mol

ΔH_{im} specific enthalpy of immersion of adsorbent, J/kg

K constant, Pa^{-1}

m saturation capacity, mol/kg

m cavity density in zeolite, Eq. (3-17), mol/kg

n specific amount adsorbed, mol/kg
n^s specific amount adsorbed at vapor pressure of adsorbate, mol/kg
n_i specific amount of i adsorbed, mol/kg
n_i° specific amount of i adsorbed at P_i°, mol/kg
n_t total specific amount adsorbed, mol/kg
n_i^e surface excess of i in adsorbed phase, mol/kg
n° amount of liquid solution before contact with adsorbent, mol/kg
N number of components
P pressure, Pa
P_i° vapor pressure of adsorbate in standard state, Pa
P^s saturation vapor pressure of liquid, Pa
q_d differential heat of adsorption, J/mol
q_{st} isosteric heat of adsorption, J/mol
q_{st}° isosteric heat at limit of zero coverage, J/mol
R gas constant, 8.3145 J/(mol-K)
s constant, Eq. (2-2); modified selectivity, Eq. (4-7)
s_0 constant, Eq. (4-8)
$s_{1,2}$ selectivity of adsorbent for comp. 1 relative to comp. 2, Eq. (3-1)
$s_{1,2}^\circ$ selectivity at limit of zero pressure, Eq. (1-26)
S_i function of ψ, Fig. 3-1
ΔS_{im} specific entropy of immersion of adsorbent, J/(kg-K)
t constant, Eq. (2-1)
T temperature, K
T_c critical temperature, K
v molar volume in saturated liquid state, m³/mol
V micropore volume of adsorbent, m³/kg
x_i mole fraction of i in liquid phase, Eqs. (1-30) to (1-34) and Chap. 4
x_i° mole fraction of i in liquid phase before contact with adsorbent, Eq. (1-30)
x_i mole fraction of i in adsorbed phase, Chap. 3 only
x_i' mole fraction of i in adsorbed phase
y_i mole fraction of i in gas phase
z variable, Eqs. (2-10), (3-14)

Greek Letters

β size ratio, m_1/m_2
γ_i activity coefficient of i in liquid phase, Chap. 1 and Chap. 4
γ_i activity coefficient of i in adsorbed phase, Chap. 3 only
γ_i' activity coefficient of i in adsorbed phase
δ Error in ψ or P_i°, Figs. 3-1, 3-2
δ_{12} $2B_{12} - B_{11} - B_{22}$, m³/mol
η $[= KP]$, dimensionless pressure, Eq. (3-15)
θ $[= n/m]$, fractional coverage
λ equilibrium enthalpy of vaporization of liquid, J/mol
Λ_{ij} constant, Eq. (3-19)
μ chemical potential, J/mol

Π	spreading pressure in adsorbed phase, N/m
σ	constant, Eq. (3-15)
σ^e	excess reciprocal amount adsorbed, Eq. (1-37), kg/mol
ϕ_i	fugacity coefficient of i in gas phase
ϕ_i°	fugacity coefficient at vapor pressure of adsorbate i, Eq. (1-35), Chap. 3
ϕ_i°	$[= \Delta G_{im}]$, specific free energy of immersion in liquid i, J/kg, Chap. 4
ψ	$[= (\Pi A / RT)]$, mol/kg
ψ_i°	$(\Pi A / RT)$ of i in standard state, mol/kg

Subscripts

i	refers to ith component

Superscripts

e	refers to excess quantity such as g^e, $(1/n)^e$ or n_i^e
s	refers to state of saturated liquid

CHAPTER 1

EQUATIONS FROM THERMODYNAMICS

In this chapter are summarized the key equations of thermodynamics for calculating quantities such as the heat of adsorption from adsorption isotherms. Equations marked with an asterisk (*) are for the case when the gas phase behaves as a perfect gas; this approximation is usually valid if the pressure does not exceed about 500 kPa.

A comprehensive reference on adsorption thermodynamics is the monograph by Young and Crowell[Yo64(62)]. Chapters 2 through 4 contain additional references and applications of the following equations to theory and experiment.

ADSORPTION OF PURE GASES

Henry Constant

$$H = \lim_{P \to 0} \left(\frac{n}{P} \right) \tag{1-1}$$

$$= \lim_{P \to 0} \left(\frac{dn}{dP} \right) \tag{1-2}$$

Adsorption Second Virial Coefficient

$$B_{1S} = RTH \tag{1-3}$$

Isosteric Heat of Adsorption

$$q_{st} = \tilde{H}_G - \bar{H}_S \tag{1-4}$$

$$= RT^2 \left[\frac{\partial \ln P}{\partial T} \right]_n = -R \left[\frac{\partial \ln P}{\partial (1/T)} \right]_n \qquad *(1-5)$$

1

Isosteric Heat at Limit of Zero Coverage

$$q_{st}^{\circ} = -RT^2 \left[\frac{d \ln H}{dT} \right] \tag{1-6}$$

$$= R \left[\frac{d \ln H}{d(1/T)} \right] \tag{1-7}$$

Differential Heat of Adsorption

$$q_d = \tilde{E}_G - \bar{E}_S \tag{1-8}$$

$$= q_{st} - RT \tag{*(1-9)}$$

Equilibrium Heat of Adsorption

$$\Delta H = \tilde{H}_G - \tilde{H}_S \tag{1-10}$$

$$= \frac{1}{n} \int_0^n q_{st} dn - \left(\frac{\Pi A}{n} \right) \tag{1-11}$$

$$= RT^2 \left[\frac{\partial \ln P}{\partial T} \right]_{\Pi} \tag{*(1-12)}$$

Integral Heat of Adsorption

$$\Delta E = \tilde{E}_G - \tilde{E}_S \tag{1-13}$$

$$= \Delta H - RT + \left(\frac{\Pi A}{n} \right) \tag{*(1-14)}$$

$$= \frac{1}{n} \int_0^n q_d \, dn \tag{1-15}$$

Spreading Pressure

$$\frac{\Pi A}{RT} = \int_0^f \frac{n}{f} \, df \tag{1-16}$$

$$= \int_0^P \frac{n}{P} \, dP \tag{*(1-17)}$$

$$= \int_0^n \frac{d \ln P}{d \ln n} \, dn \tag{*(1-18)}$$

ADSORPTION OF PURE VAPORS

Gibbs Free Energy of Immersion

$$\Delta G_{im} = -(\Pi A)^s \tag{1-19}$$

$$= -RT \int_0^{P^s} \frac{n}{P} dP \tag{*1-20}$$

$$= RT \int_0^{n^s} \ln(P/P^s) dn \tag{*1-21}$$

Enthalpy of Immersion

$$\Delta H_{im} = -\int_o^{n^s} (q_{st} - \lambda) \, dn \tag{1-22}$$

$$= R \int_0^{n^s} \left[\frac{\partial \ln(P/P^s)}{\partial (1/T)} \right]_n dn \tag{*1-23}$$

Entropy of Immersion

$$\Delta S_{im} = \frac{\Delta H_{im} - \Delta G_{im}}{T} \tag{1-24}$$

$$= -\int_0^{n^s} \left[\frac{\partial \left[RT \ln(P/P^s) \right]}{\partial T} \right]_n dn \tag{*1-25}$$

ADSORPTION OF BINARY GAS MIXTURES

Selectivity at Limit of Zero Coverage

$$s_{1,2}^{\circ} = \lim_{P \to 0} \left(\frac{n_1/y_1}{n_2/y_2} \right) \tag{1-26}$$

$$= \frac{B_{1S}}{B_{2S}} \tag{1-27}$$

Gibbs Adsorption Isotherm

$$0 = -A\,d\Pi + n_1 d\mu_1 + n_2 d\mu_2 \tag{1-28}$$

Integral Consistency Test

$$\frac{A(\Pi_1 - \Pi_2)}{RT} = \int_0^1 \left[\frac{n_1}{y_1} - \frac{n_2}{y_2}\right] dy_1 \qquad \{\text{constant } T,P\} \qquad *\text{(1-29)}$$

ADSORPTION OF BINARY LIQUID MIXTURES

Surface Excess

$$n_1^e = n^{\circ}(x_1^{\circ} - x_1) \tag{1-30}$$
$$= -n_2^e \tag{1-31}$$

Relation to Adsorption from Vapors

$$n_1^e = \lim_{P \to P^s} n(x_1' - x_1) \qquad \{\text{constant } T, y_1\} \tag{1-32}$$

Integral Consistency Test

$$\Delta G_{im_1} - \Delta G_{im_2} = -RT \int_{x_1=0}^1 \left(\frac{n_1^e}{\gamma_1 x_1 x_2}\right) d(\gamma_1 x_1) \tag{1-33}$$

Temperature Coefficient of Adsorption

$$\left[\frac{\partial n_1^e}{\partial T}\right]_{x_1} = \frac{x_1 x_2}{RT^2}\left[\frac{\partial \Delta H_{im}}{\partial x_1}\right]_T \qquad \{\text{ideal bulk liquid}\} \tag{1-34}$$

SOLUTION THERMODYNAMICS OF ADSORBED PHASE

Equations of Equilibrium

$$P y_i \phi_i = P_i^{\circ} \phi_i^{\circ} \gamma_i' x_i' \tag{1-35}$$

Excess Gibbs Free Energy

$$\frac{g^e}{RT} = \sum_{i=1}^{N} x'_i \ln \gamma'_i \qquad \{\text{constant } T, \Pi\} \qquad (1\text{-}36)$$

Excess Reciprocal Amount Adsorbed

$$\sigma^e = \left(\frac{1}{n_t}\right) - \sum_{i=1}^{N} \left(\frac{x'_i}{n_i^\circ}\right) \qquad (1\text{-}37)$$

$$= \left[\frac{\partial(g^e/RT)}{\partial(\Pi A/RT)}\right]_{T,x'_i} \qquad (1\text{-}38)$$

Activity Coefficients

$$RT \ln \gamma'_i = \left[\frac{\partial(n_t g^e)}{\partial n_i}\right]_{T,\Pi,n_j} \qquad (1\text{-}39)$$

$$\sum_{i=1}^{N} x'_i d\ln\gamma'_i = \sigma^e d\left(\frac{\Pi A}{RT}\right) \qquad \{\text{constant } T, P\} \qquad (1\text{-}40)$$

CHAPTER 2

ADSORPTION OF PURE GASES

This chapter contains adsorption isotherms of pure gases. A compilation of references to experimental data is given in Appendix D. Only those isotherms accompanied by tabulated data are included in this chapter. For each system, the experimental points are tabulated and compared with values calculated by the Toth and UNILAN equations. The constants in these equations were derived with a "least-squares" optimization routine. A graph comparing the experimental points with the equation that fits them best is provided for each adsorption isotherm.

The adsorption second virial coefficient, the free energy of immersion, and the isosteric heat of adsorption are calculated for each system. Graphs are included for the adsorption second virial coefficient as a function of temperature, and for the isosteric heat of adsorption versus amount adsorbed.

GUIDE TO TABLES

Ordering of Systems

Systems are grouped by adsorbates in alphabetical order. Gas isotherms tabulated in this chapter are indexed in Appendix A.

Heading

A sample heading is:

```
 ---------------------------------------------------------------------------
 CARBON DIOXIDE
 ---------------------------------------------------------------------------
 ACTIVATED CARBON
 ---------------------------------------------------------------------------
 ADSORBENT : BPL;PITTSBURGH CHEMICAL Co.;988 Sqm/g
 EXP.      : STATIC VOLUMETRIC
 REF.      : REICH,R.,ZIEGLER,W.T.and ROGERS,K.A.;Ind.Eng.Chem.Process Des.Dev.
             19,336(1980)
```

On the first line is the name of the gaseous adsorbate (carbon dioxide) and on the second line is the adsorbent (activated carbon).

On the third line is given information on the adsorbent: the commercial name, surface area, particle size and density (if available). In this case, the adsorbent is BPL activated carbon manufactured by Pittsburgh Chemical Co., surface area 988 m^2/g.

On the fourth line the experimental technique is described: gravimetric, volumetric, or chromatographic. In this case, a static volumetric method was used.

On the fifth line is the reference for the experimental data, which for this system is R. Reich, W. T. Ziegler and K. A. Rogers, *Ind. Eng. Chem. Process Des. Dev.* **19**, 336 (1980).

Adsorption Isotherm

A sample adsorption isotherm for the system described on page 6 is:

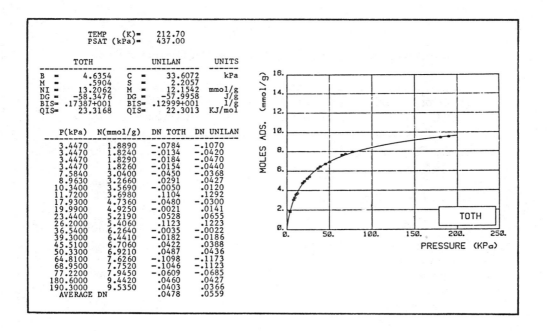

The heading is the temperature TEMP (212.7 K) of the adsorption isotherm and the vapor pressure PSAT (437.0 kPa) of saturated liquid adsorbate. PSAT is omitted if the adsorbate is above its critical temperature.

The first table for this isotherm contains constants derived for the Toth and UNILAN equations. The Toth equation [To311(71)] has three parameters $\{m, b, t\}$:

$$n = \frac{mP}{(b + P^t)^{1/t}} \tag{2-1}$$

and the UNILAN equation [Ho140(52)] also has three parameters $\{m, c, s\}$:

$$n = \frac{m}{2s} \ln \left[\frac{c + Pe^{+s}}{c + Pe^{-s}} \right] \tag{2-2}$$

where P is the pressure in kPa and n is the specific amount adsorbed in moles per kilogram of adsorbent.

Both equations reduce to the Langmuir isotherm: the Toth for $t = 1$ and the UNILAN for $s = 0$ (after application of L'Hôpital's rule for limits of the form $\frac{0}{0}$). Also for both equations, $\lim_{P \to \infty} n = m$.

Table 2-1 gives the correspondence between the symbols in the computer printout and the symbols used in the above equations.

Table 2-1. Explanation of Symbols Used in Computer Printout

Name of Variable or Constant	Symbol for Computer Printout	Symbol Used Elsewhere	Units
const., Toth eq.	B	b	$[kPa]^t$
"	M	t	–
"	NI	m	mol/kg
const., UNILAN eq.	C	c	kPa
"	S	s	–
"	M	m	mol/kg
free energy of immersion	DG	ΔG_{im}	kJ/kg
second virial coefficient	BIS	B_{1S}	m^3/kg
isosteric heat at zero coverage	QIS	q_{st}°	kJ/mol
temperature	TEMP	T	K
vapor pressure	PSAT	P^s	kPa
pressure	P	P	kPa
specific amount adsorbed	N	n	mol/kg
error	DN	Δn	mol/kg

The first three lines of the first table for the sample adsorption isotherm contain values derived for the three constants of the Toth equation {B,M,NI} and the UNILAN equation {C,S,M}. Notice that for the Toth equation, M in the computer printout is t in Eq. (2-1).

The fourth line in the first table (DG) is the free energy of immersion ΔG_{im} in J/g (= kJ/kg) calculated by numerical integration of Eq. (1-20) using both the Toth and the UNILAN equation. In most cases the two values are in excellent agreement. For supercritical adsorbates ΔG_{im} is omitted.

The fifth line in the first table for the sample adsorption isotherm is the adsorption second virial coefficient (BIS) defined by Eq. (1-3) and reported in units of liters per gram (= m^3/kg). The notation .17387+001 is shorthand for the exponential form 0.17387×10^1. The Toth equation usually gives a value for this constant that is larger than the one obtained by means of the UNILAN equation, as explained on page 12. The discrepancy is partly due to the difficulties associated with numerical differentiation of experimental data, but the main problem is the inability of any three-constant equation to provide a precise fit of the experimental data at both high and low surface coverage.

The sixth and last line in the first table (QIS) is the isosteric heat of adsorption at the limit of zero surface coverage q_{st}°, given in units of kJ per mole of adsorbate and calculated using Eq. (1-7). Agreement of the Toth and UNILAN equations for this quantity is a sensitive test of the accuracy of the second virial coefficients. If there is a large discrepancy between the two values, or if one of them is negative, the values given for the heat of adsorption and second virial coefficients should be disregarded. Value of B_{1S} and q_{st}° are reported even when they are obviously incorrect, so that the reliability of both the experimental data and the adsorption equation can be judged for each system.

The second table for the sample adsorption isotherm contains the experimental data. The first column is the pressure (P) measured in units of kPa. The second column is the specific amount adsorbed (N) in units of mmol per gram (= mol per kg). The third and fourth columns give the difference (DN) between the experimental and calculated values of the amount adsorbed:

$$DN = \Delta n = n^{calc} - n^{expt} \tag{2-3}$$

The last line of the second table is the average of the absolute value of Δn:

$$(\text{AVERAGE DN}) = \overline{\Delta n} = \frac{1}{k} \sum_{i=1}^{k} \left| n^{calc} - n^{expt} \right| \tag{2-4}$$

where k is the number of experimental points.

The graph drawn for each adsorption isotherm shows the experimental points in the same units as the second table. The solid line is either the Toth or the UNILAN equation, whichever has the smaller value of $\overline{\Delta n}$.

Supplementary Graphs

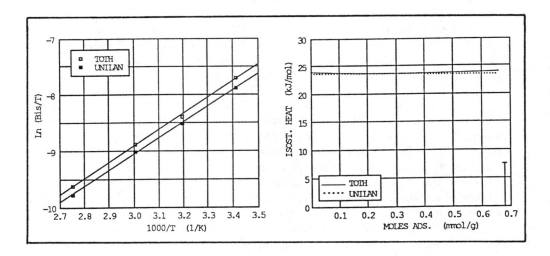

If the data set contains more than one adsorption isotherm, then two graphs are provided as shown in the above example taken from page 52 for CO_2 on activated carbon[Sz37(63)]. The first graph is the adsorption second virial coefficient as a function of temperature. Since the Henry's constant H is found empirically to vary linearly with reciprocal temperature, the coordinates of the graph are $[\ln(B_{1S}/T)]$ versus $[(10^3)/T]$. B_{1S} has units of m^3/kg and T is in K. Lines for the Toth equation and the UNILAN equation are obtained by a "least squares" regression of the points. The limiting value of the isosteric heat at zero coverage (q_{st}^o) is directly proportional to the slope of this line. If there are three or more isotherms (four in this case), then the linearity of the plotted points provides a test of the accuracy of the experimental data at low pressure. Note that the Toth equation always gives a slightly larger value of B_{1S} than the UNILAN equation, as explained on page 12.

The second graph is the isosteric heat of adsorption (q_{st}) versus the specific amount adsorbed, calculated by numerical differentiation of the adsorption equations using Eq. (1-5). The heat of adsorption is assumed to be independent of temperature in the range encompassed by the adsorption isotherms. The isosteric heat at the limit of zero coverage on this graph may be compared to q_{st}^o from Eq. (1-7), which is also tabulated under the heading QIS for each isotherm.

For vapors, the heat of vaporization (λ) of saturated liquid adsorbate at the temperature of each adsorption isotherm is given by a tick on the right-hand side of the graph for isosteric heat. The temperature corresponding to each value of λ can be deduced from the fact that λ decreases with increasing temperature. The heat of vaporization is omitted if $T > T_c$.

ADSORPTION EQUATIONS

Toth Equation

The Toth [To311(71)] equation (2-1) can be inverted to obtain P as a function of n:

$$P = \left[\frac{b}{\theta^{-t} - 1}\right]^{1/t} \qquad \{0 \le \theta < 1\} \qquad (2\text{-}5)$$

where $\theta = n/m$. The spreading pressure [My392(86)] for the Toth equation is:

$$\frac{\Pi A}{mRT} = \theta - \frac{\theta}{t}\ln(1 - \theta^t) - \sum_{j=1}^{\infty} \frac{\theta^{jt+1}}{jt(jt+1)} \qquad \{0 \le \theta < 1\} \qquad (2\text{-}6)$$

The series in Eq. (2-6) converges for $0 < t < 1$. A suitable convergence criterion is that successive estimates of the sum should differ by less than 10^{-7}.

The Henry's constant and adsorption second virial coefficient are given by:

$$H = mb^{-1/t} \qquad (2\text{-}7)$$

$$B_{1S} = RTmb^{-1/t} \qquad (2\text{-}8)$$

UNILAN Equation

The UNILAN [Ho140(52)] equation (2-2) is based on a uniform distribution of energies of adsorption (UNI for uniform distribution and LAN for the Langmuir local isotherm). Like the Toth equation, it can be inverted:

$$P = c \times \left[\frac{e^{2s\theta} - 1}{e^s - e^{s(2\theta-1)}}\right] \qquad \{0 \le \theta < 1\} \qquad (2\text{-}9)$$

where $\theta = n/m$. The spreading pressure [My365(84)] for the UNILAN equation is:

$$\frac{\Pi A}{mRT} = \frac{1}{2s}\int_{-s}^{+s} \ln\left[1 + (P/c)e^z\right]dz \qquad \{s > 0\}$$
$$= \ln\left[1 + (P/c)\right] \qquad\qquad \{s = 0\} \qquad (2\text{-}10)$$

The integral is evaluated numerically.

The Henry's constant and adsorption second virial coefficient are:

$$H = \frac{m}{c} \times \frac{\sinh(s)}{s} \qquad (2\text{-}11)$$

$$B_{1S} = \frac{RTm}{c} \times \frac{\sinh(s)}{s} \qquad (2\text{-}12)$$

Comments on Equations

Most of the adsorption isotherms contained in this data bank are for heterogeneous adsorbents, for which the shape of the isotherm is concave to the pressure axis. The isotherm on page 7 is typical: the second derivative d^2n/dP^2 is negative and its absolute value decreases monotonically with pressure.

The Toth and UNILAN equations were chosen for this data base because they are for heterogeneous adsorbents, contain only three constants and give analytical expressions for the amount adsorbed. Since both equations have temperature-dependent constants, at least two isotherms must be measured to establish the temperature coefficient of adsorption. We tried other equations but the Toth and UNILAN seem to be the best all-around equations for microporous adsorbents. Some systems exhibit the positive second derivative associated with multilayer formation at high coverage and therefore cannot be fit by either the Toth or the UNILAN equation.

A flaw of the Toth equation is that its derivatives d^kn/dP^k do not exist in the limit as $P \rightarrow 0$ for $k \geq 2$. This generates a value of B_{1S} that is too large. For the UNILAN equation, which was derived for a uniform energy distribution, these derivatives exist but the abrupt cutoff of adsorption sites on the high end of the energy distribution causes the value of B_{1S} to be underestimated. Therefore the correct value of B_{1S} usually lies between the values generated by the Toth and UNILAN equations.

PARAMETER REGRESSION

The values of constants for each adsorption equation are generated by a sequential search procedure called the Simplex algorithm [Ne308(65)]. Given k experimental points $\{P_j, n_j; \; j = 1, 2, \ldots, k\}$, the algorithm finds the set of R adjustable parameters $\mathbf{X} = \{x_1, x_2, \ldots, x_R\}$ that minimize the objective function $f(\mathbf{X})$:

$$f(\mathbf{X}) = \frac{1}{k} \sum_{j=1}^{k} [n_j^{\text{expt}}(P) - n_j^{\text{calc}}(P)]^2 \qquad (2\text{-}13)$$

In this case $R = 3$ and $\mathbf{X}$ is either the set of parameters for the UNILAN equation $\mathbf{X} = \{c, s, m\}$ or the set for the Toth equation $\mathbf{X} = \{b, t, m\}$. Every optimization was repeated several times with different initial estimates of the parameters to avoid local minima.

SAMPLE CALCULATIONS

Amount Adsorbed

For CO_2 adsorbed on activated carbon (table on page 7) at $T = 212.7$ K and at $P = 100.0$ kPa, the Toth equation (2-1) gives:

$$n = \frac{(13.2062)(100.0)}{[(4.6354) + (100.0)^{0.5904}]^{1/0.5904}} = 8.406 \, \text{mol/kg}$$

and the UNILAN equation (2-2) yields:

$$n = \frac{(12.1542)}{(2)(2.2057)} \ln \left[\frac{(33.6072) + (100.0)e^{+2.2057}}{(33.6072) + (100.0)e^{-2.2057}} \right] = 8.400 \, \text{mol/kg}$$

Temperature Coefficient of Amount Adsorbed

Interpolation is based upon the approximation that the isosteric heat of adsorption is independent of temperature, which leads to the equation:

$$\ln P = \ln P_2 + \frac{q_{st}}{R} \left[\frac{1}{T_2} - \frac{1}{T} \right] \qquad \{\text{constant } n\} \qquad (2\text{-}14)$$

where

$$\frac{q_{st}}{R} = \left[\frac{\ln(P_1/P_2)}{1/T_2 - 1/T_1} \right] \qquad (2\text{-}15)$$

For CO_2 adsorbed on activated carbon, let $n = 4.819$ mol/kg. Using the constants in the tables on pages 47 and 48, the Toth equation (2-5) gives $P_1 = 824.862$ kPa at $T_1 = 301.4$ K, and $P_2 = 19.0606$ kPa at $T_2 = 212.7$ K. From Eq. (2-15):

$$\frac{q_{st}}{R} = \left[\frac{\ln(824.862/19.0606)}{1/212.7 - 1/301.4} \right] = 2723 \, \text{K}$$

Therefore $q_{st} = (8.3145 \times 10^{-3})(2723) = 22.64$ kJ/mol. Using Eq. (2-14) to interpolate at $T = 260.2$ K:

$$\ln P = \ln(19.0606) + (2723) \times \left(\frac{1}{212.7} - \frac{1}{260.2} \right) = 5.2847$$

$$P = e^{5.2847} = 197.3 \, \text{kPa}$$

For comparison, the experimental value at $n = 4.819$ mol/kg and $T = 260.2$ K (page 47) is $P = 199.3$ kPa.

Spreading Pressure

For CO_2 adsorbed on activated carbon at 212.7 K (page 7), the amount adsorbed at $P = 100$ kPa from the Toth Eq. (2-1) is 8.4056 mol/kg. Therefore $\theta = n/m = 8.4056/13.2062 = 0.63649$ and, from Eq. (2-6):

$$\frac{\Pi A}{RT} = m\left(\theta - \frac{\theta}{t}\ln(1 - \theta^t) - \text{Sum}\right)$$

$$= 13.2062\left(0.63649 - \frac{0.63649}{0.5904}\ln[1 - (0.63649)^{0.5904}] - \text{Sum}\right)$$

$$= 13.2062(0.63649 + 1.56525 - 0.78464)$$

$$= 18.71 \text{ mol/kg}$$

This value is compared to Eq. (2-10) for the UNILAN equation: By numerical integration using the constants on page 7, the value of the integral from $s = -2.2057$ to $s = +2.2057$ is 6.7362 so:

$$\frac{\Pi A}{RT} = \frac{m}{2s} \times \text{Integral}$$

$$= \frac{12.1542}{(2)(2.2057)} \times 6.7362 = 18.56 \text{ mol/kg}$$

Entropy and Enthalpy of Immersion

The heat of immersion ΔH_{im} is defined as the isothermal change of enthalpy accompanying the immersion of unit mass of clean adsorbent in saturated liquid adsorbate. Usually this is an exothermic process and ΔH_{im} is a negative quantity. The heat of immersion is calculated from the variation of the free energy of immersion with temperature according to the Gibbs-Helmholtz relationship [My121(87)]:

$$\Delta H_{im} = \frac{d(\Delta G_{im}/T)}{d(1/T)} \tag{2-16}$$

For example, for the system on page 47 (CO_2 adsorbed on activated carbon), the free energies of immersion tabulated for the UNILAN equation at 212.7 K and 260.2 K yield:

$$\Delta H_{im} = \frac{(-55.9777/260.2) - (-57.9958/212.7)}{(1/260.2) - (1/212.7)} = -67.03 \text{ kJ/kg}$$

Assuming constant ΔH_{im}, the entropy of immersion at 212.7 K is:

$$\Delta S_{im} = \frac{\Delta H_{im} - \Delta G_{im}}{T}$$

$$= \frac{(-67.03) - (-58.00)}{212.7}$$

$$= -0.0424 \text{ kJ/(kg-K)}$$

Heats of Adsorption

Heats of adsorption [My35(86)] are defined as the energy of the adsorbate in the gas phase less the energy in the adsorbed phase and therefore are usually positive quantities.

According to Eq. (2-6), $\Pi A/RT = 5.98778$ mol/kg at $n = 4.0$ mol/kg and $T = 212.7$ K, so:

$$\frac{\Pi A}{n} = \frac{\Pi A}{RT} \times \frac{RT}{n}$$
$$= (5.98778)(8.3145 \times 10^{-3})(212.7)/(4.0)$$
$$= 2.65 \, \text{kJ/mol}$$

For the system CO_2 on activated carbon (page 47), let $T_1 = 260.2$ K and $T_2 = 212.7$ K in Eq. (2-15). Using the Toth equation to calculate P_1 and P_2, and assuming that q_{st} is independent of T in this range:

$$q_{st} = 22.56 \, \text{kJ/mol}$$

at $n = 4.0$. Next calculate q_{st} for the range $0 \le n \le 4.0$. The integral $\int_0^4 q_{st} \, dn$ is 90.77 kJ/mol. From Eq. (1-11), at 212.7 K:

$$\Delta H = \frac{1}{n} \int_0^4 q_{st} dn - \frac{\Pi A}{n}$$
$$= \frac{1}{4.0}(90.77) - 2.65$$
$$= 20.04 \, \text{kJ/mol}$$

ΔE at 212.7 K from Eq. (1-14) is:

$$\Delta E = \Delta H - RT + (\Pi A/n)$$
$$= 20.04 - 1.77 + 2.65$$
$$= 20.92 \, \text{kJ/mol}$$

q_d at 212.7 K is found from Eq. (1-9):

$$q_d = q_{st} - RT$$
$$= 22.56 - 1.77$$
$$= 20.79 \, \text{kJ/mol}$$

For heterogeneous adsorbents, $\Delta E > \Delta H$ and $\Delta E > q_d$, except at the limit $P \to 0$ where $\Delta E = \Delta H = q_d$.

Table 2-2 compares values of the various heats of adsorption for this example. The isosteric heat of adsorption is used in the differential equation for conservation of energy in a flow system. The differential and integral heats of adsorption are required for energy balances in batch processes. The equilibrium heat of adsorption is the molar change of the enthalpy of the adsorbate and arises naturally in thermodynamic formulas derived by the methods of statistical mechanics.

Table 2-2. Heats of Adsorption of Carbon Dioxide on Activated Carbon at $T = 212.7$ K, $n = 4.0$ mol/kg.

Type	Symbol	Value, kJ/mol
isosteric	q_{st}	22.56
differential	q_d	20.79
integral	ΔE	20.92
equilibrium	ΔH	20.04

Adsorption Second Virial Coefficient

For carbon dioxide adsorbed on activated carbon at 212.7 K (page 7), the adsorption second virial coefficient derived from the Toth equation is given by Eq. (2-8):

$$B_{1S} = RTmb^{-1/t} = (8.3145)(212.7)(13.2062)(4.6354)^{-1/0.5904} \times 10^{-3}$$
$$= 1.739 \, \text{m}^3/\text{kg}$$

The Henry constant from Eq. (1-3) is:

$$H = \frac{B_{1S}}{RT} = \frac{(1.739)}{(8.3145)(212.7)}$$
$$= 0.983 \times 10^{-3} \, \text{mol}/(\text{kg-Pa})$$
$$= 0.983 \, \text{mol}/(\text{kg-kPa})$$

Comparison of the equation $n = HP$ with Eq. (2-1) indicates that the former is applicable if the amount adsorbed is less than about 1% of the saturation capacity, which in this case occurs at $P < 0.1$ kPa.

ACETYLENE

ACTIVATED CARBON

ADSORBENT : BPL;PITTSBURGH ACTIVATED CARBON Co;1050-1150 Sqm/g;MESH 20X40
EXP.   : STATIC GRAVIMETRIC (McBAIN QUARTZ SPRING)
REF.   : LAUKHUF,W.L.S. and PLANK,C.A.;J.CHEM.ENG.DATA; 14,48(1969)

TEMP (K)=   303.15
PSAT (kPa)= 5422.56

	TOTH		UNILAN	UNITS
B =	5.6639	C =	955.0720	kPa
M =	.3772	S =	3.9688	
NI =	20.1893	M =	13.4985	mmol/g
DG =	-87.3009	DG =	-75.9174	J/g
BIS=	.51298+000	BIS=	.23741+000	1/g
QIS=	42.2892	QIS=	23.7201	KJ/mol

P(kPa)	N(mmol/g)	DN TOTH	DN UNILAN
7.6000	.7374	-.0781	-.1400
13.5300	.9755	.0132	-.0246
19.4800	1.3060	-.0464	-.0617
26.6700	1.4710	.0668	.0712
33.8700	1.7900	-.0106	.0058
39.3300	1.9590	-.0146	.0067
46.9300	2.1200	.0335	.0570
53.2700	2.2850	.0281	.0501
61.2700	2.4430	.0558	.0726
67.1300	2.6120	.0134	.0246
80.4000	2.9340	-.0461	-.0516
81.4000	2.9380	-.0315	-.0385
101.1000	3.2610	-.0159	-.0548
AVERAGE DN		.0349	.0507

TEMP (K)=   313.15

	TOTH		UNILAN	UNITS
B =	10.0810	C =	865.7540	kPa
M =	.4468	S =	3.3585	
NI =	19.1958	M =	13.7844	mmol/g
BIS=	.28364+000	BIS=	.17719+000	1/g
QIS=	42.2892	QIS=	23.7201	KJ/mol

P(kPa)	N(mmol/g)	DN TOTH	DN UNILAN
10.8000	.7335	-.0649	-.1055
11.4700	.7335	-.0329	-.0722
23.7300	1.2210	-.0196	-.0306
24.6700	1.2210	.0139	.0049
37.8700	1.6320	.0251	.0358
41.4700	1.7130	.0466	.0604
52.5300	2.0350	.0140	.0323
54.6700	2.0780	.0232	.0416
67.8700	2.3620	.0389	.0534
70.6000	2.4850	-.0263	-.0134
85.2700	2.7690	-.0203	-.0196
90.9300	2.8540	-.0015	-.0068
100.1000	3.0150	-.0025	-.0187
100.9000	3.0570	-.0310	-.0482
AVERAGE DN		.0258	.0388

TEMP (K)=   323.15

	TOTH		UNILAN	UNITS
B =	15.8994	C =	860.9340	kPa
M =	.5024	S =	3.0086	
NI =	17.7817	M =	13.4930	mmol/g
BIS=	.19395+000	BIS=	.14142+000	1/g
QIS=	42.2892	QIS=	23.7201	KJ/mol

P(kPa)	N(mmol/g)	DN TOTH	DN UNILAN
10.2700	.5722	-.0588	-.0881
10.4700	.5722	-.0505	-.0797
20.5300	.8987	-.0019	-.0175
23.4700	.9870	.0073	-.0040
33.9300	1.3060	.0040	.0057
36.8000	1.3860	.0035	.0080
47.4000	1.6710	-.0083	.0030
50.2000	1.7130	.0172	.0294
63.8700	1.9590	.0774	.0901
72.8000	2.2010	.0179	.0278
78.0700	2.2850	.0362	.0433
89.8700	2.5270	.0102	.0092
98.6700	2.7730	-.0849	-.0936
102.5000	2.7730	-.0217	-.0341
AVERAGE DN		.0285	.0381

DN=(NCAL-NEXP)

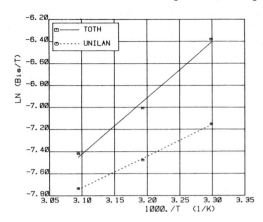
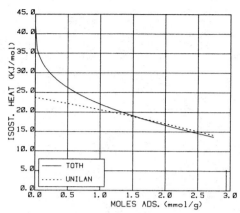

ACETYLENE

ACTIVATED CARBON

ADSORBENT : EY-51-C;PITTSBURGH COKE and CHEMICAL CO.;805 Sqm/g
EXP.   : STATIC VOLUMETRIC
REF.   : LEWIS,W.K.,GILLILAND,E.R.,CHERTOW,B. and MILLIKEN,W.;J.Am.Chem.Soc.;
            72,1157(1950)

```
                TEMP  (K)=    298.15
                PSAT (kPa)=  4846.92
```

	TOTH		UNILAN	UNITS
B =	3.7679	C =	454.8480	kPa
M =	.3256	S =	3.6394	
NI =	13.2299	M =	6.5321	mmol/g
DG =	-50.3002	DG =	-43.1537	J/g
BIS=	.55759+000	BIS=	.18605+000	1/g

P(kPa)	N(mmol/g)	DN TOTH	DN UNILAN
1.9470	.1940	-.0115	-.0586
4.0130	.3275	-.0182	-.0678
7.0130	.4710	-.0173	-.0569
12.2800	.6480	.0035	-.0142
20.4800	.8890	-.0013	.0060
26.6800	1.0330	.0004	.0187
32.8400	1.1520	.0079	.0321
39.7300	1.2810	.0044	.0312
50.2100	1.4080	.0446	.0693
66.0700	1.6720	-.0049	.0085
75.9100	1.7890	-.0058	-.0024
87.7300	1.9200	-.0104	-.0208
93.5700	1.9720	-.0043	-.0220
101.2000	2.0500	-.0102	-.0377
AVERAGE DN		.0103	.0319

DN=(NCAL-NEXP)

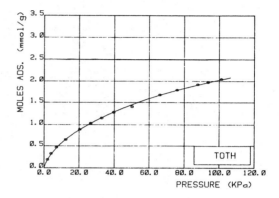

ACETYLENE

ACTIVATED CARBON

ADSORBENT : COLUMBIA GRADE L;MESH 20x40;1152 Sq m/g (N2)
EXP.   : STATIC VOLUMETRIC
REF.   : RAY,G.C.and BOX,E.O.;IND.ENG.CHEM.;42,1315(1950)

TEMP   (K)=   310.92

	TOTH		UNILAN	UNITS
B =	4.7632	C =	284918.0000	kPa
M =	.3499	S =	9.6334	
NI =	18.0210	M =	27.8921	mmol/g
BIS=	.53790+000	BIS=	.20051+000	1/g
QIS=	34.7779	QIS=	25.8353	KJ/mol

P(kPa)	N(mmol/g)	DN TOTH	DN UNILAN
8.3990	.6401	-.0262	-.1022
25.2000	1.2430	.0119	-.0059
35.6000	1.5440	-.0044	.0011
67.0600	2.1420	.0388	.0651
97.8600	2.6700	-.0298	-.0185
AVERAGE DN		.0222	.0386

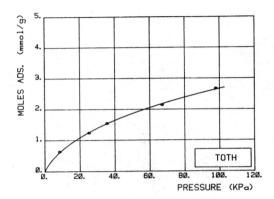

TEMP   (K)=   338.70

	TOTH		UNILAN	UNITS
B =	20.5372	C =	342169.0000	kPa
M =	.5265	S =	8.9076	
NI =	12.4925	M =	26.7166	mmol/g
BIS=	.11304+000	BIS=	.91179-001	1/g
QIS=	34.7779	QIS=	25.8353	KJ/mol

P(kPa)	N(mmol/g)	DN TOTH	DN UNILAN
19.8700	.5358	-.0015	-.0004
43.7300	.9796	.0050	.0174
68.7900	1.3660	-.0056	-.0008
92.5300	1.6580	.0022	-.0115
AVERAGE DN		.0036	.0075

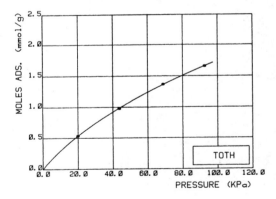

TEMP   (K)=   366.48

	TOTH		UNILAN	UNITS
B =	541.6740	C =	3400.5400	kPa
M =	.9284	S =	1.9034	
NI =	11.4343	M =	24.7901	mmol/g
BIS=	.39573-001	BIS=	.38275-001	1/g
QIS=	34.7779	QIS=	25.8353	KJ/mol

P(kPa)	N(mmol/g)	DN TOTH	DN UNILAN
26.1300	.2937	.0322	.0262
46.2600	.5799	-.0184	-.0244
76.2600	.8876	.0034	.0033
91.0600	1.0480	-.0029	.0019
AVERAGE DN		.0143	.0140

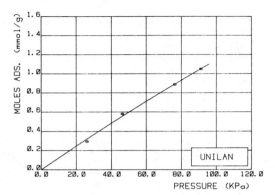

TEMP   (K)=   394.20

	TOTH		UNILAN	UNITS
B =	30.8322	C =	999920.0000	kPa
M =	.5247	S =	8.9104	
NI =	9.5910	M =	25.1492	mmol/g
BIS=	.45645-001	BIS=	.34269-001	1/g
QIS=	34.7779	QIS=	25.8353	KJ/mol

P(kPa)	N(mmol/g)	DN TOTH	DN UNILAN
24.9300	.2574	-.0022	-.0182
58.4000	.5097	.0029	-.0023
96.1300	.7505	-.0011	.0085
AVERAGE DN		.0021	.0096

DN=(NCAL-NEXP)

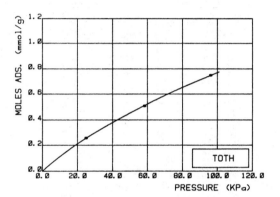

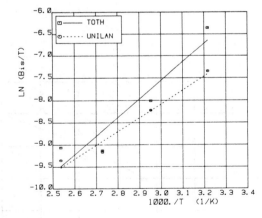

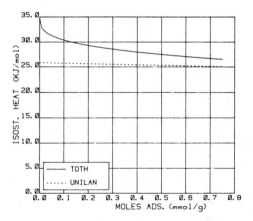

ACETYLENE

ACTIVATED CARBON

ADSORBENT : NUXIT-AL, (1.5 mm DIAMETER, 3-4 mm LONG)
EXP.   : STATIC VOLUMETRIC
REF.   : SZEPESY, L., ILLES, V., *Acta Chim. Hung.*, **35,** 37(1963)

TEMP   (K) =   293.15
PSAT (kPa) = 4320.73

TOTH		UNILAN		UNIT
B =	3.6616	C =	703.2360	kPa
M =	.3309	S =	4.1543	
NI =	18.5728	M =	11.2211	mmol/g
DG =	-72.4227	DG =	-63.7313	J/g
BIS =	.89609+00	BIS =	.29813+00	1/g
QIS =	33.3634	QIS =	24.3280	KJ/mol

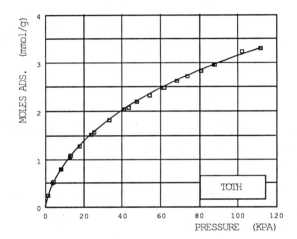

P(kPa)	N(mmol/g)	DN TOTH	DN UNILAN
1.3332	0.2427	-0.0211	-0.0888
3.7064	0.5118	-0.0409	-0.1209
4.0797	0.5220	-0.0184	-0.0974
8.0393	0.7938	0.0010	-0.0551
12.5456	1.0552	-0.0030	-0.0303
12.9322	1.0842	-0.0123	-0.0374
17.3719	1.2806	-0.0012	-0.0043
23.5447	1.5139	0.0099	0.0277
24.9579	1.5737	0.0007	0.0219
33.1172	1.8258	0.0118	0.0455
40.5566	2.0413	0.0032	0.0400
42.9697	2.0873	0.0191	0.0557
47.1160	2.2037	0.0039	0.0391
54.0621	2.3438	0.0209	0.0511
60.8348	2.4951	0.0103	0.0333
61.6214	2.5013	0.0197	0.0418
68.1542	2.6352	0.0106	0.0240
73.7937	2.7347	0.0123	0.0172
80.8865	2.8476	0.0191	0.0126
87.2459	2.9730	-0.0051	-0.0227
87.6992	2.9810	-0.0062	-0.0245
101.8313	3.2545	-0.0736	-0.1178
111.4439	3.3063	0.0030	-0.0595
AVERAGE DN		0.0142	0.0465

TEMP   (K) =   313.15

TOTH		UNILAN		UNIT
B =	5.8973	C =	1030.6200	kPa
M =	0.3759	S =	3.9625	
NI =	15.2256	M =	9.9146	mmol/g
BIS =	.353185+00	BIS =	.166149+00	1/g
QIS =	33.3634	QIS =	24.3280	KJ/mol

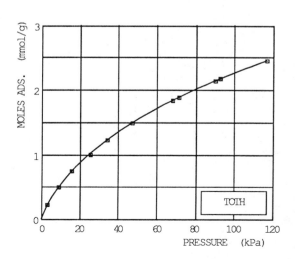

P(kPa)	N(mmol/g)	DN TOTH	DN UNILAN
2.9464	0.2276	-0.0089	-0.0524
8.6526	0.4997	-0.0030	-0.0424
15.5587	0.7576	-0.0082	-0.0269
24.9179	1.0075	0.0142	0.0183
33.8771	1.2395	-0.0019	0.0155
46.7160	1.4988	0.0002	0.0252
67.5809	1.8472	-0.0008	0.0185
71.1406	1.8990	-0.0001	0.0167
90.0990	2.1533	-0.0003	-0.0007
92.5388	2.1837	-0.0006	-0.0036
116.6568	2.4607	-0.0055	-0.0375
AVERAGE DN		0.0040	0.0234

TEMP (K) = 333.15

TOTH		UNILAN		UNIT
B =	7.4090	C =	1252.6500	kPa
M =	0.3759	S =	3.7093	
NI =	15.1724	M =	8.3411	mmol/g
BIS=.204042+00		BIS=.101440+00		l/g
QIS=	33.3634	QIS=	24.3280	KJ/mol

P(kPa)	N(mmol/g)	DN TOTH	DN UNILAN
8.3593	0.3360	-0.0295	-0.0653
20.7982	0.5925	0.0076	-0.0111
36.8369	0.8763	0.0089	0.0097
43.3163	0.9888	-0.0053	0.0001
53.7288	1.1226	0.0045	0.0140
58.6084	1.1958	-0.0063	0.0039
74.8603	1.3644	0.0149	0.0229
85.1928	1.4595	0.0288	0.0327
93.2987	1.5791	-0.0105	-0.0111
118.4433	1.8151	-0.0210	-0.0400
AVERAGE DN		0.0137	0.0211

TEMP (K) = 363.15

TOTH		UNILAN		UNIT
B =	16.5522	C =	1483.1000	kPa
M =	0.4421	S =	3.1836	
NI =	14.3226	M =	6.9867	mmol/g
BIS=7.56780-02		BIS=5.38204-02		l/g
QIS=	33.3634	QIS=	24.3280	KJ/mol

P(kPa)	N(mmol/g)	DN TOTH	DN UNILAN
18.2251	0.2820	0.0104	0.0024
20.8382	0.3074	0.0188	0.0123
26.0378	0.4243	-0.0341	-0.0375
34.9170	0.4739	0.0176	0.0188
40.8499	0.5796	-0.0252	-0.0216
57.1685	0.6862	0.0269	0.0337
69.6874	0.8344	-0.0111	-0.0048
79.9132	0.9129	-0.0054	-0.0012
108.9641	1.1226	0.0014	-0.0067
AVERAGE DN		0.0168	0.0154

DN=(NCAL-NEXP)

```
ACETYLENE
-------------------------------------------------------------------------
SILICA GEL
-------------------------------------------------------------------------
ADSORBENT : DAVISON CHEMICAL Co.;REF.GRADE;751 Sqm/g;mesh 14x20
EXP.      : STATIC VOLUMETRIC
REF.      : LEWIS,W.K.,GILLILAND,E.R.,CHERTOW,B. and MILLIKEN,W.;J.Am.Chem.Soc.;
            72,1157(1950)
```

```
              TEMP    (K)=   298.15
              PSAT (kPa)=  4846.92
```

	TOTH		UNILAN	UNITS
B =	3.4067	C =	920.8230	kPa
M =	.3107	S =	4.5192	
NI =	12.2155	M =	6.6239	mmol/g
DG =	-44.3904	DG =	-37.5420	J/g
BIS=	.58563+000	BIS=	.18100+000	1/g

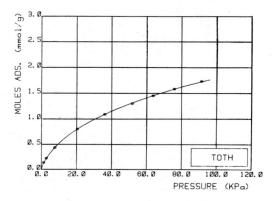

P(kPa)	N(mmol/g)	DN TOTH	DN UNILAN
1.2930	.1412	-.0156	-.0524
2.8000	.2282	-.0064	-.0479
7.6270	.4350	.0030	-.0208
20.5100	.7970	.0024	.0186
36.0000	1.0900	-.0008	.0261
51.9100	1.3010	.0136	.0325
63.6700	1.4520	.0015	.0088
75.8000	1.5830	-.0034	-.0107
91.4300	1.7310	-.0082	-.0362
AVERAGE DN		.0061	.0282

```
        DN=(NCAL-NEXP)
```

```
BENZENE
----------------------------------------------------------------------
CARBON MOLECULAR SIEVE
----------------------------------------------------------------------
ADSORBENT : MSC-5A;TAKEDA CHEMICAL Co.;0.56 cc/g;650 Sqm/g
EXP.      : STATIC GRAVIMETRIC (McBAIN QUARTZ SPRING BALANCE)
REF.      : NAKAHARA,T.,HIRATA,M.and OMORI,T.;J.Chem.Eng.Data;19,310(1974)
```

```
            TEMP  (K)=    278.55
            PSAT (kPa)=     4.86
```

	TOTH		UNILAN	UNITS
B =	.0538	C =	5232.8500	kPa
M =	.0536	S =	28.1706	
NI =	5.3450	M =	5.7736	mmol/g
DG =	-62.5720	DG =	-36.6222	J/g
BIS=	.59006+025	BIS=	.77792+008	1/g
QIS=	367.1040	QIS=	10.4911	KJ/mol

P(kPa)	N(mmol/g)	DN TOTH	DN UNILAN
.1333	1.8200	-.0171	-.0172
.3600	1.9000	.0045	.0046
.8533	1.9790	.0140	.0141
2.0000	2.0560	.0244	.0244
3.0670	2.1150	.0092	.0092
3.9070	2.1510	-.0020	-.0020
4.8000	2.2030	-.0329	-.0329
AVERAGE DN		.0149	.0149

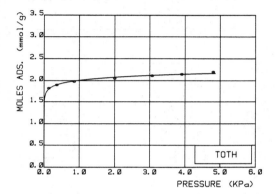

```
            TEMP  (K)=    303.15
            PSAT (kPa)=    16.03
```

	TOTH		UNILAN	UNITS
B =	.0546	C =	6622.5700	kPa
M =	.0693	S =	28.0847	
NI =	4.0705	M =	5.4962	mmol/g
DG =	-65.3569	DG =	-40.0622	J/g
BIS=	.16640+020	BIS=	.58619+008	1/g
QIS=	367.1040	QIS=	10.4911	KJ/mol

P(kPa)	N(mmol/g)	DN TOTH	DN UNILAN
.0400	1.5610	.0085	.0112
.2667	1.7720	-.0128	-.0141
.5867	1.8470	-.0097	-.0120
1.8930	1.9420	.0099	.0077
4.0930	2.0150	.0112	.0101
6.1330	2.0650	-.0003	-.0003
10.0300	2.1110	.0001	.0018
11.6300	2.1320	-.0070	-.0047
AVERAGE DN		.0074	.0077

```
            DN=(NCAL-NEXP)
```

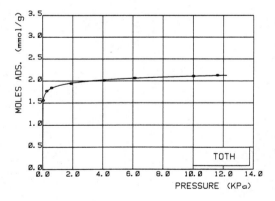

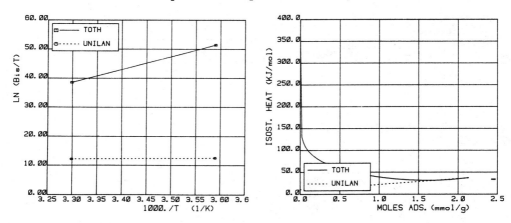

BENZENE

SILICA GEL

ADSORBENT : DAVISON CHEM.Co.
EXP.   : STATIC GRAVIMETRIC
REF.   : SIRCAR,S.and MYERS,A.L.;AIChE J.;19,159(1973)

                    TEMP (K)=   303.15
                    PSAT (kPa)=   16.03

                 TOTH   UNILAN   UNITS
         -----------------   -----------------   ------
B   =   .5271   C = 8260.2400   kPa
M   =   .5685   S =   12.1192
NI =   4.8609   M =   17.3890   mmol/g
DG =   -33.7098   DG =   -34.1236   J/g
BIS= .37785+002   BIS= .40137+002   1/g

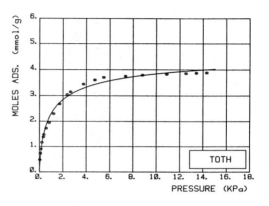

P(kPa)	N(mmol/g)	DN TOTH	DN UNILAN
.0301	.4780	-.1768	-.1106
.0693	.7320	-.1683	-.0635
.1187	.9060	-.0965	.0199
.1987	1.1700	-.0647	.0413
.3346	1.3850	.0738	.1442
.3826	1.4840	.0730	.1306
.5600	1.7310	.1169	.1325
.8133	1.9490	.1942	.1654
1.1930	2.2940	.1542	.0832
1.7200	2.6780	.0550	-.0464
2.3730	3.0430	-.0703	-.1856
2.6530	3.1510	-.0983	-.2150
3.7600	3.4460	-.1552	-.2633
4.7300	3.6080	-.1712	-.2620
5.4800	3.7080	-.1830	-.2577
7.3590	3.7570	-.0664	-.0967
8.8130	3.7940	-.0099	-.0051
10.8800	3.8320	.0539	.1074
12.5300	3.8620	.0877	.1783
13.4300	3.8720	.1078	.2179
14.3100	3.8900	.1166	.2452
AVERAGE DN		.1092	.1415

                 DN=(NCAL-NEXP)

N-BUTANE

ACTIVATED CARBON

ADSORBENT : COLUMBIA G GRADE;1157 Sq m/g;MESH 8x14
EXP.   : STATIC
REF.   : PAYNE,H.K.,STUDERVANT,G.A. and LELAND,T,W,;IND.ENG.CHEM.FUNDAM.;7,363(1968)

TEMP (K)= 313.15
PSAT (kPa)= 379.63

	TOTH		UNILAN	UNITS
B =	.7803	C =	4.8490	kPa
M =	.4294	S =	3.8924	
NI =	5.5014	M =	5.1134	mmol/g
DG =	-58.0591	DG =	-58.5524	J/g
BIS=	.25524+002	BIS=	.17284+002	1/g
QIS=	14.0408	QIS=	-35.6119	KJ/mol

P(kPa)	N(mmol/g)	DN TOTH	DN UNILAN
3.4470	2.1830	.1007	.1587
6.8950	2.7490	.0309	.0295
10.3400	3.2870	-.2256	-.2546
34.4700	3.7790	-.0327	-.0210
72.3900	4.1360	.0535	.0227
131.0000	4.3540	.0880	.0802
196.5000	4.5320	.0585	.0609
272.3000	4.6960	.0008	.0055
348.2000	4.9030	-.1335	-.1312
	AVERAGE DN	.0805	.0849

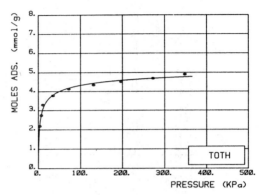

TEMP (K)= 323.15
PSAT (kPa)= 496.96

	TOTH		UNILAN	UNITS
B =	4.4951	C =	7.4358	kPa
M =	.8454	S =	1.3186	
NI =	4.6161	M =	4.5553	mmol/g
DG =	-51.4453	DG =	-51.5934	J/g
BIS=	.20958+001	BIS=	.21661+001	1/g
QIS=	14.0408	QIS=	-35.6119	KJ/mol

P(kPa)	N(mmol/g)	DN TOTH	DN UNILAN
3.4470	1.4790	.0260	.0557
6.8950	2.1120	.0773	.0903
10.3400	2.8180	-.2166	-.2132
27.5800	3.4520	-.0200	.0195
96.5300	3.9760	.1729	.1774
244.8000	4.3240	.0679	.0595
410.2000	4.6290	-.1602	-.1786
	AVERAGE DN	.1058	.1135

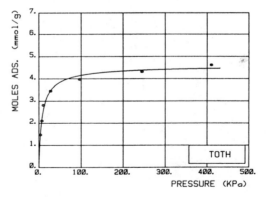

TEMP (K)= 333.15
PSAT (kPa)= 639.40

	TOTH		UNILAN	UNITS
B =	4.3262	C =	47.4772	kPa
M =	.9242	S =	6.8857	
NI =	4.3788	M =	6.7288	mmol/g
DG =	-57.6205	DG =	-61.9906	J/g
BIS=	.24863+001	BIS=	.27884+002	1/g
QIS=	14.0408	QIS=	-35.6119	KJ/mol

P(kPa)	N(mmol/g)	DN TOTH	DN UNILAN
3.4470	1.8750	-.1602	.2147
10.3400	2.7880	.0370	-.1662
55.1600	3.6100	.3155	-.1725
131.0000	3.9250	.2382	-.0659
217.2000	4.0990	.1423	.0062
282.7000	4.1950	.0753	.0383
348.2000	4.2870	.0019	.0474
417.1000	4.3710	-.0685	.0509
486.1000	4.4410	-.1286	.0550
555.0000	4.5510	-.2310	.0090
592.9000	4.6080	-.2846	-.0161
AVERAGE DN		.1530	.0765

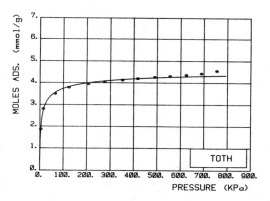

TEMP (K)= 343.15
PSAT (kPa)= 810.23

	TOTH		UNILAN	UNITS
B =	.9049	C =	44.7050	kPa
M =	.4401	S =	7.0051	
NI =	4.8315	M =	6.4375	mmol/g
DG =	-62.3007	DG =	-64.8689	J/g
BIS=	.17298+002	BIS=	.32321+002	1/g
QIS=	14.0408	QIS=	-35.6119	KJ/mol

P(kPa)	N(mmol/g)	DN TOTH	DN UNILAN
3.4470	1.8710	-.0188	.1756
13.7900	2.8160	-.0839	-.1364
65.5000	3.5000	.0614	-.1061
124.1000	3.7870	.0367	-.1001
206.8000	3.9560	.0445	-.0353
275.8000	4.0590	.0291	-.0066
351.6000	4.1360	.0196	.0272
417.1000	4.1970	.0028	.0441
489.5000	4.2620	-.0231	.0520
555.0000	4.3160	-.0480	.0551
624.0000	4.3750	-.0810	.0493
689.5000	4.4410	-.1257	.0285
755.0000	4.5570	-.2230	-.0464
AVERAGE DN		.0614	.0664

DN=(NCAL-NEXP)

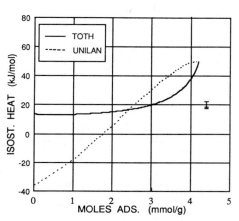

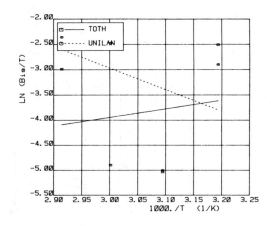

N-BUTANE

ACTIVATED CARBON

ADSORBENT : COLUMBIA GRADE L;MESH 20x40;1152 Sq m/g (N2)
EXP.   : STATIC VOLUMETRIC
REF.   : RAY,G.C.and BOX,E.O.;IND.ENG.CHEM.;42,1315(1950)

TEMP (K)= 310.92
PSAT (kPa)= 356.58

TOTH		UNILAN		UNITS
B =	.2950	C =	95911.7002	kPa
M =	.3350	S =	18.5849	
NI =	4.9477	M =	13.1634	mmol/g
DG =	-69.2430	DG =	-71.2943	J/g
BIS=	.48932+003	BIS=	.11248+004	1/g
QIS=	33.0184	QIS=	50.2660	KJ/mol

P(kPa)	N(mmol/g)	DN TOTH	DN UNILAN
5.3330	3.1120	-.0024	.0001
59.0600	3.9650	.0196	-.0013
91.5900	4.1180	-.0174	.0010
AVERAGE DN		.0131	.0008

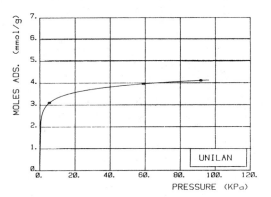

TEMP (K)= 338.70
PSAT (kPa)= 730.50

TOTH		UNILAN		UNITS
B =	12.5233	C =	1167.1900	kPa
M =	1.1518	S =	7.5155	
NI =	3.8419	M =	11.1660	mmol/g
DG =	-49.9032	DG =	-55.2170	J/g
BIS=	.12053+001	BIS=	.32912+001	1/g
QIS=	33.0184	QIS=	50.2660	KJ/mol

P(kPa)	N(mmol/g)	DN TOTH	DN UNILAN
4.2660	1.3560	-.0130	.1615
17.7300	2.7470	.0247	-.2483
71.9900	3.5260	.0364	-.0060
98.5300	3.6460	-.0035	.1054
AVERAGE DN		.0194	.1303

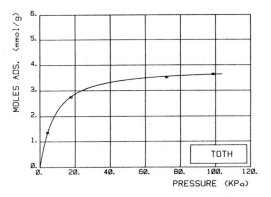

TEMP (K)= 366.48
PSAT (kPa)= 1338.13

TOTH		UNILAN		UNITS
B =	4.1489	C =	7432.6200	kPa
M =	.7689	S =	9.1630	
NI =	3.8153	M =	12.7676	mmol/g
DG =	-56.3140	DG =	-62.1305	J/g
BIS=	.18268+001	BIS=	.27239+001	1/g
QIS=	33.0184	QIS=	50.2660	KJ/mol

P(kPa)	N(mmol/g)	DN TOTH	DN UNILAN
3.2000	1.0490	.0011	.0869
19.0700	2.3990	-.0031	-.1434
73.3300	3.1580	.0135	.0154
90.6600	3.2670	-.0113	.0528
AVERAGE DN		.0072	.0746

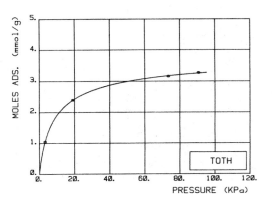

```
          TEMP   (K)=    394.20
          PSAT (kPa)=   2258.30
```

TOTH		UNILAN		UNITS
B =	1.2295	C =	70.9882	kPa
M =	.3652	S =	4.6274	
NI =	5.1408	M =	5.4612	mmol/g
DG =	-65.0754	DG =	-65.4459	J/g
BIS=	.95695+001	BIS=	.27852+001	1/g
QIS=	33.0184	QIS=	50.2660	KJ/mol

P(kPa)	N(mmol/g)	DN TOTH	DN UNILAN
3.6000	1.0620	.0143	.0129
20.6700	2.0330	-.0139	-.0127
58.2600	2.6440	-.0212	-.0277
99.0600	2.8950	.0246	.0284
AVERAGE DN		.0185	.0204

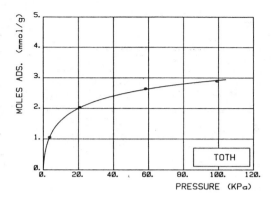

```
          TEMP   (K)=    422.00
          PSAT (kPa)=   3609.17
```

TOTH		UNILAN		UNITS
B =	1.6901	C =	93923.5000	kPa
M =	.3691	S =	10.9660	
NI =	5.1645	M =	13.1965	mmol/g
DG =	-65.1730	DG =	-65.8543	J/g
BIS=	.43720+001	BIS=	.13008+001	1/g
QIS=	33.0184	QIS=	50.2660	KJ/mol

P(kPa)	N(mmol/g)	DN TOTH	DN UNILAN
7.9990	1.0720	.0035	-.0011
26.4000	1.7110	-.0044	.0031
65.5900	2.2450	-.0036	-.0045
92.1300	2.4380	.0048	.0028
AVERAGE DN		.0041	.0029

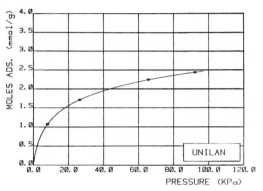

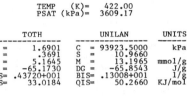
```
          TEMP   (K)=    449.80
```

TOTH		UNILAN		UNITS
B =	2.6547	C =	9294.0500	kPa
M =	.4148	S =	7.9545	
NI =	4.5585	M =	9.4129	mmol/g
BIS=	.16197+001	BIS=	.67814+000	1/g
QIS=	33.0184	QIS=	50.2660	KJ/mol

P(kPa)	N(mmol/g)	DN TOTH	DN UNILAN
11.8700	.9111	-.0014	-.0033
39.8600	1.5210	.0024	.0064
90.1300	1.9910	-.0011	-.0064
AVERAGE DN		.0017	.0054

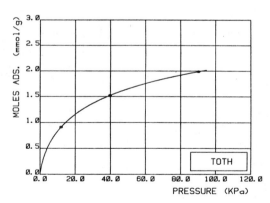

TEMP   (K)=   477.59

	TOTH		UNILAN	UNITS
B =	3.6182	C =	855340.0000	kPa
M =	.3997	S =	11.7137	
NI =	5.3818	M =	14.7899	mmol/g
BIS=	.85577+000	BIS=	.35823+000	1/g
QIS=	33.0184	QIS=	50.2660	KJ/mol

P(kPa)	N(mmol/g)	DN TOTH	DN UNILAN
12.5300	.6567	.0005	-.0088
37.2000	1.1510	-.0007	.0126
95.8600	1.7020	.0002	-.0052
AVERAGE DN		.0004	.0089

DN=(NCAL-NEXP)

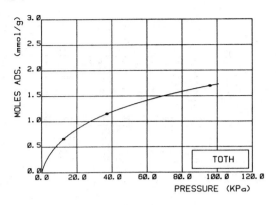

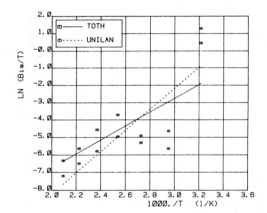

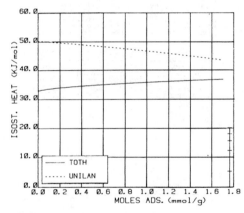

BUTANE

ACTIVATED CARBON

ADSORBENT : NUXIT-AL (1.5 mm DIAMETER, 3-4 mm LONG)
EXP.   : STATIC VOLUMETRIC
REF.   : SZEPESY, L., ILLES, V., *Acta Chim. Hung.*,**35**, 37(1963)

TEMP   (K) =   293.15
PSAT (kPa) =   208.48

TOTH		UNILAN		UNIT
B =	0.4514	C =	1113.2400	kPa
M =	0.3424	S =	10.6708	
NI =	5.2284	M =	10.5756	mmol/g
DG =	-50.3753	DG =	-50.6931	J/g
BIS =1.30066+02		BIS =.467389+02		1/g
QIS =	25.3972	QIS =	37.6670	KJ/mol

P(kPa)	N(mmol/g)	DN TOTH	DN UNILAN
0.0133	0.6006	-0.3850	-0.3944
0.3733	1.1159	0.1334	0.2405
0.6399	1.5706	-0.0489	0.0394
3.0264	2.1074	0.2743	0.2572
5.3862	2.6584	0.0435	-0.0101
10.6791	3.1590	-0.0939	-0.1727
21.7715	3.6543	-0.2410	-0.3156
26.7977	3.5976	-0.0896	-0.1561
34.4637	3.7583	-0.1400	-0.1922
61.5681	3.8734	-0.0199	-0.0199
69.4341	3.8377	0.0612	0.0753
104.8178	3.9305	0.1152	0.1866
120.2698	3.9711	0.1206	0.2141
AVERAGE DN		0.1359	0.1749

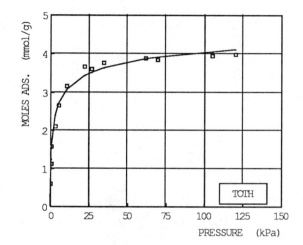

TEMP   (K) =   313.15
PSAT (kPa) =   379.61

TOTH		UNILAN		UNIT
B =	0.7374	C =	1317.4800	kPa
M =	0.5278	S =	10.1260	
NI =	4.1242	M =	10.2550	mmol/g
DG =	-52.2663	DG =	-54.0114	J/g
BIS =1.91242+01		BIS =2.50019+01		1/g
QIS =	25.3972	QIS =	37.6670	KJ/mol

P(kPa)	N(mmol/g)	DN TOTH	DN UNILAN
0.1600	0.5189	0.0156	0.1873
0.5066	1.0606	-0.0078	0.1352
1.7332	1.8013	-0.0071	-0.0176
3.9063	2.3131	-0.0074	-0.1263
11.1724	2.8645	0.0262	-0.1501
19.6517	3.1671	-0.0185	-0.1677
26.0645	3.2724	-0.0112	-0.1303
27.6377	3.2621	0.0212	-0.0904
37.9834	3.4031	-0.0080	-0.0706
64.1545	3.5615	-0.0094	0.0361
66.7543	3.5499	0.0128	0.0678
97.9517	3.6516	0.0049	0.1601
125.2693	3.7190	-0.0100	0.2172
AVERAGE DN		0.0123	0.1197

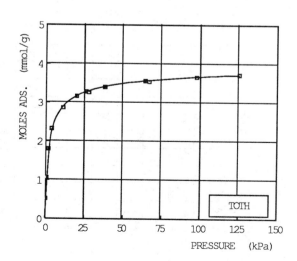

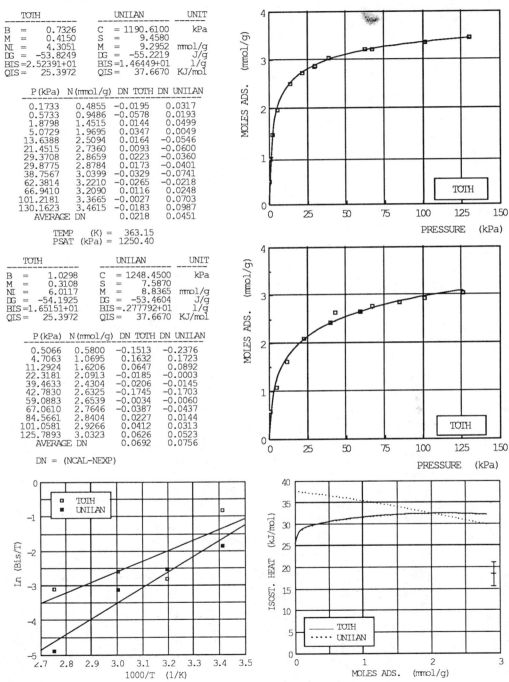

TEMP   (K) =   333.15
PSAT (kPa) =   639.40

TOTH		UNILAN		UNIT
B =	0.7326	C =	1190.6100	kPa
M =	0.4150	S =	9.4580	
NI =	4.3051	M =	9.2952	mmol/g
DG =	−53.8249	DG =	−55.2219	J/g
BIS =	2.52391+01	BIS =	1.46449+01	1/g
QIS =	25.3972	QIS =	37.6670	KJ/mol

P(kPa)	N(mmol/g)	DN TOTH	DN UNILAN
0.1733	0.4855	−0.0195	0.0317
0.5733	0.9486	−0.0578	0.0193
1.8798	1.4515	0.0144	0.0499
5.0729	1.9695	0.0347	0.0049
13.6388	2.5094	0.0164	−0.0546
21.4515	2.7360	0.0093	−0.0600
29.3708	2.8659	0.0223	−0.0360
29.8775	2.8784	0.0173	−0.0401
38.7567	3.0399	−0.0329	−0.0741
62.3814	3.2210	−0.0265	−0.0218
66.9410	3.2090	0.0116	0.0248
101.2181	3.3665	−0.0027	0.0703
130.1623	3.4615	−0.0183	0.0987
	AVERAGE DN	0.0218	0.0451

TEMP   (K) =   363.15
PSAT (kPa) =   1250.40

TOTH		UNILAN		UNIT
B =	1.0298	C =	1248.4500	kPa
M =	0.3108	S =	7.5870	
NI =	6.0117	M =	8.8365	mmol/g
DG =	−54.1925	DG =	−53.4604	J/g
BIS =	1.65151+01	BIS =	.277792+01	1/g
QIS =	25.3972	QIS =	37.6670	KJ/mol

P(kPa)	N(mmol/g)	DN TOTH	DN UNILAN
0.5066	0.5800	−0.1513	−0.2376
4.7063	1.0695	0.1632	0.1723
11.2924	1.6206	0.0647	0.0892
22.3181	2.0913	−0.0185	−0.0003
39.4633	2.4304	−0.0206	−0.0145
42.7830	2.6325	−0.1745	−0.1703
59.0883	2.6539	−0.0034	−0.0060
67.0610	2.7646	−0.0387	−0.0437
84.5661	2.8404	0.0227	0.0144
101.0581	2.9266	0.0412	0.0313
125.7893	3.0323	0.0626	0.0523
	AVERAGE DN	0.0692	0.0756

DN = (NCAL−NEXP)

BUTANE

CARBON MOLECULAR SIEVE

ADSORBENT : MSC-5A;TAKEDA CHEMICAL Co.;0.56 cc/g;650 Sqm/g
EXP.   : STATIC GRAVIMETRIC (McBAIN QUARTZ SPRING BALANCE)
REF.   : NAKAHARA,T.,HIRATA,M.and OMORI,T.;J.Chem.Eng.Data;19,310(1974)

TEMP (K)=   278.55
PSAT (kPa)=   126.63

	TOTH		UNILAN	UNITS
B =	.0944	C =	8.0316	kPa
M =	.2591	S =	16.8350	
NI =	2.2096	M =	3.5250	mmol/g
DG =	-37.9743	DG =	-33.8052	J/g
BIS=	.46288+005	BIS=	.61827+006	1/g
QIS=	62.8273	QIS=	133.9800	KJ/mol

P(kPa)	N(mmol/g)	DN TOTH	DN UNILAN
.2267	1.3260	.0127	.0630
.6400	1.5230	-.0250	-.0253
1.6000	1.6260	-.0049	-.0324
4.1330	1.7190	.0117	-.0260
9.3730	1.8000	.0113	-.0213
20.4000	1.8670	.0098	-.0069
40.5300	1.9230	.0035	.0090
60.4000	1.9540	-.0018	.0197
85.8700	1.9910	-.0179	.0196
AVERAGE DN		.0110	.0248

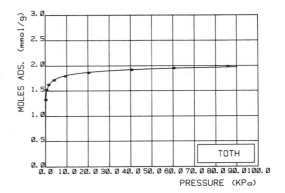

TEMP (K)=   303.15
PSAT (kPa)=   284.43

	TOTH		UNILAN	UNITS
B =	.1300	C =	9.1723	kPa
M =	.2688	S =	11.8860	
NI =	2.1559	M =	3.2368	mmol/g
DG =	-38.6321	DG =	-34.1756	J/g
BIS=	.10745+005	BIS=	.54333+004	1/g
QIS=	62.8273	QIS=	133.9800	KJ/mol

P(kPa)	N(mmol/g)	DN TOTH	DN UNILAN
.0533	.7415	.1047	.1762
.1200	.9927	.0059	.0353
.2533	1.1940	-.0582	-.0643
.5333	1.3250	-.0592	-.0940
1.6130	1.4780	-.0366	-.0963
3.9470	1.5790	-.0136	-.0754
9.3330	1.6670	.0016	-.0463
22.8000	1.7460	.0139	-.0036
40.3300	1.7890	.0212	.0310
60.4000	1.8170	.0253	.0580
86.0000	1.8440	.0243	.0791
AVERAGE DN		.0331	.0690

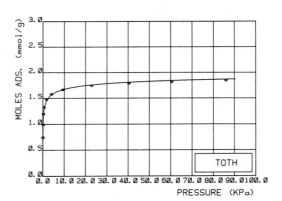

TEMP   (K)=   324.15
PSAT (kPa)=   510.02

	TOTH		UNILAN	UNITS
B =	.2224	C =	8.6261	kPa
M =	.2852	S =	8.3198	
NI =	2.1072	M =	2.7637	mmol/g
DG =	-34.3551	DG =	-32.3948	J/g
BIS=	.11043+004	BIS=	.21299+003	1/g
QIS=	62.8273	QIS=	133.9800	KJ/mol

P(kPa)	N(mmol/g)	DN TOTH	DN UNILAN
.0267	.3217	.0622	.1129
.2400	.7071	.0598	.0813
.5200	.9050	.0116	.0110
.8667	1.0560	-.0411	-.0554
1.6000	1.1870	-.0570	-.0848
2.6800	1.2770	-.0541	-.0892
3.7470	1.3020	-.0212	-.0586
6.4130	1.3800	-.0110	-.0474
8.5470	1.4040	.0096	-.0237
11.2000	1.4800	-.0261	-.0548
19.8700	1.5330	.0008	-.0126
31.0500	1.5710	.0198	.0234
71.9500	1.6410	.0449	.0928
85.6900	1.6500	.0538	.1128
AVERAGE DN		.0338	.0615

DN=(NCAL-NEXP)

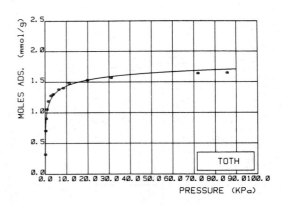

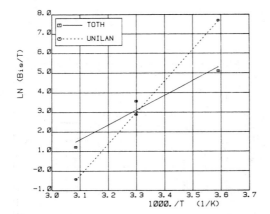

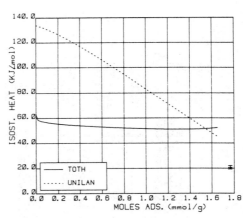

ISOBUTANE

ZEOLITE MOLECULAR SIEVE

ADSORBENT : 13X;LINDE(UNION CARBIDE);.3 cc/g;525 Sqm/g (N2)
EXP.   : STATIC VOLUMETRIC
REF.   : HYUN,S.H.and DANNER,R.P.;J.Chem.Eng.Data;27,196(1982)

TEMP (K)= 298.15
PSAT (kPa)= 351.86

	TOTH		UNILAN	UNITS
B =	.1993	C =	.3087	kPa
M =	.3169	S =	5.0003	
NI =	2.1244	M =	1.9132	mmol/g
DG =	-33.3426	DG =	-31.8390	J/g
BIS=	.85467+003	BIS=	.22803+003	1/g
QIS=	82.1562	QIS=	66.9532	KJ/mol

P(kPa)	N(mmol/g)	DN TOTH	DN UNILAN
.2000	.7500	.1100	.1247
.2400	.9900	-.0909	-.0809
.4700	1.0400	-.0023	-.0041
1.5200	1.3600	-.0812	-.1044
5.2300	1.4800	.0143	-.0026
7.3800	1.4900	.0569	.0453
12.3700	1.6200	-.0004	-.0030
13.1200	1.6300	-.0026	-.0042
23.0000	1.6300	-.0671	.0735
23.9700	1.7100	-.0081	-.0012
26.3700	1.7100	.0028	-.0105
40.9000	1.7900	-.0301	-.0206
57.4600	1.8400	-.0468	-.0390
76.6400	1.8200	-.0008	.0035
90.5000	1.8100	.0234	-.0248
137.8000	1.8800	-.0133	-.0217
AVERAGE DN		.0344	.0353

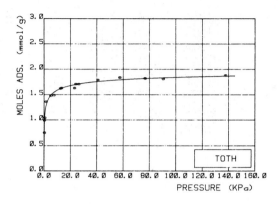

TEMP (K)= 323.15
PSAT (kPa)= 684.55

	TOTH		UNILAN	UNITS
B =	.3764	C =	.6062	kPa
M =	.5454	S =	2.9487	
NI =	1.6959	M =	1.6278	mmol/g
DG =	-30.6668	DG =	-30.2499	J/g
BIS=	.27335+002	BIS=	.23278+002	1/g
QIS=	82.1562	QIS=	66.9532	KJ/mol

P(kPa)	N(mmol/g)	DN TOTH	DN UNILAN
.4000	.7000	.0000	.0109
.7800	.8600	.0191	.0165
1.4800	1.0000	.0426	.0329
2.4300	1.1900	-.0330	-.0419
2.6600	1.2500	-.0735	-.0817
8.1300	1.3700	.0077	.0147
17.2000	1.4300	.0434	.0564
17.4300	1.4600	.0148	.0278
29.1300	1.5200	.0045	.0158
43.9700	1.5500	.0068	.0136
45.4500	1.5500	.0091	.0154
75.0200	1.5900	.0002	-.0016
81.8000	1.5900	.0049	.0014
110.6000	1.6000	.0096	.0005
130.2000	1.6400	-.0233	-.0356
137.8000	1.6500	-.0310	-.0443
AVERAGE DN		.0202	.0257

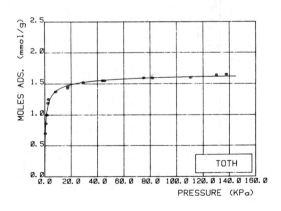

TEMP (K)=   373.15
PSAT (kPa)= 1981.92

	TOTH		UNILAN	UNITS
B =	3.6025	C =	3.5366	kPa
M =	1.0088	S =1.0031-007		
NI =	1.3706	M =	1.3732	mmol/g
DG =	-26.9462	DG =	-26.8757	J/g
BIS=	.11935+001	BIS=	.12046+001	1/g
QIS=	82.1562	QIS=	66.9532	KJ/mol

P(kPa)	N(mmol/g)	DN TOTH	DN UNILAN
.6400	.1700	.0395	.0395
1.0100	.2300	.0741	.0744
1.6900	.4600	-.0166	-.0169
1.7200	.4700	-.0213	-.0215
3.2400	.6900	-.0333	-.0338
4.5200	.8200	-.0489	-.0501
10.4200	1.0300	-.0036	-.0050
14.3300	1.0800	.0225	.0202
14.8000	1.1000	.0094	.0073
16.0800	1.0800	.0466	.0444
27.6500	1.2100	.0079	.0075
32.6600	1.2300	.0092	.0080
40.6600	1.2600	.0032	.0031
64.5700	1.3300	-.0288	-.0293
80.5700	1.3300	-.0154	-.0150
107.2000	1.3300	-.0018	-.0015
137.8000	1.3400	-.0026	-.0017
AVERAGE DN		.0226	.0223

DN=(NCAL-NEXP)

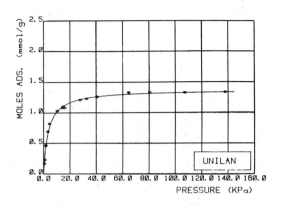

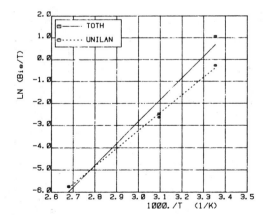

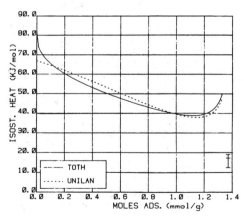

CARBON DIOXIDE

ACTIVATED CARBON

ADSORBENT : Fiber Carbon. KF-1500, Toyobo Co., Ltd.
EXP.   : Gravimetric
REF.   : Kuro-Oka, M., Suzuki, T., Nitta, T., and
            Katayama, T., J. Chem. Eng. Japan, **17**, 588
            (1984)

TEMP(K)   =   273.15
PSAT(kPa) =   3484.90

TOTH		UNILAN		UNIT
B =	3.4812	C =	576.2396	kPa
M =	.2056	S =	3.3673	
NI =	265.6045	M =	15.2317	mmol/g
DG =	-108.7401	DG =	-75.9167	J/g
BIS =	.139829+01	BIS =	.258183+00	1/g
QIS =	44.9976	QIS =	21.2224	KJ/mol

P(kPa)	N(mmol/g)	DN TOTH	DN UNILAN
0.0906	0.0260	-0.0006	-0.0157
0.1480	0.0410	-0.0026	-0.0242
0.4280	0.0890	0.0030	-0.0409
0.5550	0.1140	-0.0007	-0.0518
0.8070	0.1610	-0.0087	-0.0711
1.4950	0.2380	0.0073	-0.0741
2.4180	0.3480	0.0044	-0.0886
3.2910	0.4460	-0.0037	-0.0998
4.0600	0.5090	0.0063	-0.0890
6.6860	0.7320	0.0033	-0.0769
10.8700	1.0280	0.0009	-0.0426
16.5700	1.3770	-0.0116	-0.0076
22.4800	1.6720	-0.0048	0.0365
28.9700	1.9590	0.0027	0.0711
39.8300	2.3900	0.0057	0.0928
54.7700	2.9100	0.0016	0.0759
73.0800	3.4600	-0.0009	0.0196
88.3800	3.8700	-0.0036	-0.0478
96.6800	4.0700	0.0025	-0.0818
AVERAGE DN		0.0039	0.0583

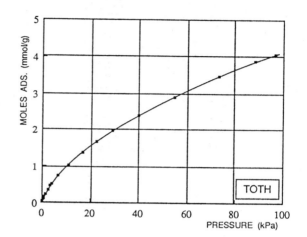

TEMP(K)   =   298.15
PSAT(kPa) =   6439.81

TOTH		UNILAN		UNIT
B =	9.2698	C =	671.5916	kPa
M =	.3953	S =	3.0570	
NI =	25.0000	M =	10.4226	mmol/g
DG =	-83.4438	DG =	-64.9057	J/g
BIS =	.221709+00	BIS =	.133496+00	1/g
QIS =	44.9976	QIS =	21.2224	KJ/mol

P(kPa)	N(mmol/g)	DN TOTH	DN UNILAN
0.2040	0.0250	-0.0092	-0.0140
0.8370	0.0670	-0.0082	-0.0225
1.3980	0.0990	-0.0058	-0.0253
2.0120	0.1410	-0.0124	-0.0360
2.8690	0.1810	-0.0061	-0.0331
3.9050	0.2370	-0.0096	-0.0387
5.0540	0.2880	-0.0058	-0.0356
6.4440	0.3460	-0.0013	-0.0303
9.1890	0.4630	-0.0042	-0.0288
14.6500	0.6600	-0.0004	-0.0121
19.9700	0.8310	0.0014	0.0019
26.7700	1.0220	0.0083	0.0213
34.4000	1.2210	0.0090	0.0309
46.0000	1.4910	0.0093	0.0356
58.4500	1.7530	0.0045	0.0258
68.9500	1.9570	-0.0027	0.0086
83.2700	2.2100	-0.0106	-0.0191
96.6000	2.4100	-0.0019	-0.0335
AVERAGE DN		0.0061	0.0252

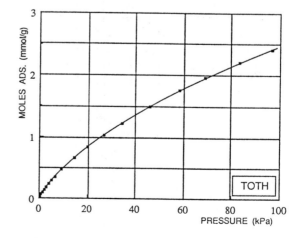

TEMP (K)   =   323.15

TOTH		UNILAN		UNIT
B =	42.2911	C =	999.5480	kPa
M =	.6973	S =	2.9371	
NI =	6.2643	M =	8.3417	mmol/g
BIS =.783249-01		BIS =.717893-01		1/g
QIS =	44.9976	QIS =	21.2224	KJ/mol

P (kPa)	N (mmol/g)	DN TOTH	DN UNILAN
0.7060	0.0260	-0.0060	-0.0073
0.8970	0.0360	-0.0107	-0.0122
1.9830	0.0710	-0.0162	-0.0190
3.2900	0.1020	-0.0131	-0.0167
4.2640	0.1310	-0.0174	-0.0214
6.6950	0.1870	-0.0143	-0.0186
8.7360	0.2340	-0.0140	-0.0180
10.8000	0.2760	-0.0099	-0.0134
13.9300	0.3380	-0.0051	-0.0077
19.8900	0.4510	0.0006	0.0000
26.4000	0.5620	0.0087	0.0100
32.5900	0.6680	0.0074	0.0102
40.0500	0.7810	0.0116	0.0153
53.5300	0.9730	0.0107	0.0144
68.5400	1.1650	0.0067	0.0085
83.1900	1.3410	-0.0058	-0.0070
94.2600	1.4590	-0.0109	-0.0148
AVERAGE DN		0.0099	0.0126

DN= (NCAL-NEX

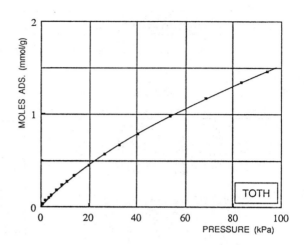

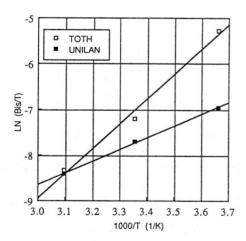

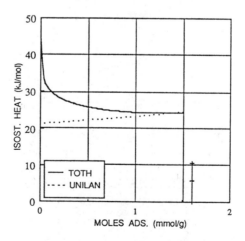

CARBON DIOXIDE

ACTIVATED CARBON

ADSORBENT : BPL;PITTSBURGH COKE & CHEM.Co;1050-1150 Sq m/g;MESH 12x30
EXP.   : STATIC GRAVIMETRIC
REF.   : MEREDITH,J.M.and PLANK,C.A.;J.CHEM.ENG.DATA; 12,259(1967)

TEMP   (K)=   303.15
PSAT (kPa)=   7212.60

	TOTH		UNILAN	UNITS
B =	11.2027	C =	83996.2002	kPa
M =	.4238	S =	7.7469	
NI =	14.7171	M =	17.9553	mmol/g
DG =	-56.4919	DG =	-45.6484	J/g
BIS=	.12403+000	BIS=	.80477-001	1/g
QIS=	13.5500	QIS=	6.5397	KJ/mol

P(kPa)	N(mmol/g)	DN TOTH	DN UNILAN
3.8660	.1272	.0073	-.0099
7.1330	.2522	-.0262	-.0443
10.5300	.3363	-.0256	-.0411
18.8700	.5067	-.0156	-.0215
19.4700	.5135	-.0106	-.0158
22.8000	.5885	-.0219	-.0236
35.2400	.7544	.0244	.0319
43.7300	.8634	.0429	.0529
44.4600	.8839	.0329	.0429
61.4600	1.1380	.0034	.0102
70.3300	1.2520	-.0052	-.0031
75.7300	1.3020	.0058	.0041
87.9900	1.4770	-.0387	-.0505
AVERAGE DN		.0200	.0271

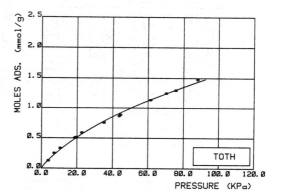

TEMP   (K)=   313.15

	TOTH		UNILAN	UNITS
B =	10.2138	C =	185680.0000	kPa
M =	.4425	S =	8.7230	
NI =	9.4740	M =	16.5206	mmol/g
BIS=	.12920+000	BIS=	.81563-001	1/g
QIS=	13.5500	QIS=	6.5397	KJ/mol

P(kPa)	N(mmol/g)	DN TOTH	DN UNILAN
8.1990	.2658	-.0194	-.0385
18.3300	.4522	.0059	-.0033
26.6600	.5953	.0051	.0035
35.3300	.7385	-.0090	-.0054
40.0000	.7998	-.0068	-.0015
51.7300	.9248	.0133	.0197
55.0600	.9657	.0104	.0166
64.1300	1.0680	.0060	.0100
73.8600	1.1700	.0008	.0009
92.7900	1.3540	-.0141	-.0249
AVERAGE DN		.0091	.0124

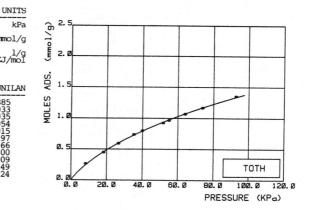

TEMP   (K)=   323.15

	TOTH		UNILAN	UNITS
B =	16.5924	C =	221844.0000	kPa
M =	.5394	S =	8.7713	
NI =	6.4194	M =	16.3938	mmol/g
BIS=	.94426-001	BIS=	.72958-001	1/g
QIS=	13.5500	QIS=	6.5397	KJ/mol

P(kPa)	N(mmol/g)	DN TOTH	DN UNILAN
3.3330	.0841	.0116	.0023
7.9990	.2249	-.0197	-.0296
11.0700	.2613	.0076	-.0007
14.5300	.3272	.0078	.0020
21.4700	.4317	.0224	.0214
28.2600	.5749	-.0174	-.0147
34.2600	.6567	-.0161	-.0110
45.6000	.7794	.0022	.0091
54.8000	.8816	.0020	.0086
67.4600	1.0040	.0062	.0102
80.6600	1.1290	-.0004	-.0009
101.6000	1.2950	-.0008	-.0107
	AVERAGE DN	.0095	.0101

DN=(NCAL-NEXP)

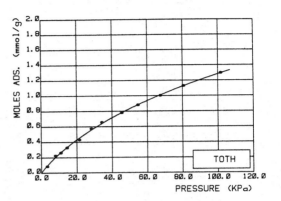

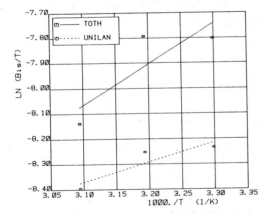

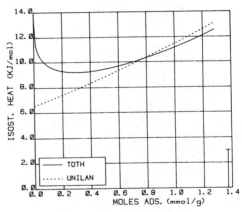

CARBON DIOXIDE

ACTIVATED CARBON

ADSORBENT : ASC(METAL INPREG.BPL);PITTSBURGH COKE & CHEM.Co;800-850 Sq m/g;MESH 12x30
EXP.   : STATIC GRAVIMETRIC
REF.   : MEREDITH,J.M.and PLANK,C.A.;J.CHEM.ENG.DATA; 12,259(1967)

TEMP   (K)=   303.15
PSAT (kPa)= 7212.60

	TOTH		UNILAN	UNITS
B =	6.8702	C =	726744.0000	kPa
M =	.3523	S =	10.2529	
NI =	18.6182	M =	20.6549	mmol/g
DG =	-57.9210	DG =	-44.4986	J/g
BIS=	.19740+000	BIS=	.99086-001	1/g
QIS=	35.8498	QIS=	21.6877	KJ/mol

P(kPa)	N(mmol/g)	DN TOTH	DN UNILAN
5.3330	.2295	-.0140	-.0390
5.8660	.2658	-.0335	-.0581
9.5330	.3431	-.0051	-.0245
13.8700	.4953	-.0486	-.0595
14.6700	.4385	.0269	.0175
21.8600	.6476	-.0291	-.0262
22.7300	.6090	.0265	.0307
37.6000	.8748	.0185	.0350
47.0000	1.0290	.0033	.0204
48.8000	.9907	.0666	.0834
70.7900	1.3340	-.0021	.0008
80.2600	1.4500	-.0140	-.0208
89.7300	1.5630	-.0293	-.0475
AVERAGE DN		.0244	.0356

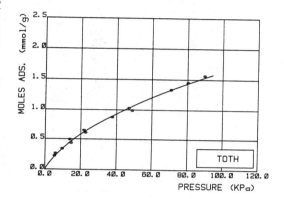

TEMP   (K)=   313.15

	TOTH		UNILAN	UNITS
B =	10.2354	C =	841236.0000	kPa
M =	.4777	S =	10.4785	
NI =	6.0560	M =	15.4721	mmol/g
BTS=	.12108+000	BIS=	.81212-001	1/g
QIS=	35.8498	QIS=	21.6877	KJ/mol

P(kPa)	N(mmol/g)	DN TOTH	DN UNILAN
6.0000	.1522	.0287	.0146
11.4000	.3045	-.0044	-.0142
18.8700	.4590	-.0235	-.0262
23.0000	.4772	.0236	.0240
23.7300	.5340	-.0222	-.0213
28.6000	.5726	.0088	.0123
38.8600	.7248	-.0136	-.0076
45.2600	.7816	.0015	.0076
47.2000	.7998	.0040	.0099
54.6000	.8953	-.0167	-.0123
59.6000	.9157	.0101	.0130
66.7900	.9725	.0172	.0174
74.5300	1.0470	.0068	.0035
90.1300	1.1820	-.0110	-.0227
AVERAGE DN		.0137	.0147

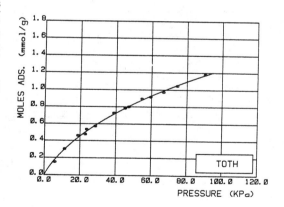

TEMP   (K)=   323.15

	TOTH		UNILAN	UNITS
B =	12.7430	C =	867068.0000	kPa
M =	.4885	S =	10.2365	
NI =	5.9516	M =	14.6698	mmol/g
BIS=	.87377-001	BIS=	.61953-001	1/g
QIS=	35.8498	QIS=	21.6877	KJ/mol

P(kPa)	N(mmol/g)	DN TOTH	DN UNILAN
4.3330	.0954	.0085	-.0019
9.3990	.1886	.0100	.0008
12.6000	.2613	-.0103	-.0174
22.7300	.3795	.0139	.0139
25.0600	.4272	-.0048	-.0035
35.0600	.5453	-.0094	-.0041
46.0000	.6385	.0056	.0124
53.4000	.7112	-.0011	.0054
63.2000	.8044	-.0139	-.0092
77.8600	.8998	-.0010	-.0010
88.5900	.9702	.0004	-.0041
94.0600	.9952	.0099	.0030
AVERAGE DN		.0074	.0064

DN=(NCAL-NEXP)

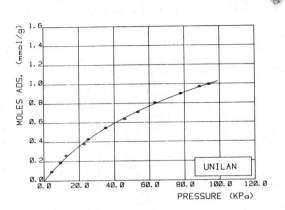

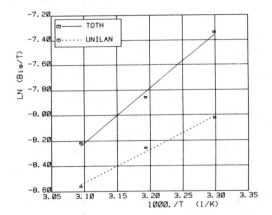

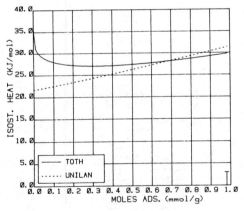

CARBON DIOXIDE

ACTIVATED CARBON

ADSORBENT : COLUMBIA GRADE L;MESH 20x40;1152 Sq m/g (N2)
EXP.   : STATIC VOLUMETRIC
REF.   : RAY,G.C.and BOX,E.O.;IND.ENG.CHEM.;42,1315(1950)

TEMP   (K)=   310.92

	TOTH		UNILAN	UNITS
B =	9.8363	C =	588929.0000	kPa
M =	.4376	S =	9.9060	
NI =	11.4157	M =	22.3413	mmol/g
BIS=	.15891+000	BIS=	.99241-001	1/g
QIS=	29.3978	QIS=	25.4931	KJ/mol

P(kPa)	N(mmol/g)	DN TOTH	DN UNILAN
16.9300	.5180	.0057	-.0048
32.1400	.8387	-.0120	-.0052
71.8600	1.3720	.0153	.0233
95.0700	1.6440	-.0087	-.0159
AVERAGE DN		.0104	.0123

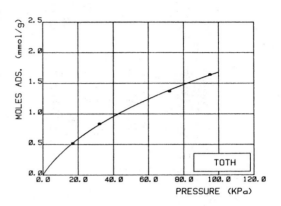

TEMP   (K)=   338.70

	TOTH		UNILAN	UNITS
B =	34.9607	C =	81720.9004	kPa
M =	.6130	S =	7.1312	
NI =	6.7191	M =	16.4457	mmol/g
BIS=	.57394-001	BIS=	.49680-001	1/g
QIS=	29.3978	QIS=	25.4931	KJ/mol

P(kPa)	N(mmol/g)	DN TOTH	DN UNILAN
21.1900	.3098	.0172	.0140
42.8000	.6079	-.0292	-.0271
68.4000	.8034	.0196	.0224
95.8600	1.0480	-.0047	-.0069
AVERAGE DN		.0177	.0176

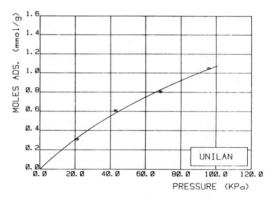

TEMP   (K)=   366.48

	TOTH		UNILAN	UNITS
B =	23.6498	C =	999988.0000	kPa
M =	.5127	S =	9.1687	
NI =	6.0860	M =	17.1189	mmol/g
BIS=	.38763-001	BIS=	.27285-001	1/g
QIS=	29.3978	QIS=	25.4931	KJ/mol

P(kPa)	N(mmol/g)	DN TOTH	DN UNILAN
34.2700	.2710	.0073	-.0057
60.2600	.4394	-.0098	-.0135
98.9300	.6096	.0036	.0134
AVERAGE DN		.0069	.0109

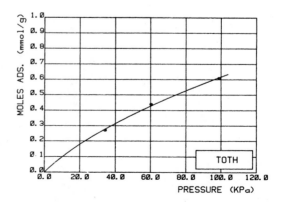

TEMP   (K)=   394.20

	TOTH		UNILAN	UNITS
B =	135.4960	C =	894781.0000	kPa
M =	.6757	S =	8.2267	
NI =	7.1194	M =	18.6900	mmol/g
BIS=	.16330-001	BIS=	.15558-001	l/g
QIS=	29.3978	QIS=	25.4931	KJ/mol

P(kPa)	N(mmol/g)	DN TOTH	DN UNILAN
45.7300	.1963	.0022	.0024
80.0000	.3320	-.0047	-.0044
98.2600	.3884	.0027	.0024
AVERAGE DN		.0032	.0030

DN=(NCAL−NEXP)

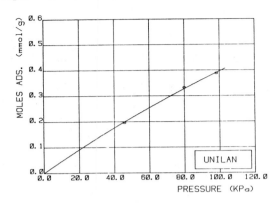

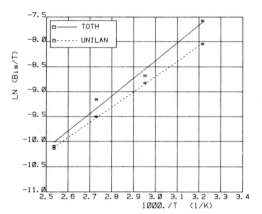

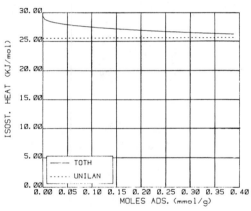

```
CARBON DIOXIDE
---------------------------------------------------------------------------
ACTIVATED CARBON
---------------------------------------------------------------------------
ADSORBENT : BPL;PITTSBURGH CHEMICAL Co.;988 Sqm/g
EXP.      : STATIC VOLUMETRIC
REF.      : REICH,R.,ZIEGLER,W.T.and ROGERS,K.A.;Ind.Eng.Chem.Process Des.Dev.
            19,336(1980)
```

```
            TEMP   (K)=    212.70
            PSAT (kPa)=    437.00
```

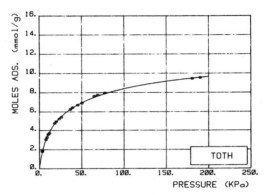

	TOTH		UNILAN	UNITS
B =	4.6354	C =	33.6072	kPa
M =	.5904	S =	2.2057	
NI =	13.2062	M =	12.1542	mmol/g
DG =	-58.3476	DG =	-57.9958	J/g
BIS=	.17387+001	BIS=	.12999+001	1/g
QIS=	23.3168	QIS=	22.3013	KJ/mol

P(kPa)	N(mmol/g)	DN TOTH	DN UNILAN
3.4470	1.8890	-.0784	-.1070
3.4470	1.8240	-.0134	-.0420
3.4470	1.8290	-.0184	-.0470
3.4470	1.8260	-.0154	-.0440
7.5840	3.0400	-.0450	-.0368
8.9630	3.2660	.0291	.0427
10.3400	3.5690	-.0050	.0120
11.7200	3.6980	.1104	.1292
17.9300	4.7360	-.0480	-.0300
19.9900	4.9250	-.0021	.0141
23.4400	5.2190	.0528	.0655
26.2000	5.4060	.1123	.1223
36.5400	6.2640	-.0035	-.0022
39.3000	6.4410	-.0182	-.0186
45.5100	6.7060	.0422	.0388
50.3300	6.9210	.0487	.0436
64.8100	7.6260	-.1098	-.1173
68.9500	7.7520	-.1046	-.1123
77.2200	7.9450	-.0609	-.0685
180.6000	9.4420	.0460	.0427
190.3000	9.5350	.0403	.0366
AVERAGE DN		.0478	.0559

```
            TEMP   (K)=    260.20
            PSAT (kPa)=   2430.00
```

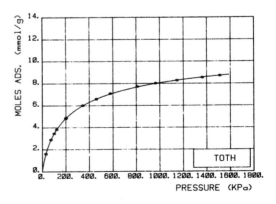

	TOTH		UNILAN	UNITS
B =	24.0405	C =	262.7560	kPa
M =	.6502	S =	1.9062	
NI =	11.6400	M =	10.8573	mmol/g
DG =	-56.2747	DG =	-55.9777	J/g
BIS=	.18935+000	BIS=	.15425+000	1/g
QIS=	23.3168	QIS=	22.3013	KJ/mol

P(kPa)	N(mmol/g)	DN TOTH	DN UNILAN
28.9600	1.6080	-.0481	-.0745
73.0800	2.8860	.0007	.0023
100.7000	3.4530	.0133	.0199
122.0000	3.8180	.0190	.0266
199.3000	4.8190	.0235	.0279
203.4000	4.8770	.0084	.0125
344.7000	6.0020	-.0002	-.0020
459.5000	6.6600	-.0009	-.0036
579.2000	7.0910	-.0307	-.0326
808.7000	7.7230	-.0317	-.0313
965.3000	8.0180	-.0127	-.0116
1147.0000	8.2900	.0059	.0067
1363.0000	8.5500	.0208	.0196
1512.0000	8.7120	.0163	.0128
AVERAGE DN		.0166	.0203

TEMP   (K)=   301.40
PSAT (kPa)= 6930.00

	TOTH		UNILAN	UNITS
B =	88.9193	C =	762.7610	kPa
M =	.7261	S =	1.5876	
NI =	9.8321	M =	9.3342	mmol/g
DG =	-55.1355	DG =	-54.9588	J/g
BIS=	.50967-001	BIS=	.45271-001	1/g
QIS=	23.3168	QIS=	22.3013	KJ/mol

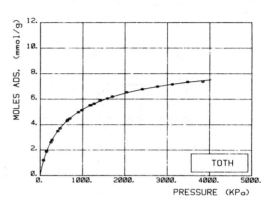

P(kPa)	N(mmol/g)	DN TOTH	DN UNILAN
77.9100	1.1900	-.0446	-.0594
79.9800	1.2100	-.0406	-.0551
149.6000	1.8700	-.0049	-.0081
151.7000	1.8870	-.0037	-.0066
252.3000	2.6210	.0113	.0160
279.2000	2.7830	.0163	.0218
417.8000	3.4870	.0248	.0309
468.2000	3.7000	.0244	.0299
630.2000	4.2720	.0220	.0250
650.2000	4.3690	-.0144	-.0116
698.4000	4.4680	.0258	.0280
897.7000	4.9720	.0108	.0110
979.7000	5.1330	.0190	.0187
1171.0000	5.5140	-.0208	-.0216
1193.0000	5.5250	.0034	.0026
1268.0000	5.6340	.0091	.0082
1411.0000	5.8970	-.0556	-.0564
1587.0000	6.0520	.0031	.0026
1706.0000	6.2040	-.0200	-.0203
2040.0000	6.5380	-.0448	-.0445
2408.0000	6.7780	-.0114	-.0107
2766.0000	6.9890	-.0045	-.0041
3117.0000	7.1620	.0021	.0017
3482.0000	7.3480	-.0243	-.0261
3840.0000	7.3790	.0800	.0764
AVERAGE DN		.0217	.0239

DN=(NCAL-NEXP)

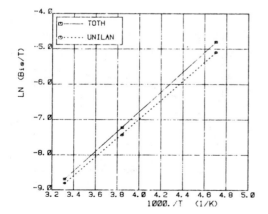

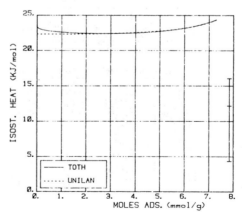

CARBON DIOXIDE

ACTIVATED CARBON

ADSORBENT : PCB, Calgon Corp.
EXP.   : Static
REF.   : Ritter, J. A. and R. T. Yang, *Ind. Eng. Chem.*
            Res. **26**, 1679 (1987).

TEMP(K)   =   296.
PSAT(kPa) =   6128.64

TOTH		UNILAN		UNIT
B =	90.0520	C =	348.6813	kPa
M =	.8093	S =	1.3007	
NI =	10.2968	M =	10.0274	mmol/g
DG =	-72.5419	DG =	-72.4804	J/g
BIS=.974503-01		BIS=.924852-01		1/g
QIS=	19.8640	QIS=	20.9051	KJ/mol

P(kPa)	N(mmol/g)	DN TOTH	DN UNILAN
111.0047	2.6857	-0.0266	-0.0218
510.8973	5.7998	0.0568	0.0492
1723.6750	8.1643	-0.0806	-0.0714
3626.6123	8.9228	0.0436	0.0388
AVERAGE DN		0.0519	0.0453

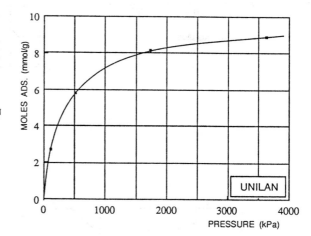

TEMP(K)   =   373.

TOTH		UNILAN		UNIT
B =	142.8367	C =	2581.4320	kPa
M =	.693525	S =	1.9630	
NI =	9.6416	M =	9.6529	mmol/g
BIS=.233672-01		BIS=.206174-01		1/g
QIS=	19.8640	QIS=	20.9051	KJ/mol

P(kPa)	N(mmol/g)	DN TOTH	DN UNILAN
200.6358	1.0618	0.0012	-0.0055
645.3439	2.4225	-0.0017	0.0067
1792.6221	4.1580	0.0014	-0.0041
3378.4031	5.3224	-0.0005	0.0013
AVERAGE DN		0.0012	0.0044

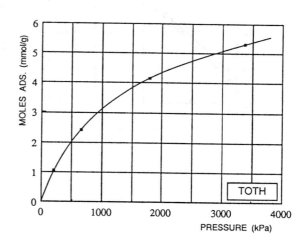

TEMP (K)   =   480.

TOTH		UNILAN		UNIT
B =	158.4412	C =	30817.0500	kPa
M =	.5795	S =	3.1486	
NI =	11.2187	M =	12.0669	mmol/g
BIS =	.715923−02	BIS =	.577201−02	1/g
QIS =	19.8640	QIS =	20.9051	KJ/mol

P (kPa)	N (mmol/g)	DN TOTH	DN UNILAN
348.1823	0.4684	−0.0042	−0.0215
1089.3625	1.1510	−0.0062	−0.0025
2268.3562	1.8827	0.0155	0.0254
3674.8752	2.5519	−0.0042	−0.0142
AVERAGE DN		0.0075	0.0159

DN= (NCAL−NEXP)

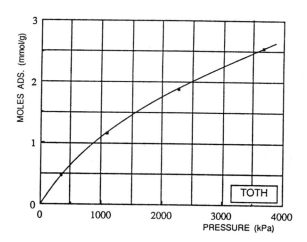

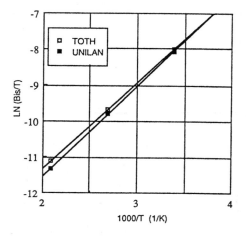

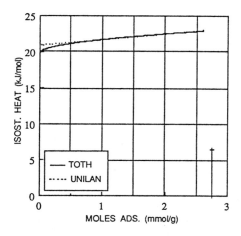

CARBON DIOXIDE

ACTIVATED CARBON

ADSORBENT : NUXIT-AL, (1.5 mm DIAMETER, 3-4 mm LONG)
EXP.   : STATIC VOLUMETRIC
REF.   : SZEPESY, L., ILLES, V., *Acta Chim. Hung.*, **35**, 37(1963)

TEMP (K) =   293.15
PSAT (kPa) = 5734.21

TOTH	UNILAN	UNIT
B = 22.8755	C = 843.8980	kPa
M = 0.6286	S = 3.1675	
NI = 7.9348	M = 10.2582	mmol/g
DG = -54.2454	DG = -56.3996	J/g
BIS =.133022+00	BIS =.110869+00	1/g
QIS = 24.0131	QIS = 23.7155	KJ/mol

P(kPa)	N(mmol/g)	DN TOTH	DN UNILAN
7.0394	0.3409	-0.0329	-0.0488
20.1450	0.6242	0.1102	0.1014
25.0645	0.8817	-0.0139	-0.0194
31.6640	1.0347	-0.0038	-0.0056
39.3300	1.1953	0.0069	0.0081
42.31164	1.3078	-0.0433	-0.0412
49.5691	1.3908	0.0159	0.0194
56.3019	1.5608	-0.0319	-0.0278
60.0349	1.6157	-0.0226	-0.0185
73.9404	1.8401	-0.0274	-0.0241
75.6069	1.8062	0.0311	0.0342
90.9789	2.0721	-0.0239	-0.0232
97.6050	2.0949	0.0368	0.0361
111.0839	2.2872	0.0023	-0.0011
		0.0288	0.0292

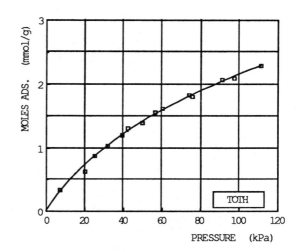

TEMP (K) =   313.15

TOTH	UNILAN	UNIT
B = 40.5174	C = 1160.9900	kPa
M = 0.6746	S = 2.9663	
NI = 6.5506	M = 8.5711	mmol/g
BIS =7.05926-02	BIS =6.27539-02	1/g
QIS = 24.0131	QIS = 23.7155	KJ/mol

P(kPa)	N(mmol/g)	DN TOTH	DN UNILAN
8.5859	0.2151	-0.0144	-0.0217
20.9049	0.4243	0.0126	0.0076
39.1033	0.7032	0.0214	0.0214
46.7827	0.8625	-0.0317	-0.0304
57.6351	0.9557	0.0134	0.0158
66.7677	1.0954	-0.0189	-0.0162
78.2467	1.1869	0.0147	0.0170
95.6852	1.3890	-0.0150	-0.0143
97.0717	1.3752	0.0118	0.0122
124.0028	1.6188	-0.0005	-0.0042
		0.0154	0.0161

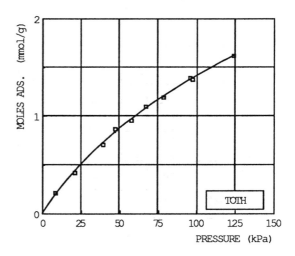

TEMP   (K) =   333.15

TOTH		UNILAN		UNIT
B =	47.6393	C = 1499.9900		kPa
M =	0.6504	S =	2.8571	
NI =	6.3518	M =	7.2791	mmol/g
BIS=4.62882−02		BIS=4.08216−02		1/g
QIS=	24.0131	QIS=	23.7155	KJ/mol

P(kPa)	N(mmol/g)	DN TOTH	DN UNILAN
30.0641	0.3802	0.0033	−0.0003
33.1838	0.4324	−0.0159	−0.0189
44.0229	0.5171	0.0074	0.0065
52.1289	0.6073	−0.0074	−0.0071
60.7415	0.6657	0.0097	0.0110
79.9665	0.8268	0.0035	0.0057
88.7258	0.8991	−0.0036	−0.0015
96.4851	0.9490	0.0017	0.0034
121.8296	1.1213	−0.0037	−0.0045
		0.0063	0.0066

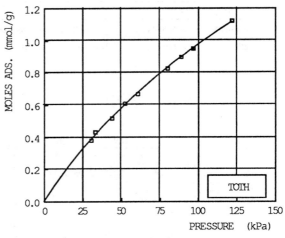

TEMP   (K) =   363.15

TOTH		UNILAN		UNIT
B =	124.4030	C = 1500.0000		kPa
M =	0.7833	S =	1.7240	
NI =	3.7783	M =	6.5436	mmol/g
BIS=2.41462−02		BIS=2.07376−02		1/g
QIS=	24.0131	QIS=	23.7155	KJ/mol

P(kPa)	N(mmol/g)	DN TOTH	DN UNILAN
14.9721	0.1178	−0.0076	−0.0178
24.2646	0.1660	0.0063	−0.0067
25.8378	0.1861	−0.0037	−0.0169
38.7434	0.2624	−0.0005	−0.0144
48.1159	0.3065	0.0094	−0.0035
67.4983	0.4176	0.0021	−0.0058
75.2869	0.4498	0.0090	0.0039
100.6048	0.5751	0.0017	0.0081
132.5887	0.7224	−0.0127	0.0119
		0.0059	0.0099

DN = (NCAL−NEXP)

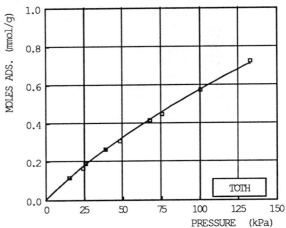

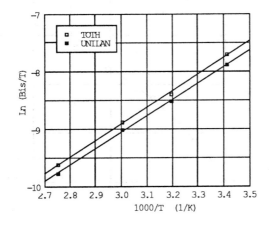

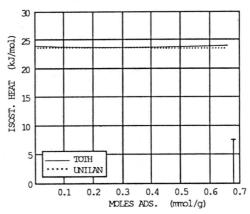

CARBON DIOXIDE
```
------------------------------------------------------------
```
ACTIVATED CARBON
```
------------------------------------------------------------
```
ADSORBENT : NUXIT-AL, (1.5 mm DIAMETER, 3-4 mm LONG)
EXP.   : STATIC VOLUMETRIC
REF.   : SZEPESY, L., ILLES, V., *Acta Chim. Hung.*,**35**, 53(1963)

TEMP (K) = 293.15
PSAT (kPa) = 5710.00

TOTH		UNILAN		UNIT
B =	26.2845	C =	1366.1400	kPa
M =	0.5831	S =	2.9747	
NI =	12.2269	M =	14.0438	mmol/g
DG =	-63.3026	DG =	-64.0370	J/g
BIS=	1.09520-01	BIS=	.822619-01	1/g
QIS=	20.1547	QIS=	20.6391	KJ/mol

P(kPa)	N(mmol/g)	DN TOTH	DN UNILAN
104.2634	2.1645	-0.0067	-0.0158
125.0350	2.4197	-0.0030	-0.0067
164.1465	2.8342	0.0053	0.0073
196.2665	3.1323	0.0063	0.0102
222.4084	3.3536	0.0036	0.0078
250.3741	3.5700	0.0008	0.0048
277.5292	3.7636	-0.0024	0.0009
308.9399	3.9689	-0.0050	-0.0027
336.6017	4.1277	0.0010	0.0025
376.1184	4.3508	-0.0053	-0.0050
411.5821	4.5275	-0.0036	-0.0041
449.1737	4.7037	-0.0051	-0.0061
489.9064	4.8773	-0.0037	-0.0048
542.0887	5.0772	0.0019	0.0012
584.6452	5.2289	0.0043	0.0046
AVERAGE DN		0.0039	0.0056

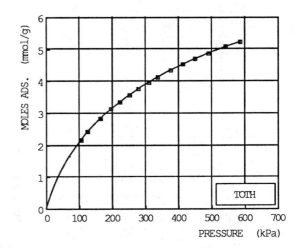

TEMP (K) = 313.15

TOTH		UNILAN		UNIT
B =	18.0310	C =	2965.5701	kPa
M =	0.4999	S =	3.5813	
NI =	11.1780	M =	12.5886	mmol/g
BIS=	8.94138-02	BIS=	5.53838-02	1/g
QIS=	20.1547	QIS=	20.6391	KJ/mol

P(kPa)	N(mmol/g)	DN TOTH	DN UNILAN
127.3655	1.6286	0.0263	0.0095
158.3710	1.8852	0.0017	-0.0055
189.2751	2.0980	-0.0063	-0.0067
228.8932	2.3371	-0.0124	-0.0074
271.1457	2.5750	-0.0304	-0.0223
320.3896	2.7994	-0.0277	-0.0185
367.8098	2.9810	-0.0140	-0.0053
422.5252	3.1796	-0.0100	-0.0032
511.5899	3.3504	0.1076	0.1098
557.3888	3.5646	0.0259	0.0254
609.9765	3.7511	-0.0194	-0.0232
743.8268	4.0951	-0.0471	-0.0588
AVERAGE DN		0.0274	0.0246

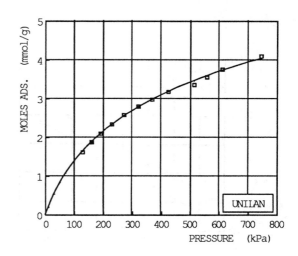

TEMP   (K) =   333.15

TOTH		UNILAN		UNIT
B =	31.9312	C =	2995.7900	kPa
M =	0.5363	S =	3.0179	
NI =	11.2923	M =	11.1885	mmol/g
BIS =	4.90287−02	BIS =	3.49633−02	1/g
QIS =	20.1547	QIS =	20.6391	KJ/mol

P (kPa)	N (mmol/g)	DN TOTH	DN UNILAN
102.3382	1.0575	−0.0569	−0.0786
139.4232	1.1547	0.0918	0.0806
167.5916	1.4233	−0.0100	−0.0147
205.5884	1.6188	−0.0015	0.0004
240.6469	1.7883	−0.0002	0.0057
280.3663	1.9820	−0.0168	−0.0081
322.1122	2.1386	−0.0028	0.0070
364.5674	2.3010	−0.0055	0.0041
407.8331	2.4514	−0.0051	0.0030
460.6234	2.6214	−0.0052	0.0001
521.4184	2.7958	−0.0002	0.0007
594.4738	2.9935	−0.0013	−0.0067
652.5330	3.1300	0.0058	−0.0052
AVERAGE DN		0.0156	0.0165

TEMP   (K) =   363.15

TOTH		UNILAN		UNIT
B =	32.4048	C =	2870.4299	kPa
M =	0.4827	S =	2.4858	
NI =	13.1062	M =	8.0186	mmol/g
BIS =	2.93693−02	BIS =	2.02357−02	1/g
QIS =	20.1547	QIS =	20.6391	KJ/mol

P (kPa)	N (mmol/g)	DN TOTH	DN UNILAN
106.7965	0.6024	0.0065	−0.0115
143.4762	0.7429	0.0191	0.0088
165.6664	0.8491	−0.0015	−0.0075
208.9321	1.0026	−0.0004	0.0009
244.3959	1.1619	−0.0429	−0.0372
297.5915	1.2792	0.0014	0.0114
354.7388	1.4399	−0.0005	0.0114
418.0669	1.5996	0.0010	0.0121
482.5096	1.7227	0.0291	0.0369
603.7957	2.0239	−0.0160	−0.0192
690.0232	2.1738	−0.0020	−0.0160
AVERAGE DN		0.0109	0.0157

DN = (NCAL−NEXP)

```
CARBON DIOXIDE
-------------------------------------------------------------------------------
ZEOLITE MOLECULAR SIEVE
-------------------------------------------------------------------------------
ADSORBENT : 13X;LINDE(UNION CARBIDE);.3 cc/g;525 Sqm/g (N2)
EXP.      : STATIC VOLUMETRIC
REF.      : HYUN,S.H.and DANNER,R.P.;J.Chem.Eng.Data;27,196(1982)
```

```
              TEMP   (K)=   298.15
              PSAT (kPa)=  6439.81
```

	TOTH		UNILAN	UNITS
B =	1.2054	C =	11.5219	kPa
M =	.4166	S =	3.0918	
NI =	5.7976	M =	4.9063	mmol/g
DG =	-79.6692	DG =	-76.2191	J/g
BIS=	.91782+001	BIS=	.37506+001	1/g
QIS=	47.1689	QIS=	37.6840	KJ/mol

P(kPa)	N(mmol/g)	DN TOTH	DN UNILAN
.2400	.4400	-.0804	-.1412
.3700	.6100	-.1303	-.1869
.7800	.7400	.0157	-.0184
.8800	.9300	-.1209	-.1501
2.1600	1.1900	.0928	.1002
2.3900	1.2800	.0641	.0748
2.8000	1.4000	.0429	.0580
3.7800	1.5700	.0689	.0900
7.2500	2.0200	.0751	.0985
8.2600	2.1600	.0299	.0524
15.9100	2.6800	-.0071	.0066
22.1900	2.9900	-.0735	-.0649
23.1600	3.0200	-.0725	-.0645
29.7400	3.1900	-.0632	-.0586
45.7900	3.4400	-.0143	-.0147
54.8600	3.6000	-.0540	-.0565
72.7600	3.7100	.0172	.0110
87.2500	3.8400	-.0009	-.0103
102.3000	3.9200	.0140	.0010
137.8000	4.0200	.0834	.0608
AVERAGE DN		.0561	.0660

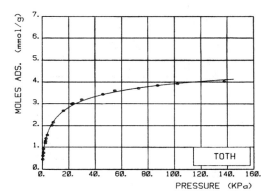

```
              TEMP   (K)=   323.15
```

	TOTH		UNILAN	UNITS
B =	2.3913	C =	38.1007	kPa
M =	.4683	S =	3.0781	
NI =	5.4655	M =	5.0517	mmol/g
BIS=	.22824+001	BIS=	.12540+001	1/g
QIS=	47.1689	QIS=	37.6840	KJ/mol

P(kPa)	N(mmol/g)	DN TOTH	DN UNILAN
.4400	.2100	.0089	-.0268
1.1100	.4900	-.0566	-.0888
1.5900	.6000	-.0473	-.0722
2.5000	.7600	-.0237	-.0356
4.6800	.9700	.0848	.0918
7.7600	1.3400	.0235	.0397
8.6300	1.4100	.0236	.0406
15.1100	1.8400	-.0125	.0022
17.9700	1.9700	-.0142	-.0020
23.3700	2.1400	.0142	.0213
28.3600	2.2900	.0121	.0152
47.6100	2.7100	-.0113	-.0175
51.9300	2.8000	-.0355	-.0427
57.2200	2.8700	-.0325	-.0405
94.0600	3.2200	-.0189	-.0258
113.1000	3.3000	.0301	.0262
137.8000	3.4400	.0240	.0248
AVERAGE DN		.0279	.0361

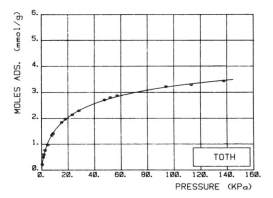

```
              DN=(NCAL-NEXP)
```

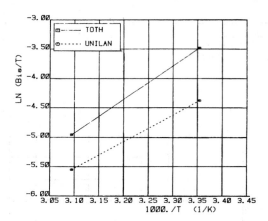

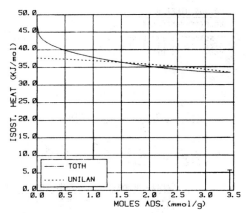

CARBON DIOXIDE

H-MORDENITE

ADSORBENT : Norton Co.; Type Z-900H; 1/16 in Pellets
EXP.   : Static Volumetric
REF.   : Talu, O., Zwiebel, I.; *AIChE J.* **32**, 1263 (1986)

TEMP (K)   =   283.15
PSAT (kPa) =   4504.34

TOTH		UNILAN		UNIT
B =	.6750	C =	554.0828	kPa
M =	.1475	S =	7.1599	
NI =	19.8267	M =	6.5592	mmol/g
DG =	-56.7899	DG =	-48.0814	J/g
BIS=.67044+003		BIS=.25043+001		1/g
QIS=	70.4400	QIS=	40.1533	KJ/mol

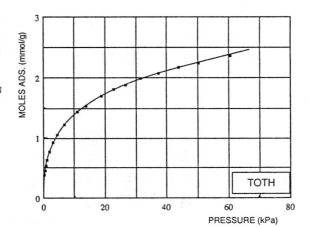

P (kPa)	N (mmol/g)	DN TOTH	DN UNILAN
.32	.38	-.0093	-.1255
.47	.45	-.0115	-.1120
.75	.53	.0040	-.0680
1.12	.63	-.0016	-.0431
1.91	.77	.0037	.0055
3.03	.92	-.0013	.0346
4.48	1.04	.0166	.0749
6.85	1.21	.0128	.0853
10.97	1.43	-.0021	.0707
13.68	1.53	.0017	.0684
18.85	1.70	-.0085	.0414
22.72	1.80	-.0105	.0251
26.59	1.88	-.0052	.0159
31.48	1.98	-.0107	-.0079
37.43	2.07	-.0007	-.0196
43.67	2.16	.0010	-.0397
50.12	2.24	.0050	-.0572
60.76	2.36	.0057	-.0897
AVERAGE DN		.0062	.0547

TEMP (K)   =   303.15
PSAT (kPa) =   7212.62

TOTH		UNILAN		UNIT
B =	1.0065	C =	4459.4580	kPa
M =	.2080	S =	8.0694	
NI =	10.5220	M =	8.1360	mmol/g
DG =	-53.8775	DG =	-48.5616	J/g
BIS=.25707+002		BIS=.91042+000		1/g
QIS=	70.4400	QIS=	40.1533	KJ/mol

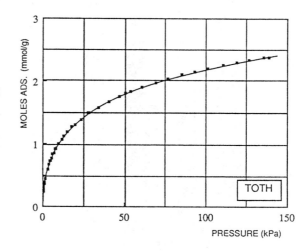

P (kPa)	N (mmol/g)	DN TOTH	DN UNILAN
.37	.25	-.0311	-.1315
.57	.29	-.0133	-.1174
.63	.31	-.0183	-.1222
.72	.32	-.0072	-.1103
.94	.36	-.0015	-.1004
1.20	.40	.0050	-.0872
1.57	.45	.0114	-.0700
2.97	.60	.0192	-.0251
3.63	.68	-.0036	-.0341
4.24	.73	-.0068	-.0264
4.56	.73	.0160	.0015
5.08	.78	.0006	-.0063
6.02	.84	-.0026	.0020
6.88	.86	.0240	.0373
7.51	.92	-.0044	.0143
8.22	.92	.0290	.0530
9.59	1.01	-.0023	.0301
11.74	1.07	.0185	.0602
12.71	1.13	-.0087	.0361

14.67	1.20	−.0178	.0317
17.71	1.28	−.0146	.0391
19.87	1.30	.0179	.0731
23.20	1.39	.0008	.0564
27.50	1.49	−.0166	.0377
28.29	1.49	−.0025	.0513
34.13	1.58	.0025	.0518
40.29	1.67	−.0007	.0424
46.72	1.75	−.0010	.0347
50.46	1.81	−.0188	.0125
53.63	1.83	−.0050	.0224
60.84	1.90	−.0040	.0144
68.94	1.97	−.0023	.0061
76.62	2.04	−.0106	−.0116
85.06	2.10	−.0087	−.0199
93.19	2.15	−.0038	−.0245
101.19	2.20	−.0038	−.0336
110.63	2.25	.0010	−.0392
119.18	2.29	.0072	−.0422
126.52	2.33	.0046	−.0524
135.71	2.37	.0088	−.0574
138.67	2.38	.0124	−.0566
AVERAGE DN		.0095	.0458

TEMP (K)   =   323.15

TOTH		UNILAN		UNIT
B =	1.1926	C = 3250.3440		kPa
M =	.1736	S =	6.6265	
NI =	20.5374	M =	7.3357	mmol/g
BIS=.20000+002		BIS=.34537+000		l/g
QIS =	70.4400	QIS =	40.1533	KJ/mol

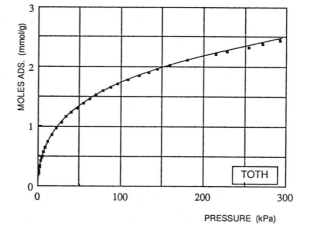

P (kPa)	N (mmol/g)	DN TOTH	DN UNILAN
.63	.21	−.0373	−.1344
1.14	.26	−.0205	−.1300
1.71	.31	−.0131	−.1249
2.17	.34	−.0043	−.1141
2.25	.34	.0019	−.1073
3.53	.42	.0081	−.0886
4.46	.47	.0094	−.0765
4.99	.49	.0158	−.0640
6.74	.56	.0221	−.0386
8.80	.64	.0173	−.0239
11.78	.73	.0181	−.0005
16.48	.84	.0244	.0314
22.06	.96	.0161	.0430
27.92	1.06	.0140	.0541
34.07	1.15	.0123	.0608
40.20	1.22	.0196	.0728
46.91	1.30	.0150	.0705
54.37	1.38	.0101	.0658
62.06	1.45	.0100	.0639
69.91	1.52	.0050	.0558
78.40	1.59	−.0006	.0458
87.41	1.65	.0023	.0430
96.00	1.70	.0078	.0426
108.74	1.78	.0035	.0287
122.19	1.84	.0164	.0310
133.44	1.90	.0127	.0181
143.73	1.96	.0011	−.0020
159.03	2.03	−.0017	−.0176
180.13	2.11	.0030	−.0303
214.47	2.21	.0255	−.0358
228.05	2.25	.0296	−.0425
254.73	2.33	.0305	−.0623
271.62	2.38	.0282	−.0774
292.86	2.44	.0250	−.0963
AVERAGE DN		.0142	.0587

DN= (NCAL−NEXP)

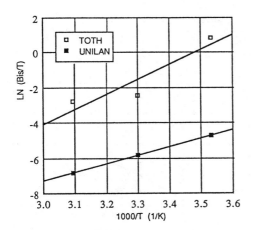

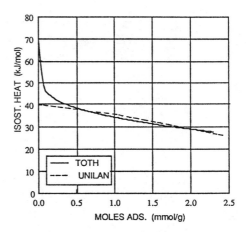

CARBON MONOXIDE

ACTIVATED CARBON

ADSORBENT : COLUMBIA GRADE L;MESH 20x40;1152 Sq m/g (N2)
EXP.   : STATIC VOLUMETRIC
REF.   : RAY,G.C.and BOX,E.O.;IND.ENG.CHEM.;42,1315(1950)

TEMP   (K)=   310.92

	TOTH		UNILAN	UNITS
B =	34.4345	C =	30241.0000	kPa
M =	.3869	S =	3.4922	
NI =	45.8216	M =	18.8938	mmol/g
BIS=	.12621-001	BIS=	.75911-002	1/g
QIS=	15.4365	QIS=	12.3672	KJ/mol

P(kPa)	N(mmol/g)	DN TOTH	DN UNILAN
27.4700	.1148	-.0111	-.0353
59.7400	.2352	-.0279	-.0652
99.9700	.3738	-.0503	-.0951
100.7000	.3788	-.0533	-.0982
1148.0000	2.0430	.1273	.1444
1793.0000	3.0050	-.0739	-.0851
AVERAGE DN		.0573	.0872

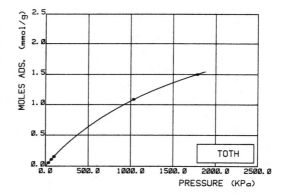

TEMP   (K)=   366.48

	TOTH		UNILAN	UNITS
B =	792.1250	C =	803663.0000	kPa
M =	.8664	S =	7.0348	
NI =	3.7467	M =	16.8211	mmol/g
BIS=	.51478-002	BIS=	.51466-002	1/g
QIS=	15.4365	QIS=	12.3672	KJ/mol

P(kPa)	N(mmol/g)	DN TOTH	DN UNILAN
33.7400	.0545	.0008	.0011
69.8700	.1078	.0038	.0048
99.9700	.1536	.0029	.0044
101.5000	.1545	.0043	.0057
1034.0000	1.0850	-.0045	-.0084
1786.0000	1.5010	.0021	.0047
AVERAGE DN		.0031	.0048

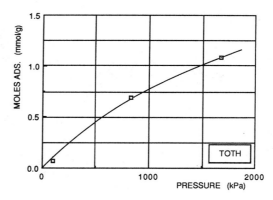

TEMP   (K)=   422.00

	TOTH		UNILAN	UNITS
B =	1668.6800	C =	999993.0000	kPa
M =	.9465	S =	6.6826	
NI =	2.8376	M =	17.3129	mmol/g
BIS=	.39217-002	BIS=	.36285-002	1/g
QIS=	15.4365	QIS=	12.3672	KJ/mol

P(kPa)	N(mmol/g)	DN TOTH	DN UNILAN
99.9700	.0665	.0399	.0329
827.4000	.6912	-.0156	-.0342
1675.0000	1.0800	.0059	.0197
AVERAGE DN		.0205	.0290

TEMP   (K)=   477.59

	TOTH		UNILAN	UNITS
B =	646.8830	C =	999769.0000	kPa
M =	.6848	S =	5.9585	
NI =	7.1149	M =	16.1004	mmol/g
BIS=	.22193-002	BIS=	.20768-002	1/g
QIS=	15.4365	QIS=	12.3672	KJ/mol

P(kPa)	N(mmol/g)	DN TOTH	DN UNILAN
99.9700	.0461	.0070	.0052
675.7000	.3174	-.0031	-.0035
1400.0000	.5837	.0012	.0014
AVERAGE DN		.0037	.0034

DN=(NCAL−NEXP)

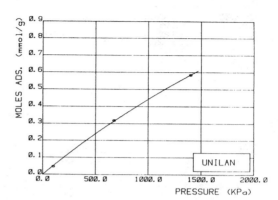

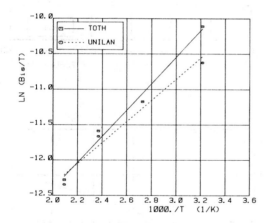

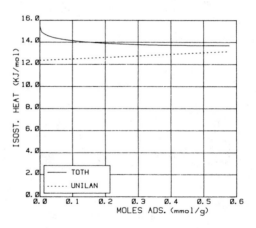

CARBON MONOXIDE

ACTIVATED CARBON

ADSORBENT : PCB, Calgon Corp.
EXP.   : Static
REF.   : Ritter, J. A. and R. T. Yang, *Ind. Eng. Chem.*
            Res. **26**, 1679 (1987).

TEMP(K)   =   296.

TOTH		UNILAN		UNIT
B =	86.3496	C =	1776.0890	kPa
M =	.6533	S =	1.8809	
NI =	7.2357	M =	6.7243	mmol/g
BIS=	.193548-01	BIS=	.158686-01	1/g
QIS=	15.8519	QIS=	15.0655	KJ/mol

P(kPa)	N(mmol/g)	DN TOTH	DN UNILAN
231.6619	1.1064	-0.0256	-0.0366
806.6799	2.3110	0.0299	0.0381
1971.8842	3.4799	0.0173	0.0198
3633.5068	4.3543	-0.0680	-0.0693
5446.8130	4.7291	0.0398	0.0374
AVERAGE DN		0.0361	0.0402

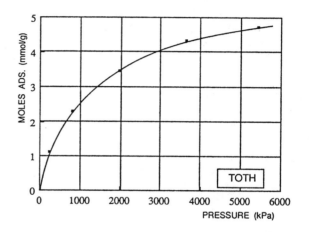

TEMP(K)   =   373.

TOTH		UNILAN		UNIT
B =	703.0723	C =	4598.2060	kPa
M =	.80734	S =	1.3263	
NI =	5.5368	M =	5.4329	mmol/g
BIS=	.510980-02	BIS=	.483681-02	1/g
QIS=	15.8519	QIS=	15.0655	KJ/mol

P(kPa)	N(mmol/g)	DN TOTH	DN UNILAN
383.3453	0.5309	-0.0127	-0.0164
1054.8892	1.1377	0.0158	0.0169
1944.3054	1.7355	-0.0034	-0.0020
3075.0361	2.2485	-0.0058	-0.0062
5667.4434	2.9624	0.0025	0.0021
AVERAGE DN		0.0081	0.0087

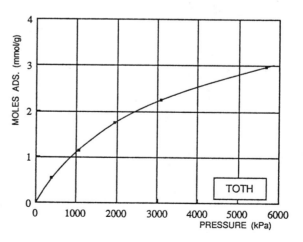

TEMP(K)   =   473.

TOTH	UNILAN	UNIT
B = 717.2849	C = 16420.5700	kPa
M = .7363	S = 2.0197	
NI = 5.0949	M = 5.5734	mmol/g
BIS = .269013-02	BIS = .248253-02	1/g
QIS = 15.8519	QIS = 15.0655	KJ/mol

P(kPa)	N(mmol/g)	DN TOTH	DN UNILAN
479.1816	0.2806	-0.0074	-0.0117
1130.7308	0.5577	0.0072	0.0065
1978.7789	0.8700	-0.0025	-0.0004
3288.7720	1.2313	-0.0001	0.0019
5812.2319	1.7310	0.0008	-0.0014
AVERAGE DN		0.0036	0.0044

DN= (NCAL-NEXP)

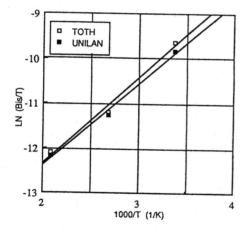

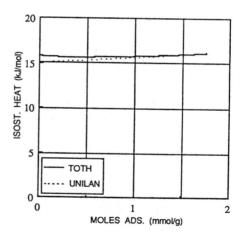

CARBON MONOXIDE ————————————————————————————————————

ZEOLITE ————————————————————————————

ADSORBENT : 10X Linde; 1/16 in. pellets; 672 m^2/gm
EXP.   : Static Volumetric
REF.   : Nolan, J. T., T. W. McKeehan and R. P. Danner,
             J. Chem. Eng. Data, **26**, 112 (1981)

TEMP (K)   =   172.04

TOTH		UNILAN		UNIT
B =	.6299	C =	43.2510	kPa
M =	.27455	S =	5.5801	
NI =	7.5810	M =	7.2396	mmol/g
BIS=.583855+02		BIS=.568704+01		1/g
QIS=	11.2405	QIS=	17.5968	KJ/mol

P (kPa)	N (mmol/g)	DN TOTH	DN UNILAN
0.5733	0.9592	0.0622	0.0185
2.4933	1.8738	-0.1005	-0.0640
10.4266	2.6456	0.0306	0.0608
26.3733	3.2657	0.0346	0.0357
50.6532	3.7297	0.0065	-0.0082
81.8798	4.0554	-0.0067	-0.0249
118.2930	4.2918	-0.0109	-0.0251
132.9997	4.3632	-0.0098	-0.0214
162.6796	4.4837	-0.0076	-0.0130
202.9728	4.6086	-0.0007	0.0033
231.7061	4.6800	0.0053	0.0161
264.1727	4.7558	0.0049	0.0235
	AVERAGE DN	0.0234	0.0262

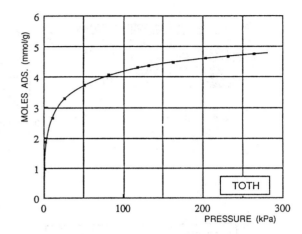

TEMP (K)   =   227.59

TOTH		UNILAN		UNIT
B =	.9959	C =	736.5909	kPa
M =	.13956	S =	5.4744	
NI =	40.7024	M =	6.3019	mmol/g
BIS=.793213+02		BIS=.352643+00		1/g
QIS=	11.2405	QIS=	17.5968	KJ/mol

P (kPa)	N (mmol/g)	DN TOTH	DN UNILAN
0.3333	0.0491	0.1138	0.0099
0.4400	0.1294	0.0595	-0.0527
0.6400	0.2275	0.0021	-0.1191
0.6800	0.3257	-0.0888	-0.2111
0.9467	0.3257	-0.0456	-0.1718
1.6533	0.4194	-0.0511	-0.1726
2.1867	0.4640	-0.0432	-0.1559
2.9067	0.5086	-0.0280	-0.1268
3.9067	0.5755	-0.0255	-0.1050
4.6267	0.5978	-0.0044	-0.0709
5.4000	0.6380	-0.0025	-0.0561
6.4400	0.6826	0.0038	-0.0341
7.4933	0.7183	0.0145	-0.0095
9.8800	0.7986	0.0254	0.0273
12.4400	0.8789	0.0280	0.0506
15.6800	0.9726	0.0241	0.0660
19.5866	1.0618	0.0277	0.0856
23.4399	1.1421	0.0272	0.0956
26.7066	1.2090	0.0211	0.0955
33.2132	1.3206	0.0167	0.0977
39.8266	1.4232	0.0086	0.0914
45.2532	1.4946	0.0067	0.0885
53.1065	1.6061	-0.0141	0.0636
59.7598	1.6596	0.0018	0.0744
66.4265	1.7399	-0.0143	0.0521
73.0398	1.7935	-0.0087	0.0509
79.6798	1.8515	-0.0111	0.0410

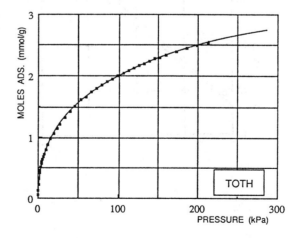

86.2931	1.9050	−0.0126	0.0317
92.5998	1.9496	−0.0103	0.0263
99.3464	2.0032	−0.0162	0.0120
106.4264	2.0433	−0.0089	0.0103
112.9331	2.0879	−0.0120	−0.0012
119.4130	2.1236	−0.0081	−0.0056
126.0530	2.1638	−0.0092	−0.0154
132.7597	2.1995	−0.0071	−0.0220
139.3063	2.2352	−0.0073	−0.0306
146.0530	2.2753	−0.0121	−0.0441
152.6530	2.2931	0.0034	−0.0371
159.3329	2.3333	−0.0040	−0.0531
172.7596	2.3868	0.0050	−0.0609
185.9062	2.4404	0.0093	−0.0730
199.1195	2.4894	0.0152	−0.0832
212.5195	2.5341	0.0236	−0.0910
AVERAGE DN		0.0205	0.0668

TEMP (K)   =   273.15

TOTH	UNILAN	UNIT
B = 1.6826	C = 5108.1920	kPa
M = .16596	S = 6.2494	
NI = 32.2401	M = 5.2322	mmol/g
BIS =.318398+01	BIS =.963471−01	1/g
QIS= 11.2405	QIS = 17.5968	KJ/mol

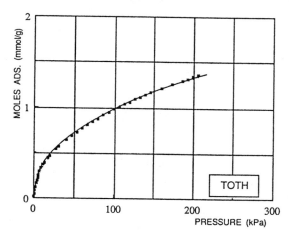

P (kPa)	N (mmol/g)	DN TOTH	DN UNILAN
0.2667	0.0170	0.0187	−0.0058
0.5467	0.0442	0.0132	−0.0216
1.0400	0.0794	0.0070	−0.0375
1.8267	0.1213	0.0009	−0.0502
3.0400	0.1660	−0.0006	−0.0535
3.9733	0.1914	0.0016	−0.0497
5.0267	0.2534	−0.0328	−0.0811
5.1200	0.2222	0.0007	−0.0472
7.0533	0.2945	−0.0282	−0.0687
9.2000	0.3324	−0.0247	−0.0566
11.1600	0.3658	−0.0248	−0.0491
13.3200	0.3930	−0.0187	−0.0354
17.2000	0.4412	−0.0140	−0.0188
19.4533	0.4671	−0.0122	−0.0112
26.8799	0.5367	−0.0019	0.0137
30.4133	0.5644	0.0039	0.0245
39.8266	0.6384	0.0088	0.0383
46.3065	0.6835	0.0117	0.0446
53.0799	0.7254	0.0157	0.0503
59.5865	0.7691	0.0127	0.0478
66.2265	0.8022	0.0185	0.0530
72.2932	0.8401	0.0139	0.0471
79.6131	0.8726	0.0193	0.0503
86.0531	0.9124	0.0111	0.0397
92.9731	0.9405	0.0153	0.0407
99.5864	0.9770	0.0081	0.0302
106.2931	1.0002	0.0134	0.0319
112.8264	1.0346	0.0058	0.0205
119.7464	1.0587	0.0088	0.0194
126.1197	1.0895	0.0022	0.0087
132.8663	1.1104	0.0059	0.0080
139.3730	1.1426	−0.0032	−0.0055
146.0130	1.1658	−0.0037	−0.0105
159.3063	1.2117	−0.0061	−0.0222
172.5862	1.2559	−0.0091	−0.0347
186.3062	1.3009	−0.0139	−0.0494
192.4529	1.3170	−0.0126	−0.0525
199.1595	1.3406	−0.0177	−0.0625
205.5728	1.3567	−0.0164	−0.0660
AVERAGE DN		0.0117	0.0374

DN= (NCAL−NEXP)

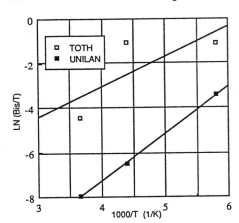 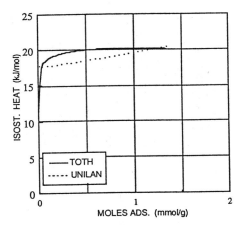

CYCLOHEXANE
```
-----------------------------------------------------------------------
CARBON MOLECULAR SIEVE
-----------------------------------------------------------------------
ADSORBENT : MSC-5A;TAKEDA CHEMICAL Co.;0.56 cc/g;650 Sqm/g
EXP.      : STATIC GRAVIMETRIC (McBAIN QUARTZ SPRING BALANCE)
REF.      : NAKAHARA,T.,HIRATA,M.and OMORI,T.;J.Chem.Eng.Data;19,310(1974)
```

```
            TEMP   (K)=    303.15
            PSAT (kPa)=     16.30
```

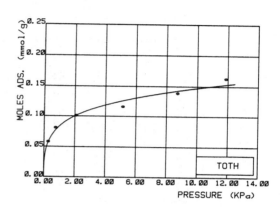

	TOTH		UNILAN	UNITS
B =	.3679	C =	24.6790	kPa
M =	.0988	S =	6.4282	
NI =	1.9895	M =	.3394	mmol/g
DG =	-1.4949	DG =	-1.3104	J/g
BIS=	.12470+006	BIS=	.16688+001	1/g

P(kPa)	N(mmol/g)	DN TOTH	DN UNILAN
.2667	.0582	-.0008	-.0044
.7333	.0808	-.0041	-.0025
2.0800	.1010	.0001	.0039
5.1330	.1164	.0099	.0120
8.7070	.1390	.0039	.0033
11.8700	.1628	-.0095	-.0124
AVERAGE DN		.0047	.0064

```
            DN=(NCAL-NEXP)
```

CYCLOHEXANE

SILICA GEL

ADSORBENT : DAVISON CHEM.Co.
EXP.   : STATIC GRAVIMETRIC
REF.   : SIRCAR,S.and MYERS,A.L.;AIChE J.;19,159(1973)

TEMP (K)=   303.15
PSAT (kPa)=   16.30

	TOTH		UNILAN	UNITS
B =	.8307	C =	95.2152	kPa
M =	.1893	S =	5.6451	
NI =	29.9744	M =	10.1398	mmol/g
DG =	−21.9327	DG =	−20.7077	J/g
BIS=	.20122+003	BIS=	.67256+001	1/g

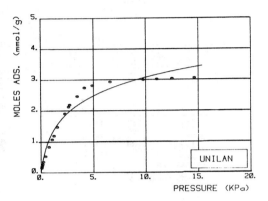

P(kPa)	N(mmol/g)	DN TOTH	DN UNILAN
.0667	.1490	.1536	.0133
.0920	.1860	.1780	.0310
.1240	.2300	.2002	.0517
.1533	.2690	.2141	.0681
.2253	.3480	.2452	.1123
.4400	.5290	.3047	.2219
.7746	.8300	.2632	.2427
1.0670	1.0700	.1960	.2125
1.2470	1.1900	.1673	.2008
1.5600	1.4690	.0283	.0838
2.2800	1.9050	−.1462	−.0632
2.6400	2.1270	−.2587	−.1694
2.7330	2.1910	−.2961	−.2058
3.4530	2.4670	−.3854	−.2926
4.1460	2.7410	−.5046	−.4158
4.8930	2.8110	−.4275	−.3474
6.6130	2.9320	−.2641	−.2131
9.7460	2.9920	.0744	.0610
11.0500	3.0120	.1913	.1502
12.5100	3.0280	.3144	.2425
14.6400	3.0520	.4720	.3562
AVERAGE DN		.2517	.1788

DN=(NCAL−NEXP)

ETHANE

ACTIVATED CARBON

ADSORBENT : Fiber Carbon, KF-1500, Toyobo Co., Ltd.
EXP.   : Gravimetric
REF.   : Kuro-Oka, M., Suzuki, T., Nitta, T., and
            Katayama, T., J. Chem. Eng. Japan, **17**, 588
            (1984)

            TEMP (K)   =   273.15
            PSAT (kPa) =   2386.54

TOTH		UNILAN		UNIT
B =	1.5373	C =	4205.1630	kPa
M =	.3086	S =	7.8692	
NI =	16.2473	M =	21.6139	mmol/g
DG =	-92.5162	DG =	-88.2932	J/g
BIS =	.915869+01	BIS =	.193977+01	1/g
QIS =	32.8373	QIS =	26.7117	KJ/mol

P (kPa)	N (mmol/g)	DN TOTH	DN UNILAN
0.0213	0.0600	-0.0122	-0.0419
0.0381	0.0910	-0.0139	-0.0588
0.0632	0.1360	-0.0207	-0.0830
0.0840	0.1640	-0.0203	-0.0941
0.1450	0.2380	-0.0213	-0.1194
0.2440	0.3380	-0.0227	-0.1440
0.4100	0.4720	-0.0218	-0.1601
0.5960	0.5890	-0.0134	-0.1560
1.0280	0.8160	-0.0074	-0.1371
1.8810	1.1380	0.0109	-0.0741
2.6310	1.3580	0.0218	-0.0267
4.2060	1.7240	0.0327	0.0412
7.6370	2.3000	0.0303	0.1022
11.9100	2.8200	0.0059	0.1040
16.5500	3.2200	0.0085	0.1093
22.6100	3.6400	-0.0037	0.0848
30.3200	4.0800	-0.0391	0.0243
39.8200	4.4700	-0.0372	-0.0080
66.5900	5.2200	-0.0130	-0.0734
99.1900	5.7900	0.0398	-0.1069
AVERAGE DN		0.0198	0.0875

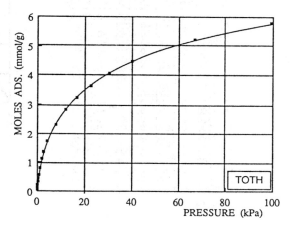

            TEMP (K)   =   298.15
            PSAT (kPa) =   4191.42

TOTH		UNILAN		UNIT
B =	2.3460	C =	3958.5620	kPa
M =	.3110	S =	6.8540	
NI =	18.3913	M =	18.2628	mmol/g
DG =	-93.3327	DG =	-84.1156	J/g
BIS =	.293841+01	BIS =	.790596+00	1/g
QIS =	32.8373	QIS =	26.7117	KJ/mol

P (kPa)	N (mmol/g)	DN TOTH	DN UNILAN
0.0316	0.0300	-0.0058	-0.0200
0.1150	0.0830	-0.0106	-0.0468
0.2490	0.1430	-0.0084	-0.0659
0.3720	0.1960	-0.0124	-0.0823
0.5280	0.2460	-0.0072	-0.0874
0.8620	0.3370	0.0039	-0.0871
1.3640	0.4760	-0.0069	-0.0995
1.8350	0.5740	-0.0017	-0.0889
2.2810	0.6550	0.0046	-0.0745
2.7010	0.7430	-0.0082	-0.0786
3.9000	0.9180	0.0045	-0.0395
5.2550	1.0970	0.0045	-0.0119
8.2410	1.4150	0.0061	0.0365
11.7100	1.6980	0.0173	0.0817
18.4700	2.1500	0.0083	0.1021
26.4300	2.5600	-0.0035	0.0933
38.7500	3.0400	-0.0102	0.0640
51.8400	3.4400	-0.0180	0.0186
79.6600	4.0600	-0.0115	-0.0642
100.0200	4.3800	0.0214	-0.0946
AVERAGE DN		0.0087	0.0669

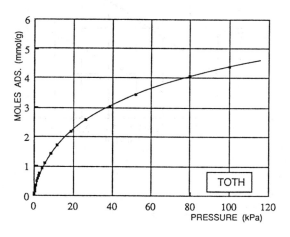

TEMP (K)   =   323.15

TOTH		UNILAN		UNIT
B =	3.3731	C =	3733.3940	kPa
M =	.3089	S =	6.0188	
NI =	22.0645	M =	15.1300	mmol/g
BIS =	.115756+01	BIS =	.371825+00	1/g
QIS =	32.8373	QIS =	26.7117	KJ/mol

P (kPa)	N (mmol/g)	DN TOTH	DN UNILAN
0.0007	0.0030	-0.0027	-0.0029
0.1020	0.0320	-0.0038	-0.0180
0.1810	0.0400	0.0063	-0.0152
0.2680	0.0680	-0.0036	-0.0314
0.5700	0.1150	0.0044	-0.0385
0.7400	0.1500	-0.0030	-0.0515
1.0280	0.1950	-0.0052	-0.0602
1.2800	0.2270	-0.0026	-0.0613
1.8210	0.2910	0.0011	-0.0613
2.8020	0.4000	-0.0010	-0.0620
3.4410	0.4610	0.0001	-0.0572
5.3390	0.6220	0.0003	-0.0408
7.9560	0.8070	0.0008	-0.0162
10.6100	0.9660	0.0023	0.0068
14.5000	1.1700	0.0007	0.0293
19.8500	1.4030	0.0036	0.0533
28.5700	1.7270	-0.0029	0.0604
39.6500	2.0500	0.0028	0.0617
66.1100	2.6600	-0.0093	-0.0032
96.7600	3.1600	0.0052	-0.0735
	AVERAGE DN	0.0031	0.0402

DN= (NCAL-NEXP)

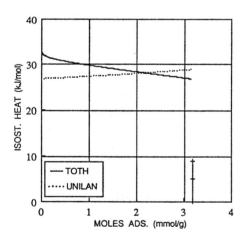

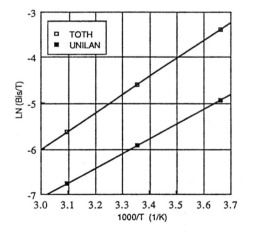

ETHANE

ACTIVATED CARBON

ADSORBENT : ATRITION RESISTANT BEADS; TAIYO KAKEN Co.,
            JAPAN; 800–1200 g/m2
EXP.   : STATIC VOLUMETRIC
REF.   : KAUL, B. K., *IND. ENG. CHEM. RES.* **26**, 928 (1987)

                 TEMP (K)   =   310.95

TOTH		UNILAN		UNIT
B =	3.6201	C =	299.9969	kPa
M =	.3820	S =	3.4782	
NI =	10.4034	M =	8.4169	mmol/g
BIS =	.92699+000	BIS =	.33752+000	1/g
QIS =	33.8425	QIS =	25.5038	KJ/mol

P(kPa)	N(mmol/g)	DN TOTH	DN UNILAN
0.5516	0.1651	-0.0476	-0.0951
0.6205	0.1655	-0.0361	-0.0871
2.8268	0.4702	-0.0585	-0.1482
2.8958	0.4823	-0.0635	-0.1534
2.8958	0.4238	-0.0050	-0.0949
3.2405	0.4341	0.0192	-0.0714
11.8589	1.0181	0.0252	-0.0219
13.0999	1.0614	0.0439	0.0038
14.4789	1.2041	-0.0339	-0.0668
15.7199	1.2590	-0.0334	-0.0601
19.1673	1.3393	0.0272	0.0156
21.5804	1.4062	0.0496	0.0469
37.9898	1.9461	-0.0131	0.0213
39.4377	1.9336	0.0338	0.0701
45.7808	2.0629	0.0453	0.0881
53.9166	2.2650	0.0038	0.0518
59.1565	2.3552	0.0074	0.0574
64.2586	2.4306	0.0173	0.0684
101.3521	2.9619	-0.0198	0.0268
127.0004	3.2305	-0.0304	0.0074
170.9886	3.6052	-0.0545	-0.0342
203.3247	3.8109	-0.0516	-0.0442
223.7330	3.8881	-0.0127	-0.0130
254.9660	4.0576	-0.0226	-0.0341
281.9933	4.1308	0.0279	0.0073
305.0905	4.2771	-0.0216	-0.0495
339.2192	4.3338	0.0524	0.0145
396.4453	4.5408	0.0376	-0.0154
AVERAGE DN		0.0320	0.0525

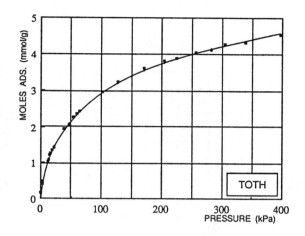

                 TEMP (K)   =   422.05

TOTH		UNILAN		UNIT
B =	53.6899	C =	1446.1960	kPa
M =	.6401	S =	2.4524	
NI =	5.7624	M =	5.9841	mmol/g
BIS =	.40111-001	BIS =	.34132-001	1/g
QIS =	33.8425	QIS =	25.5038	KJ/mol

P(kPa)	N(mmol/g)	DN TOTH	DN UNILAN
10.5489	0.1155	-0.0093	-0.0171
13.3068	0.1423	-0.0108	-0.0194
28.8198	0.2659	-0.0047	-0.0139
31.8535	0.2931	-0.0085	-0.0175
41.7819	0.3640	-0.0063	-0.0140
57.5018	0.4693	-0.0050	-0.0102
69.2228	0.5282	0.0095	0.0062
70.1191	0.5412	0.0020	-0.0012
89.8379	0.6563	0.0003	0.0000
100.5247	0.6991	0.0148	0.0158
123.0704	0.8209	0.0057	0.0089
143.6166	0.8963	0.0248	0.0293
174.5049	1.0297	0.0213	0.0267
201.0495	1.1292	0.0237	0.0291
225.9393	1.3036	-0.0622	-0.0573
249.6571	1.3005	0.0200	0.0240

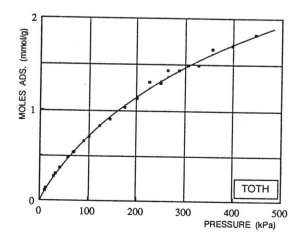

```
264.4117  1.4357  -0.0684  -0.0650
287.6469  1.4361   0.0014   0.0037
304.4699  1.4932  -0.0071  -0.0058
326.8088  1.4928   0.0550   0.0550
355.7665  1.6663  -0.0429  -0.0448
397.8242  1.7060   0.0197   0.0150
445.3976  1.8207   0.0113   0.0033
    AVERAGE DN      0.0189   0.0210
```

DN= (NCAL−NEXP)

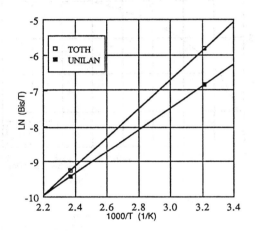

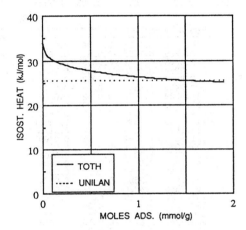

ETHANE

ACTIVATED CARBON

ADSORBENT : BPL;PITTSBURGH ACTIVATED CARBON Co;1050-1150 Sqm/g;MESH 20X40
EXP.   : STATIC GRAVIMETRIC (McBAIN QUARTZ SPRING)
REF.   : LAUKHUF,W.L.S. and PLANK,C.A.;J.CHEM.ENG.DATA; 14,48(1969)

TEMP   (K)=   303.15
PSAT (kPa)=   4652.80

TOTH		UNILAN		UNITS
B =	2.0101	C =	6780.2500	kPa
M =	.2887	S =	7.2869	
NI =	14.4656	M =	15.2659	mmol/g
DG =	-73.4384	DG =	-66.5030	J/g
BIS=	.32479+001	BIS=	.56888+000	1/g
QIS=	34.2299	QIS=	18.1663	KJ/mol

P(kPa)	N(mmol/g)	DN TOTH	DN UNILAN
4.2000	.8646	-.0899	-.1895
8.8000	1.1310	.0275	-.0171
12.6000	1.4300	-.0415	-.0553
21.0000	1.7290	.0393	.0615
23.0700	1.8620	-.0170	.0098
33.2000	2.0620	.0997	.1362
34.8700	2.2280	-.0212	.0154
44.7300	2.3940	.0501	.0828
48.3300	2.5270	-.0062	.0235
58.4700	2.7270	-.0085	.0103
64.2700	2.7930	.0212	.0330
72.6700	2.9600	-.0140	-.0135
77.0000	3.0260	-.0169	-.0224
88.4000	3.1930	-.0306	-.0525
101.4000	3.3260	-.0076	-.0485
103.7000	3.3590	-.0148	-.0590
AVERAGE DN		.0316	.0519

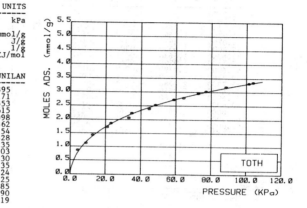

TEMP   (K)=   313.15

TOTH		UNILAN		UNITS
B =	2.2592	C =	6664.3600	kPa
M =	.3024	S =	7.1227	
NI =	13.2376	M =	14.3700	mmol/g
BIS=	.23273+001	BIS=	.48858+000	1/g
QIS=	34.2299	QIS=	18.1663	KJ/mol

P(kPa)	N(mmol/g)	DN TOTH	DN UNILAN
4.8670	.7649	-.0330	-.1146
7.0000	.8646	.0358	-.0234
14.6000	1.3300	-.0018	-.0058
17.8700	1.4630	.0044	.0141
24.8000	1.6960	.0177	.0443
31.3300	1.9290	-.0242	.0089
36.2000	2.0290	.0004	.0346
44.0700	2.2280	-.0214	.0104
49.5300	2.2610	-.0549	.0831
56.3300	2.4610	-.0213	.0009
61.5300	2.5270	-.0002	.0165
74.0700	2.7270	-.0120	-.0102
77.0000	2.7600	-.0047	-.0067
88.2700	2.9270	-.0277	-.0443
88.8000	2.9270	-.0213	-.0386
102.7000	3.0260	.0371	.0012
103.9000	3.0600	.0158	-.0216
AVERAGE DN		.0196	.0282

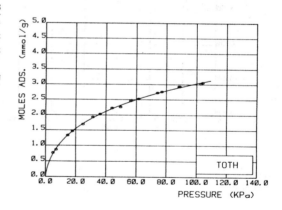

TEMP   (K)=   323.15

	TOTH		UNILAN	UNITS
B =	2.5916	C =	7733.7300	kPa
M =	.3226	S =	7.1455	
NI =	10.6268	M =	12.5779	mmol/g
BIS=	.14916+001	BIS=	.38780+000	1/g
QIS=	34.2299	QIS=	18.1663	KJ/mol

P(kPa)	N(mmol/g)	DN TOTH	DN UNILAN
4.5470	.5653	-.0090	-.0749
7.0670	.7316	-.0078	-.0542
12.3300	.9644	.0228	.0090
21.5300	1.2640	.0500	.0658
24.0700	1.3970	-.0098	.0102
35.4000	1.6300	.0304	.0579
35.6000	1.6630	.0016	.0291
47.7300	1.9290	-.0360	-.0122
49.6000	1.9290	-.0048	.0179
59.2700	2.0950	-.0226	-.0072
65.7300	2.2280	-.0667	-.0573
70.6700	2.2610	-.0363	-.0317
76.6000	2.3280	-.0317	-.0332
88.4000	2.3940	.0323	.0182
88.8000	2.3940	.0365	.0219
100.9000	2.5270	.0224	-.0055
102.0000	2.5270	.0326	.0035
AVERAGE DN		.0267	.0300

DN=(NCAL-NEXP)

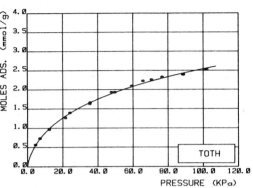

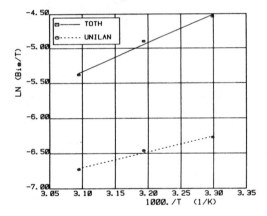

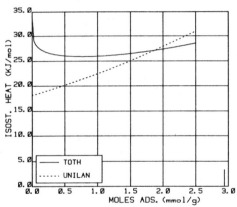

ETHANE
```
------------------------------------------------------------------------
```
ACTIVATED CARBON
```
------------------------------------------------------------------------
```
ADSORBENT : COLUMBIA GRADE L;MESH 20x40;1152 Sq m/g (N2)
EXP.   : STATIC VOLUMETRIC
REF.   : RAY,G.C.and BOX,E.O.;IND.ENG.CHEM.;42,1315(1950)

TEMP   (K)=   310.92

	TOTH		UNILAN		UNITS
B =	4.0447	C =	96.3086		kPa
M =	.4706	S =	2.8078		
NI =	7.7259	M =	6.8212		mmol/g
BIS=	.10255+001	BIS=	.53836+000		1/g
QIS=	30.1551	QIS=	28.5112		KJ/mol

P(kPa)	N(mmol/g)	DN TOTH	DN UNILAN
6.6670	1.0570	-.0876	-.1338
27.2000	2.0470	.0328	.0425
58.0100	2.8300	.0225	.0374
94.6700	3.3710	.0115	.0211
99.9700	3.4050	.0368	.0457
310.3000	4.6090	.0268	.0295
555.0000	5.1610	.0220	.0272
775.7000	5.7400	-.2709	-.2674
1155.0000	5.7500	.0294	.0230
1469.0000	5.8060	.1448	.1265
AVERAGE DN		.0685	.0754

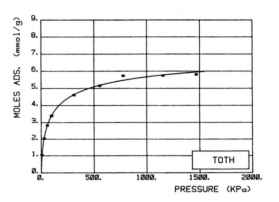

TEMP   (K)=   338.70

	TOTH		UNILAN		UNITS
B =	6.5217	C =	85689.7998		kPa
M =	.5023	S =	9.2185		
NI =	6.6289	M =	17.6783		mmol/g
BIS=	.44661+000	BIS=	.31769+000		1/g
QIS=	30.1551	QIS=	28.5112		KJ/mol

P(kPa)	N(mmol/g)	DN TOTH	DN UNILAN
8.4010	.6698	-.0309	-.0107
28.5400	1.3500	.0158	.0614
59.0700	1.9440	.0241	.0439
98.5400	2.4500	-.0036	-.0210
99.9700	2.4730	-.0128	-.0313
448.2000	3.9070	.0030	-.0869
834.3000	4.4590	-.0134	-.0514
1162.0000	4.7200	-.0170	.0025
1493.0000	4.8590	.0244	.1023
AVERAGE DN		.0161	.0457

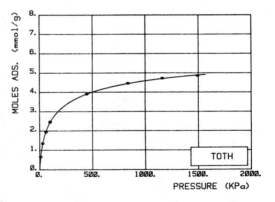

TEMP   (K)=   366.48

TOTH		UNILAN		UNITS
B =	12.3766	C =	9839.1899	kPa
M =	.5457	S =	6.1158	
NI =	6.2915	M =	12.5999	mmol/g
BIS=	.19072+000	BIS=	.14449+000	1/g
QIS=	30.1551	QIS=	28.5112	KJ/mol

P(kPa)	N(mmol/g)	DN TOTH	DN UNILAN
7.5980	.3439	-.0253	-.0349
34.5400	.9744	-.0154	.0059
65.7300	1.4290	-.0179	.0057
99.0700	1.7520	.0017	.0154
99.9700	1.7860	-.0243	-.0110
265.4000	2.6420	.0516	.0172
493.0000	3.2510	.0584	.0095
775.7000	3.7580	-.0164	-.0466
1107.0000	4.0960	-.0362	-.0268
1458.0000	4.3050	-.0148	.0430
AVERAGE DN		.0262	.0216

TEMP   (K)=   394.20

TOTH		UNILAN		UNITS
B =	10.2724	C =	=694262.0000	kPa
M =	.4576	S =	9.7051	
NI =	7.4878	M =	20.5500	mmol/g
BIS=	.15107+000	BIS=	.81973-001	1/g
QIS=	30.1551	QIS=	28.5112	KJ/mol

P(kPa)	N(mmol/g)	DN TOTH	DN UNILAN
9.4660	.2531	.0047	-.0394
25.0600	.5816	-.0490	-.0893
69.8700	1.0270	.0100	.0050
99.9700	1.2590	.0154	.0247
100.0000	1.2620	.0126	.0219
258.6000	2.0380	.0139	.0386
506.8000	2.7130	-.0102	.0004
803.2000	3.1840	-.0106	-.0137
1127.0000	3.5590	-.0365	-.0454
1493.0000	3.7730	.0368	.0289
AVERAGE DN		.0200	.0307

TEMP   (K)=   422.00

TOTH		UNILAN		UNITS
B =	82.4663	C =	9447.7000	kPa
M =	.7447	S =	4.6661	
NI =	5.3722	M =	11.7363	mmol/g
BIS=	.50345-001	BIS=	.49633-001	1/g
QIS=	30.1551	QIS=	28.5112	KJ/mol

P(kPa)	N(mmol/g)	DN TOTH	DN UNILAN
20.0000	.2745	-.0259	-.0193
43.4700	.5149	-.0273	-.0143
71.7400	.7147	.0149	.0293
97.8700	.8845	.0373	.0490
99.9700	.8909	.0453	.0567
396.4000	2.2470	-.0684	-.1129
792.9000	2.9290	-.0023	-.0444
1117.0000	3.2260	.0568	.0519
1493.0000	3.5940	-.0285	.0247
AVERAGE DN		.0341	.0447

TEMP (K)= 449.80

	TOTH		UNILAN	UNITS
B =	27.7999	C =	703092.0000	kPa
M =	.5210	S =	8.5047	
NI =	7.2037	M =	19.8887	mmol/g
BIS=	.45569-001	BIS=	.30711-001	1/g
QIS=	30.1551	QIS=	28.5112	KJ/mol

P(kPa)	N(mmol/g)	DN TOTH	DN UNILAN
99.9700	.6400	.0020	-.0181
444.7000	1.6540	-.0110	.0025
761.9000	2.1300	.0229	.0316
1117.0000	2.5700	-.0201	-.0211
1493.0000	2.8570	.0064	-.0027
	AVERAGE DN	.0125	.0152

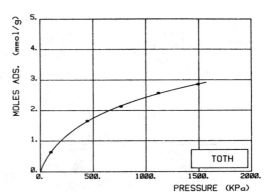

TEMP (K)= 477.59

	TOTH		UNILAN	UNITS
B =	37.3320	C =	699638.0000	kPa
M =	.5021	S =	7.7870	
NI =	9.1193	M =	20.5649	mmol/g
BIS=	.26790-001	BIS=	.18054-001	1/g
QIS=	30.1551	QIS=	28.5112	KJ/mol

P(kPa)	N(mmol/g)	DN TOTH	DN UNILAN
99.9700	.4199	-.0012	-.0293
441.3000	1.2080	.0045	.0122
761.9000	1.7000	-.0133	.0000
1127.0000	2.0730	.0167	.0203
1493.0000	2.4160	-.0073	-.0194
	AVERAGE DN	.0086	.0162

DN=(NCAL-NEXP)

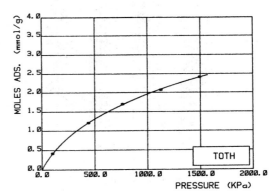

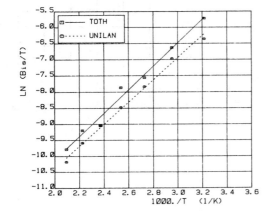

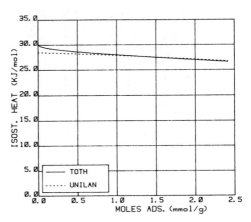

ETHANE
```
--------------------------------------------------------------------------------
```
ACTIVATED CARBON
```
--------------------------------------------------------------------------------
```
ADSORBENT : BPL;PITTSBURGH CHEMICAL Co.;988 Sqm/g
EXP.   : STATIC VOLUMETRIC
REF.   : REICH,R.,ZIEGLER,W.T.and ROGERS,K.A.;Ind.Eng.Chem.Process Des.Dev.
              19,336(1980)

                    TEMP (K)=   212.70
                    PSAT (kPa)= 373.40
```

|          | TOTH |          | UNILAN | UNITS |
|----------|---------|----------|-----------|--------|
| B  =     | .6040   | C  =     | 5.2000    | kPa    |
| M  =     | .3600   | S  =     | 4.0182    |        |
| NI =     | 8.7700  | M  =     | 7.8000    | mmol/g |
| DG =     | -60.9399 | DG =    | -59.8563  | J/g    |
| BIS= | .62923+002 | BIS= | .18346+002 | 1/g |
| QIS= | 31.1949 | QIS= | 26.5089 | KJ/mol |

| P(kPa) | N(mmol/g) | DN TOTH | DN UNILAN |
|--------|-----------|---------|-----------|
| .6890  | 1.8690 | .1704  | .1905  |
| .6890  | 1.8570 | .1824  | .2025  |
| .6890  | 1.8590 | .1804  | .2005  |
| 1.3780 | 2.9600 | -.3081 | -.2898 |
| 3.4470 | 3.6880 | -.1524 | -.1726 |
| 3.4470 | 3.6950 | -.1594 | -.1796 |
| 11.0300 | 4.7210 | -.0495 | -.1193 |
| 13.1000 | 4.8500 | -.0167 | -.0893 |
| 24.8200 | 5.3460 | .0625  | -.0053 |
| 47.5700 | 5.8440 | .0988  | .0585  |
| 73.7700 | 6.1260 | .1439  | .1289  |
| 108.2000 | 6.3700 | .1624 | .1683  |
| 140.7000 | 6.5630 | .1370 | .1537  |
| 199.9000 | 6.8270 | .0814 | .1052  |
| 239.2000 | 7.0100 | -.0018 | .0213 |
| 292.3000 | 7.2020 | -.0875 | -.0692 |
| 296.5000 | 7.2540 | -.1322 | -.1143 |
| 310.3000 | 7.3170 | -.1719 | -.1558 |
| AVERAGE DN |     | .1277  | .1347  |

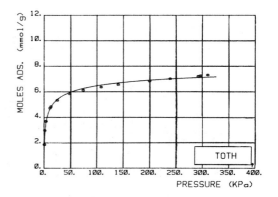

```
                    TEMP  (K)=   260.20
                    PSAT (kPa)=  1720.00
```

	TOTH		UNILAN	UNITS
B =	1.6400	C =	37.7920	kPa
M =	.3970	S =	3.4180	
NI =	7.8490	M =	6.8747	mmol/g
DG =	-58.7293	DG =	-57.9446	J/g
BIS=	.48839+001	BIS=	.17544+001	1/g
QIS=	31.1949	QIS=	26.5089	KJ/mol

P(kPa)	N(mmol/g)	DN TOTH	DN UNILAN
6.2050	1.7960	.0034	.0020
6.8900	1.8370	.0468	.0490
19.3100	2.8320	-.0350	-.0241
35.8500	3.4080	-.0207	-.0203
74.4600	4.1100	-.0273	-.0370
107.6000	4.4220	-.0014	-.0106
157.9000	4.7550	.0031	-.0009
281.3000	5.2360	-.0049	.0049
382.7000	5.4600	.0036	.0206
570.9000	5.7380	.0059	.0275
710.8000	5.8630	.0239	.0440
873.6000	6.0150	-.0004	.0149
1064.0000	6.1240	.0065	.0136
1189.0000	6.2250	-.0318	-.0310
1365.0000	6.3060	-.0375	-.0462
AVERAGE DN		.0168	.0231

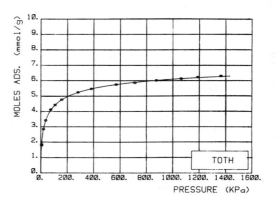
```

TEMP   (K)=   301.40
PSAT (kPa)=   4484.00

TOTH		UNILAN		UNITS
B =	6.7940	C =	119.3310	kPa
M =	.5360	S =	2.5155	
NI =	6.5170	M =	5.9514	mmol/g
DG =	-55.2703	DG =	-54.8879	J/g
BIS=	.45764+000	BIS=	.30534+000	1/g
QIS=	31.1949	QIS=	26.5089	KJ/mol

P(kPa)	N(mmol/g)	DN TOTH	DN UNILAN
12.4100	1.0130	-.0334	-.0445
31.0300	1.6200	.0454	.0579
64.8100	2.3880	-.0346	-.0211
96.5200	2.7490	.0054	.0135
142.0000	3.1760	-.0279	-.0254
170.3000	3.3290	.0030	.0035
326.1000	3.9650	-.0016	-.0017
437.1000	4.2260	.0016	.0045
541.9000	4.4170	-.0058	-.0003
684.0000	4.5980	.0014	.0094
768.1000	4.7080	-.0191	-.0104
938.4000	4.8390	-.0026	-.0064
1038.0000	4.9250	-.0177	-.0093
1211.0000	5.0140	-.0026	.0038
1373.0000	5.1000	-.0077	-.0040
1711.0000	5.2180	.0079	.0041
AVERAGE DN		.0136	.0137

DN=(NCAL-NEXP)

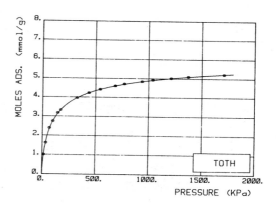

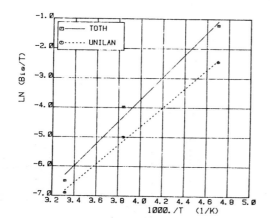

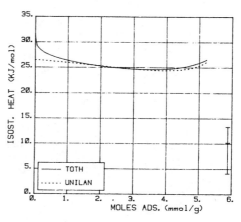

ETHANE

ACTIVATED CARBON

ADSORBENT : NUXIT-AL, (1.5 mm DIAMETER, 3-4 mm LONG)
EXP.   : STATIC VOLUMETRIC
REF.   : SZEPESY, L., ILLES, V., *Acta Chim. Hung.*, **35**, 37(1963)

          TEMP   (K) =   293.15
          PSAT (kPa) = 3766.7

TOTH		UNILAN		UNIT
B =	2.4909	C =	864.1530	kPa
M =	0.3548	S =	5.3523	
NI =	9.8668	M =	10.5617	mmol/g
DG =	-62.2830	DG =	-59.8679	J/g
BIS=1.83643+00		BIS=.587434+00		1/g
QIS=	35.6024	QIS=	26.1186	KJ/mol

P(kPa)	N(mmol/g)	DN TOTH	DN UNILAN
0.5600	0.2432	-0.0530	-0.1167
2.3465	0.5470	-0.0267	-0.1000
2.5331	0.4850	0.0617	-0.0098
4.5729	0.8049	-0.0178	-0.0649
4.7329	0.8018	0.0015	-0.0438
7.7060	1.0642	-0.0045	-0.0197
9.5325	1.1690	0.0190	0.0174
11.7723	1.3238	0.0016	0.0127
15.1987	1.4979	0.0075	0.0316
17.2385	1.6063	-0.0069	0.0224
24.3446	1.8785	-0.0036	0.0338
28.4642	1.9802	0.0277	0.0655
32.0373	2.1167	-0.0050	0.0317
37.0635	2.2399	0.0035	0.0371
43.9563	2.4108	-0.0084	0.0189
50.0624	2.5402	-0.0132	0.0075
54.2621	2.6169	-0.0114	0.0044
60.4615	2.7200	-0.0074	0.0010
68.3275	2.8458	-0.0103	-0.0115
72.5938	2.8967	0.0005	-0.0059
85.0728	3.0658	-0.0051	-0.0263
85.2327	3.0497	0.0130	-0.0085
101.9913	3.2393	0.0117	-0.0286
113.2570	3.3576	0.0049	-0.0473
AVERAGE DN		0.0136	0.0320

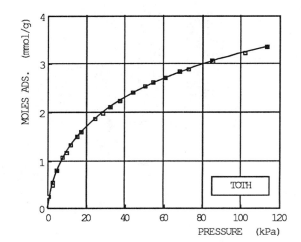

          TEMP   (K) =   313.15

TOTH		UNILAN		UNIT
B =	3.0056	C =	1133.8199	kPa
M =	0.3405	S =	4.9383	
NI =	10.4184	M =	9.4832	mmol/g
BIS=.107095+01		BIS=.307638+00		1/g
QIS=	35.6024	QIS=	26.1186	KJ/mol

P(kPa)	N(mmol/g)	DN TOTH	DN UNILAN
13.3322	0.9821	-0.0121	-0.0498
19.6650	1.1958	-0.0002	-0.0152
29.0642	1.4537	0.0026	0.0065
30.5974	1.4880	0.0051	0.0110
36.7969	1.6246	0.0056	0.0170
37.7301	1.6442	0.0052	0.0171
49.0625	1.8570	0.0017	0.0162
55.1953	1.9548	0.0026	0.0163
65.3944	2.1060	-0.0016	0.0084
80.3932	2.2925	-0.0016	0.0000
93.4587	2.4326	-0.0006	-0.0080
94.1253	2.4407	-0.0019	-0.0097
114.7902	2.6312	-0.0003	-0.0236
AVERAGE DN		0.0032	0.0153

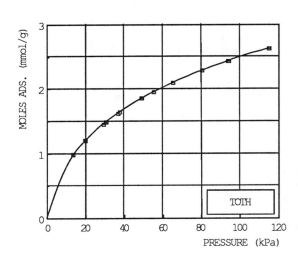

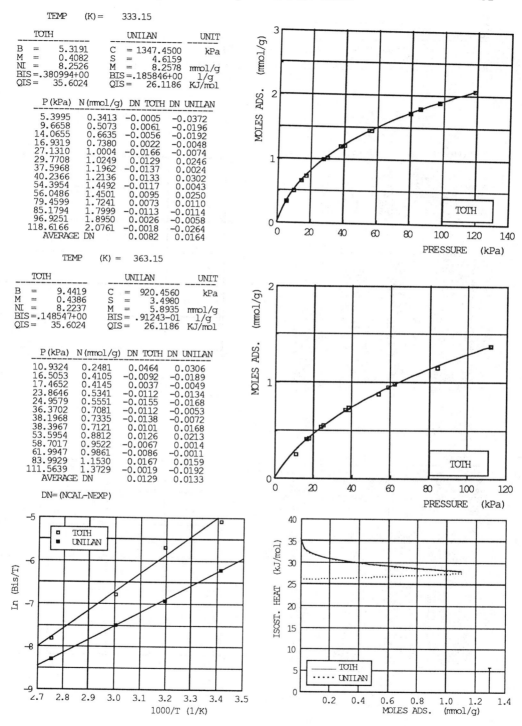

TEMP (K) = 333.15

TOTH		UNILAN		UNIT
B =	5.3191	C =	1347.4500	kPa
M =	0.4082	S =	4.6159	
NI =	8.2526	M =	8.2578	mmol/g
BIS =	.380994+00	BIS =	.185846+00	1/g
QIS =	35.6024	QIS =	26.1186	KJ/mol

P (kPa)	N (mmol/g)	DN TOTH	DN UNILAN
5.3995	0.3413	−0.0005	−0.0372
9.6658	0.5073	0.0061	−0.0196
14.0655	0.6635	−0.0056	−0.0192
16.9319	0.7380	0.0022	−0.0048
27.1310	1.0004	−0.0166	−0.0074
29.7708	1.0249	0.0129	0.0246
37.5968	1.1962	−0.0137	0.0024
40.2366	1.2136	0.0133	0.0302
54.3954	1.4492	−0.0117	0.0043
56.0486	1.4501	0.0095	0.0250
79.4599	1.7241	0.0073	0.0110
85.1794	1.7999	−0.0113	−0.0114
96.9251	1.8950	0.0026	−0.0058
118.6166	2.0761	−0.0018	−0.0264
AVERAGE DN		0.0082	0.0164

TEMP (K) = 363.15

TOTH		UNILAN		UNIT
B =	9.4419	C =	920.4560	kPa
M =	0.4386	S =	3.4980	
NI =	8.2237	M =	5.8935	mmol/g
BIS =	.148547+00	BIS =	.91243−01	1/g
QIS =	35.6024	QIS =	26.1186	KJ/mol

P (kPa)	N (mmol/g)	DN TOTH	DN UNILAN
10.9324	0.2481	0.0464	0.0306
16.5053	0.4105	−0.0092	−0.0189
17.4652	0.4145	0.0037	−0.0049
23.8646	0.5341	−0.0112	−0.0134
24.9579	0.5551	−0.0155	−0.0168
36.3702	0.7081	−0.0112	−0.0053
38.1968	0.7335	−0.0138	−0.0072
38.3967	0.7121	0.0101	0.0168
53.5954	0.8812	0.0126	0.0213
58.7017	0.9522	−0.0067	0.0014
61.9947	0.9861	−0.0086	−0.0011
83.9929	1.1530	0.0167	0.0159
111.5639	1.3729	−0.0019	−0.0192
AVERAGE DN		0.0129	0.0133

DN= (NCAL−NEXP)

ETHANE

ACTIVATED CARBON

ADSORBENT : NUXIT–AL, (1.5 mm DIAMETER, 3–4 mm LONG)
EXP.   : STATIC VOLUMETRIC
REF.   : SZEPESY, L., ILLES, V., *Acta Chim. Hung.*, **35,** 53(1963)

TEMP   (K) =   293.15
PSAT (kPa) = 3766.70

TOTH			UNILAN			UNIT
B	=	4.7773	C	=	877.3510	kPa
M	=	0.4914	S	=	5.3680	
NI	=	7.3022	M	=	10.7923	mmol/g
DG	=	−58.6737	DG	=	−61.0069	J/g
BIS	=.738307+00		BIS	=.598831+00		1/g
QIS	=	26.0453	QIS	=	30.4603	KJ/mol

P(kPa)	N(mmol/g)	DN TOTH	DN UNILAN
108.0124	3.2438	0.0510	0.0835
118.2463	3.4165	−0.0252	−0.0014
137.7007	3.5892	−0.0360	−0.0258
158.3710	3.7101	−0.0094	−0.0101
175.1909	3.8096	−0.0032	−0.0106
194.3413	3.9140	0.0001	−0.0131
223.1176	4.0577	−0.0020	−0.0208
241.9641	4.1362	0.0016	−0.0193
272.7669	4.2522	0.0054	−0.0169
303.2657	4.3526	0.0093	−0.0124
362.0342	4.5221	0.0101	−0.0064
426.1729	4.6756	0.0087	0.0020
487.4745	4.8189	−0.0128	−0.0076
555.2610	4.9201	0.0005	0.0206
631.7614	5.0388	−0.0078	0.0303
	AVERAGE DN	0.0122	0.0187

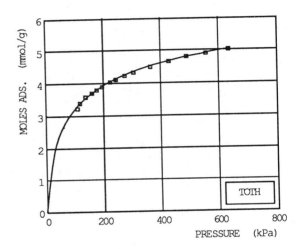

TEMP   (K) =   313.15

TOTH			UNILAN			UNIT
B	=	6.0052	C	=	506.8200	kPa
M	=	0.4755	S	=	3.8245	
NI	=	7.5622	M	=	8.5655	mmol/g
BIS	=.453889+00		BIS	=.263409+00		1/g
QIS	=	26.0453	QIS	=	30.4603	KJ/mol

P(kPa)	N(mmol/g)	DN TOTH	DN UNILAN
96.7654	2.5482	−0.0173	−0.0031
110.5456	2.6575	0.0111	0.0216
127.0615	2.8097	0.0051	0.0112
147.4279	2.9690	0.0035	0.0050
196.3678	3.2733	0.0064	−0.0003
215.8223	3.3821	−0.0009	−0.0097
248.7529	3.5338	−0.0001	−0.0112
282.8994	3.6895	−0.0181	−0.0303
331.0287	3.8386	−0.0001	−0.0118
384.5284	3.9970	−0.0009	−0.0101
446.1339	4.1491	0.0012	−0.0031
507.4356	4.2763	0.0057	0.0078
576.8432	4.4151	−0.0041	0.0064
658.2072	4.5418	−0.0005	0.0210
	AVERAGE DN	0.0054	0.0109

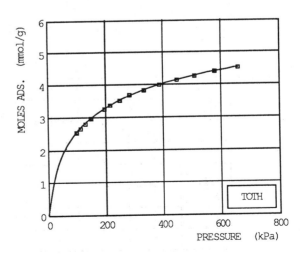

TEMP   (K) =   333.15

TOTH		UNILAN		UNIT
B =	8.1114	C =	1042.5000	kPa
M =	0.4653	S =	3.7029	
NI =	8.0199	M =	8.8835	mmol/g
BIS =.247089+00		BIS =.129209+00		1/g
QIS =	26.0453	QIS =	30.4603	KJ/mol

P(kPa)	N(mmol/g)	DN TOTH	DN UNILAN
99.9064	1.9052	−0.0005	−0.0048
120.4754	2.0837	−0.0011	−0.0019
141.4497	2.2434	−0.0025	−0.0014
164.1465	2.3916	0.0007	0.0025
183.1956	2.5005	0.0059	0.0078
208.9321	2.6437	0.0016	0.0032
238.8230	2.7909	−0.0019	−0.0010
269.6258	2.9225	−0.0014	−0.0014
305.3935	3.0573	0.0010	−0.0000
346.7341	3.1992	−0.0001	−0.0019
390.5066	3.3295	0.0023	−0.0002
477.9500	3.5570	0.0010	−0.0014
544.1152	3.7096	−0.0065	−0.0078
603.2890	3.8243	−0.0059	−0.0056
670.3662	3.9287	0.0070	0.0096
AVERAGE DN		0.0026	0.0034

TEMP   (K) =   363.15

TOTH		UNILAN		UNIT
B =	12.7185	C =	2888.2100	kPa
M =	0.4785	S =	3.9753	
NI =	8.0416	M =	9.6694	mmol/g
BIS =1.19437−01		BIS =.676995−01		1/g
QIS =	26.0453	QIS =	30.4603	KJ/mol

P(kPa)	N(mmol/g)	DN TOTH	DN UNILAN
98.0826	1.2716	−0.0004	−0.0164
119.8675	1.4175	0.0090	0.0002
147.0226	1.5889	0.0075	0.0049
174.4816	1.7531	−0.0055	−0.0040
204.0685	1.9012	−0.0086	−0.0044
239.8363	2.0543	−0.0058	0.0000
275.4013	2.2033	−0.0167	−0.0104
325.8612	2.3688	−0.0088	−0.0031
373.7879	2.5134	−0.0080	−0.0036
421.5120	2.5433	0.0920	0.0947
499.8362	2.8364	−0.0134	−0.0138
557.8954	2.9649	−0.0190	−0.0215
615.8533	3.0814	−0.0239	−0.0284
AVERAGE DN		0.0168	0.0158

DN = (NCAL−NEXP)

ETHANE

CARBON MOLECULAR SIEVE

ADSORBENT : MSC-5A;TAKEDA CHEMICAL Co.;0.56 cc/g;650 Sqm/g
EXP.   : STATIC GRAVIMETRIC (McBAIN QUARTZ SPRING BALANCE)
REF.   : NAKAHARA,T.,HIRATA,M.and OMORI,T.;J.Chem.Eng.Data;19,310(1974)

TEMP (K)=   278.65
PSAT (kPa)=   2720.02

	TOTH		UNILAN	UNITS
B =	.8492	C =	4.4289	kPa
M =	.4770	S =	3.4036	
NI =	3.0635	M =	2.9324	mmol/g
DG =	-43.3799	DG =	-42.5702	J/g
BIS=	.99991+001	BIS=	.67685+001	1/g
QIS=	37.1403	QIS=	37.9935	KJ/mol

P(kPa)	N(mmol/g)	DN TOTH	DN UNILAN
.9600	.8680	-.0395	-.0021
2.8930	1.2940	-.0058	.0009
6.5330	1.6200	.0207	.0026
10.6700	1.8120	.0302	.0057
15.7300	1.9650	.0261	.0031
23.8000	2.1450	-.0074	-.0226
37.6000	2.2810	.0020	.0011
50.1300	2.3610	.0045	.0141
66.9300	2.4540	-.0127	.0075
83.8700	2.5340	-.0382	-.0105
AVERAGE DN		.0187	.0070

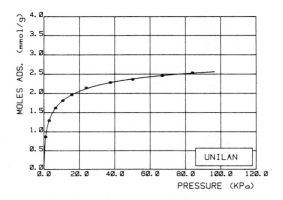

TEMP (K)=   303.15
PSAT (kPa)=   4652.80

	TOTH		UNILAN	UNITS
B =	1.4965	C =	9.5596	kPa
M =	.4956	S =	2.6986	
NI =	2.9177	M =	2.6151	mmol/g
DG =	-41.7844	DG =	-40.2486	J/g
BIS=	.32602+001	BIS=	.18896+001	1/g
QIS=	37.1403	QIS=	37.9935	KJ/mol

P(kPa)	N(mmol/g)	DN TOTH	DN UNILAN
1.4670	.5653	.0091	.0053
3.0670	.8580	-.0228	-.0194
5.7330	1.0810	.0079	.0112
9.7330	1.3100	.0047	.0052
15.1300	1.4930	.0097	.0082
21.4700	1.6490	-.0013	-.0033
27.8700	1.7460	.0058	.0043
36.1300	1.8620	-.0109	-.0115
45.1300	1.9390	-.0066	-.0063
58.5300	2.0250	-.0025	-.0014
83.0000	2.1280	.0067	.0075
AVERAGE DN		.0080	.0076

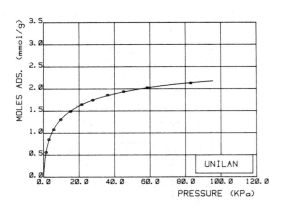

TEMP (K)=   323.15

	TOTH		UNILAN	UNITS
B =	2.5547	C =	30.5974	kPa
M =	.5149	S =	2.9792	
NI =	2.9011	M =	2.8496	mmol/g
BIS=	.12609+001	BIS=	.82395+000	1/g
QIS=	37.1403	QIS=	37.9935	KJ/mol

P(kPa)	N(mmol/g)	DN TOTH	DN UNILAN
.6400	.1530	.0245	.0113
1.7330	.3691	-.0082	-.0124
4.1330	.6119	-.0010	.0050
6.4000	.7715	-.0034	.0040
10.1300	.9644	-.0129	-.0079
13.6000	1.0710	.0052	.0071
18.8000	1.2110	.0062	.0042
27.0700	1.3830	-.0055	-.0111
40.5300	1.5500	.0026	-.0039
54.9300	1.6790	.0013	-.0028
73.3300	1.7690	.0276	.0289
89.6000	1.8990	-.0251	-.0185
AVERAGE DN		.0103	.0098

DN=(NCAL-NEXP)

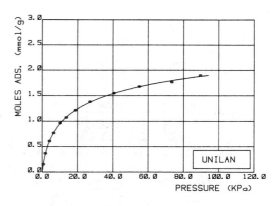

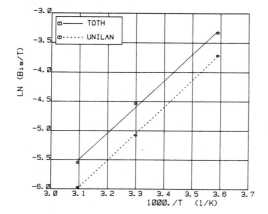

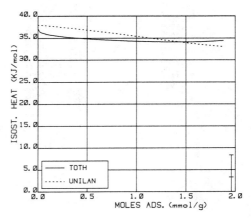

ETHANE

ZEOLITE MOLECULAR SIEVE

ADSORBENT : 13X;LINDE(UNION CARBIDE);.3 cc/g;525 Sqm/g (N2)
EXP.   : STATIC VOLUMETRIC
REF.   : HYUN,S.H.and DANNER,R.P.;J.Chem.Eng.Data;27,196(1982)

TEMP   (K)=   273.15
PSAT (kPa)=   2386.54

	TOTH		UNILAN	UNITS
B =	45.1947	C =	9.1204	kPa
M =	1.5259	S =	2.0101-009	
NI =	2.6264	M =	2.8794	mmol/g
DG =	-34.4851	DG =	-36.3466	J/g
BIS=	.49083+000	BIS=	.71806+000	1/g
QIS=	26.9694	QIS=	29.7908	KJ/mol

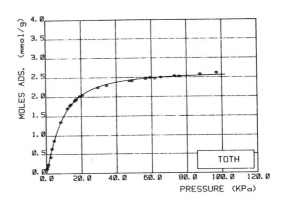

P(kPa)	N(mmol/g)	DN TOTH	DN UNILAN
.4000	.1200	-.0339	-.0068
.8400	.1900	-.0104	.0516
1.1500	.2300	.0142	.0871
2.4600	.4200	.0833	.1839
3.2000	.6400	-.0018	.0998
4.5200	.8600	-.0029	.0912
8.3000	1.3500	-.0090	.0239
12.1100	1.7000	-.0353	-.0543
13.5200	1.7700	-.0149	-.0488
14.2300	1.8100	-.0140	-.0586
16.2200	1.9000	-.0036	-.0580
16.8900	1.9300	-.0038	-.0654
17.0300	1.9400	-.0079	-.0603
19.1500	2.0200	-.0061	-.0723
20.6400	2.0500	.0129	-.0495
29.6100	2.2400	.0209	-.0357
34.5300	2.2900	.0366	-.0102
47.5400	2.4100	.0217	.0057
48.9300	2.4100	.0293	.0208
56.6300	2.4600	.0140	.0237
59.2800	2.4900	-.0064	.0088
61.8100	2.4700	.0218	.0363
65.2500	2.5000	.0019	.0214
73.1400	2.5300	-.0092	.0292
75.9000	2.5200	.0063	.0543
87.7700	2.5800	-.0346	.0245
97.0400	2.6100	-.0535	.0247
AVERAGE DN		.0189	.0484

TEMP   (K)=   298.15
PSAT (kPa)=   4191.41

	TOTH		UNILAN	UNITS
B =	246.5450	C =	27.7920	kPa
M =	1.5955	S =	3.1061-009	
NI =	2.4226	M =	2.8664	mmol/g
DG =	-32.6182	DG =	-35.6250	J/g
BIS=	.19028+000	BIS=	.25604+000	1/g
QIS=	26.9694	QIS=	29.7908	KJ/mol

P(kPa)	N(mmol/g)	DN TOTH	DN UNILAN
.5400	.0400	.0014	.0088
1.0500	.0900	-.0096	.0075
1.2100	.1100	-.0174	.0085
3.5400	.2700	-.0033	.0505
7.5400	.5100	.0346	.0962
9.7100	.6600	.0219	.0856
10.0500	.6900	.0125	.0695
10.4000	.7000	.0235	.0804
13.5500	.9100	-.0099	.0306
14.0000	.9900	-.0664	-.0354
16.7200	1.0400	.0171	.0330
20.0000	1.2000	-.0007	-.0015
20.9400	1.2400	-.0035	-.0067
23.6400	1.3400	-.0043	-.0231
24.8200	1.3900	-.0145	-.0383
33.0900	1.6200	-.0142	-.0592
40.9400	1.7400	.0232	-.0329
43.0500	1.7900	.0079	-.0481
46.5300	1.8500	-.0005	-.0523
50.1400	1.9000	-.0034	-.0605
50.7500	1.9000	.0040	-.0466
54.4600	1.9700	-.0243	-.0678
63.6900	2.0200	.0097	-.0272
68.3800	2.0700	-.0060	-.0284
71.1800	2.0600	.0223	.0025
80.0500	2.1000	.0316	.0252
84.2600	2.1500	.0012	.0100
94.2100	2.2000	-.0099	.0157
110.8000	2.2500	-.0120	.0354
124.7000	2.2800	-.0129	.0681
137.8000	2.2900	-.0016	.0930
AVERAGE DN		.0137	.0403

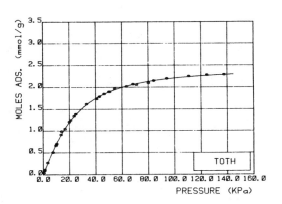

TEMP   (K)=   323.15

	TOTH		UNILAN	UNITS
B =	1028.5900	C =	74.5424	kPa
M =	1.6235	S =	2.0071-009	
NI =	2.2727	M =	2.9683	mmol/g
BIS=	.85182-001	BIS=	.10708+000	1/g
QIS=	26.9694	QIS=	29.7908	KJ/mol

P(kPa)	N(mmol/g)	DN TOTH	DN UNILAN
2.1200	.0800	-.0129	-.0018
4.2500	.1500	-.0161	.0095
5.2600	.1600	.0053	.0339
6.2400	.1900	.0056	.0383
14.2300	.4400	-.0079	.0354
28.2900	.7900	.0031	.0231
32.5100	.8700	.0166	.0307
34.0600	.9200	-.0007	.0120
52.1600	1.2600	-.0204	-.0372
55.0000	1.2800	.0007	-.0165
55.2300	1.2800	.0040	-.0165
63.2200	1.3900	-.0018	-.0296
72.7600	1.4900	.0040	-.0201
78.4300	1.5500	-.0014	-.0270
92.4900	1.6600	.0026	-.0150
104.3000	1.7300	.0092	-.0006
107.5000	1.7600	-.0025	-.0087
126.8000	1.8600	-.0096	.0102
134.9000	1.8800	.0020	.0308
137.8000	1.8900	.0025	.0365
AVERAGE DN		.0064	.0217

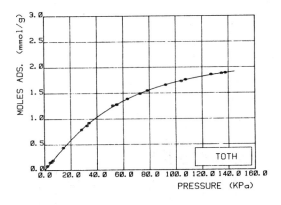

TEMP   (K)=   373.15

	TOTH		UNILAN	UNITS
B =	1715.1600	C =	433.1760	kPa
M =	1.3584	S =	.6350	
NI =	2.1651	M =	3.8207	mmol/g
BIS=	.27934-001	BIS=	.29239-001	1/g
QIS=	26.9694	QIS=	29.7908	KJ/mol

P(kPa)	N(mmol/g)	DN TOTH	DN UNILAN
3.3000	.0300	-.0004	.0008
9.8800	.0800	.0081	.0106
11.7000	.1100	-.0059	-.0032
16.3200	.1500	-.0058	-.0029
17.8400	.1600	-.0027	.0002
18.2400	.1600	.0007	.0036
25.5900	.2200	.0026	.0052
32.7700	.2800	.0014	.0032
37.0600	.3200	-.0044	-.0032
37.1900	.3200	-.0034	-.0021
39.3500	.3400	-.0065	-.0055
46.8700	.3900	.0012	.0012
51.1800	.4200	.0033	.0028
55.5700	.4500	.0053	.0043
56.2400	.4600	.0001	-.0009
56.9800	.4700	-.0046	-.0057
61.0000	.5000	-.0061	-.0076
66.1600	.5200	.0096	.0077
74.5800	.5900	-.0042	-.0063
80.5200	.6100	.0139	.0119
85.2400	.6600	-.0066	-.0084
87.4000	.6700	-.0034	-.0050
88.6500	.6700	.0042	.0026
100.4000	.7400	.0030	.0030
121.4000	.8600	-.0046	.0005
AVERAGE DN		.0045	.0043

DN=(NCAL-NEXP)

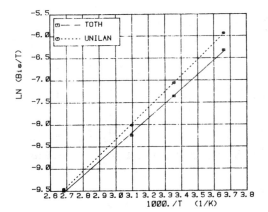

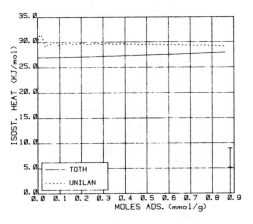

ETHANE

ZEOLITE

ADSORBENT : 13X; UNION CARBIDE; 1/16 in PELLETS
EXP.   : STATIC VOLUMETRIC
REF.   : KAUL, B. K., *IND. ENG. CHEM. RES.* **26**, 928 (1987)

TEMP (K) = 323.15

TOTH	UNILAN	UNIT
B =26806.5100	C = 84.8868	kPa
M = 2.2646	S = 1.2986-04	
NI = 2.2573	M = 3.0031	mmol/g
BIS =.67206-001	BIS =.95040-001	1/g
QIS = 22.9165	QIS = 26.6759	KJ/mol

P (kPa)	N (mmol/g)	DN TOTH	DN UNILAN
0.6205	0.1093	-0.0938	-0.0873
7.5842	0.1343	0.0551	0.1124
11.0315	0.2320	0.0429	0.1126
12.2726	0.2454	0.0601	0.1336
13.4447	0.2989	0.0354	0.1118
14.4789	0.3480	0.0117	0.0903
20.6841	0.4595	0.0499	0.1290
22.0630	0.4908	0.0514	0.1281
24.8209	0.5354	0.0713	0.1441
24.1315	0.5309	0.0597	0.1334
25.5104	0.5577	0.0649	0.1356
31.0261	0.7138	0.0335	0.0911
33.7840	0.7495	0.0581	0.1050
34.4735	0.8030	0.0194	0.0639
35.1630	0.8700	-0.0328	0.0107
38.6103	0.9280	-0.0187	0.0120
41.3682	1.0484	-0.0833	-0.0644
46.8840	1.0930	-0.0216	-0.0249
47.5734	1.1600	-0.0757	-0.0808
51.7103	1.2403	-0.0817	-0.1032
59.2944	1.3607	-0.0768	-0.1245
64.8102	1.4009	-0.0344	-0.0998
67.5681	1.4499	-0.0448	-0.1186
77.9101	1.5526	-0.0176	-0.1151
86.8732	1.6195	0.0106	-0.1007
100.6626	1.7266	0.0230	-0.0975
108.2468	1.7756	0.0279	-0.0929
114.4520	1.8069	0.0356	-0.0828
135.8256	1.8916	0.0564	-0.0435
155.1308	1.9675	0.0475	-0.0270
195.8095	2.0612	0.0425	0.0336
235.7988	2.1281	0.0245	0.0797
293.0247	2.2396	-0.0483	0.0881
361.9717	2.2931	-0.0775	0.1393
AVERAGE DN		0.0467	0.0917

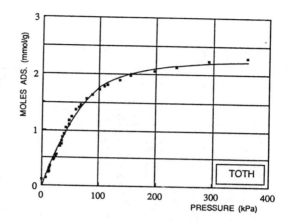

TEMP (K) = 373.15

TOTH	UNILAN	UNIT
B = 2776.1550	C = 254.9327	kPa
M = 1.4695	S = 3.7850-03	
NI = 1.9263	M = 2.5638	mmol/g
BIS =.27112-001	BIS =.31200-001	1/g
QIS = 22.9165	QIS = 26.6759	KJ/mol

P (kPa)	N (mmol/g)	DN TOTH	DN UNILAN
14.4789	0.1428	-0.0178	-0.0050
17.9262	0.1651	-0.0110	0.0033
24.8209	0.2097	0.0015	0.0178
28.9577	0.2543	-0.0096	0.0072
43.4366	0.3703	-0.0127	0.0030
49.6418	0.3926	0.0111	0.0253
51.7103	0.4283	-0.0096	0.0040
56.5365	0.4595	-0.0063	0.0058
64.1207	0.5086	-0.0029	0.0067
80.3232	0.6112	-0.0004	0.0031
92.3890	0.6915	-0.0084	-0.0096
97.9047	0.7004	0.0140	0.0110

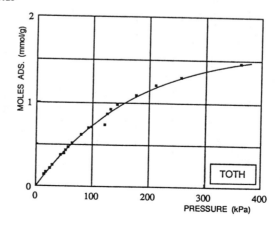

121.3467	0.7317	0.1054	0.0952
125.4835	0.8566	0.0003	-0.0109
132.3783	0.9146	-0.0256	-0.0383
143.4098	0.9681	-0.0304	-0.0451
176.5043	1.0797	-0.0137	-0.0308
210.9778	1.2001	-0.0246	-0.0391
257.1723	1.2938	-0.0011	-0.0063
363.3507	1.4633	0.0121	0.0434
AVERAGE DN		0.0159	0.0205

TEMP (K)   =   423.15

TOTH		UNILAN		UNIT
B = 1706.4130		C = 1253.6810		kPa
M = 1.1110		S = .8225		
NI = 2.6806		M = 3.7754		mmol/g
BIS =.11625-001		BIS =.11830-001		1/g
QIS = 22.9165		QIS = 26.6759		KJ/mol

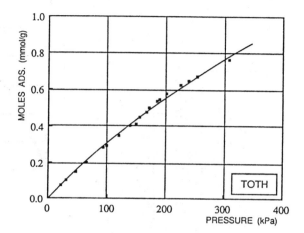

P (kPa)	N (mmol/g)	DN TOTH	DN UNILAN
22.7525	0.0758	-0.0019	-0.0012
31.7156	0.1026	-0.0003	0.0005
46.8840	0.1472	0.0020	0.0029
65.4996	0.2008	0.0044	0.0050
93.7679	0.2811	0.0054	0.0056
99.2837	0.2900	0.0118	0.0119
119.2783	0.3480	0.0082	0.0079
138.5835	0.4015	0.0053	0.0047
148.9255	0.4104	0.0227	0.0221
155.1308	0.4506	-0.0018	-0.0026
167.5412	0.4751	0.0042	0.0035
172.3675	0.4997	-0.0086	-0.0094
184.7780	0.5354	-0.0146	-0.0154
190.2937	0.5443	-0.0106	-0.0113
202.0147	0.5800	-0.0192	-0.0198
225.1120	0.6255	-0.0129	-0.0133
239.2461	0.6469	-0.0037	-0.0038
253.0355	0.6692	0.0031	0.0034
307.8484	0.7629	0.0183	0.0210
AVERAGE DN		0.0084	0.0087

DN= (NCAL-NEXP)

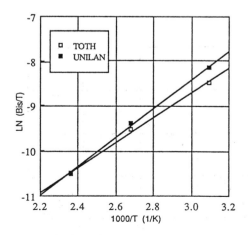

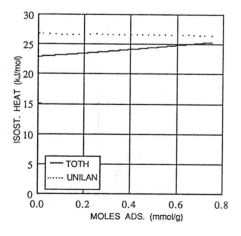

ETHYLENE

ACTIVATED CARBON

ADSORBENT : ATRITION RESISTANT BEADS; TAIYO KAKEN Co.,
            JAPAN; 800-1200 g/m2
EXP.   : STATIC VOLUMETRIC
REF.   : KAUL, B. K.,*IND. ENG. CHEM. RES.* **26**, 928 (1987)

TEMP (K)   =   310.95

TOTH	UNILAN	UNIT
B = 3.7219	C = 6924.3630	kPa
M = .3433	S = 6.0432	
NI = 13.4725	M = 16.1464	mmol/g
BIS =.75751+000	BIS =.21010+000	1/g
QIS = 32.8983	QIS = 23.2516	KJ/mol

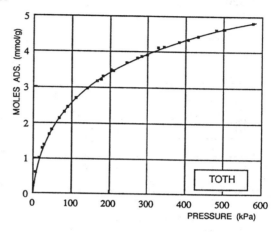

P (kPa)	N(mmol/g)	DN TOTH	DN UNILAN
2.2753	0.2851	-0.0107	-0.1119
2.3442	0.2945	-0.0140	-0.1164
6.4121	0.5871	-0.0198	-0.1471
15.3752	0.9922	-0.0094	-0.1102
24.6830	1.2925	-0.0034	-0.0676
27.9235	1.3786	0.0007	-0.0524
41.0235	1.6802	0.0086	-0.0083
47.9182	1.8136	0.0125	0.0097
66.4649	2.1209	0.0173	0.0406
80.3232	2.3181	0.0149	0.0499
89.7690	2.4368	0.0152	0.0556
111.2115	2.6849	0.0059	0.0533
140.9277	2.9628	0.0058	0.0549
166.4380	3.1738	-0.0019	0.0445
177.2627	3.2082	0.0424	0.0870
182.3648	3.2983	-0.0120	0.0316
204.8415	3.4915	-0.0571	-0.0185
210.1505	3.4687	-0.0014	0.0360
246.5545	3.6802	-0.0040	0.0235
272.9612	3.8270	-0.0153	0.0045
285.7164	3.8711	0.0019	0.0180
299.1610	3.9251	0.0101	0.0221
327.4983	4.1067	-0.0481	-0.0446
342.6666	4.1366	-0.0158	-0.0168
382.6559	4.2829	-0.0093	-0.0221
407.4768	4.3539	0.0076	-0.0125
432.9872	4.4324	0.0143	-0.0129
480.5606	4.6046	-0.0106	-0.0507
499.8658	4.6492	0.0007	-0.0445
501.9342	4.6287	0.0271	-0.0186
575.0180	4.8285	0.0215	-0.0423
AVERAGE DN		0.0142	0.0461

TEMP (K)   =   422.05

TOTH	UNILAN	UNIT
B = 31.3106	C = 5338.4090	kPa
M = .5159	S = 3.6170	
NI = 8.1530	M = 7.9062	mmol/g
BIS =.36084-001	BIS =.26723-001	1/g
QIS = 32.8983	QIS = 23.2516	KJ/mol

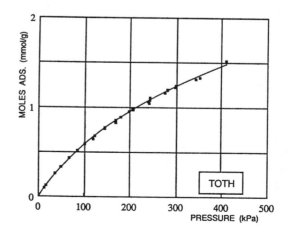

P (kPa)	N(mmol/g)	DN TOTH	DN UNILAN
11.9278	0.0955	0.0039	-0.0082
15.5820	0.1227	0.0034	-0.0100
36.1972	0.2614	-0.0014	-0.0156
48.4697	0.3355	-0.0051	-0.0175
67.5681	0.4332	-0.0029	-0.0116
86.0459	0.5193	-0.0010	-0.0061
101.1452	0.5889	-0.0037	-0.0061
119.6920	0.6398	0.0225	0.0228
122.7946	0.6830	-0.0084	-0.0077
145.0645	0.7647	-0.0047	-0.0016
145.6850	0.7696	-0.0073	-0.0042
168.9891	0.8262	0.0192	0.0238
169.2649	0.8508	-0.0044	0.0002
179.1243	0.8874	-0.0075	-0.0025
196.8437	0.9498	-0.0118	-0.0066
205.1863	0.9873	-0.0229	-0.0177

207.3236	0.9659	0.0052	0.0104
239.0392	1.0591	0.0074	0.0116
240.3492	1.0449	0.0254	0.0296
242.0040	1.0743	0.0008	0.0049
242.9692	1.1020	-0.0241	-0.0201
276.0638	1.1546	0.0151	0.0169
280.0627	1.1907	-0.0104	-0.0088
297.9200	1.2184	0.0086	0.0085
299.5747	1.2394	-0.0081	-0.0084
342.6666	1.3121	0.0252	0.0201
350.9402	1.3348	0.0219	0.0158
409.5452	1.5227	-0.0356	-0.0499
	AVERAGE DN	0.0114	0.0131

DN= (NCAL-NEXP)

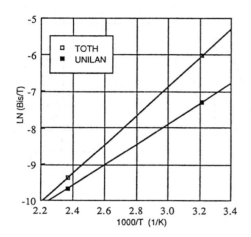
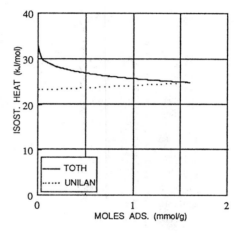

ETHYLENE

ACTIVATED CARBON

ADSORBENT : EY-51-C;PITTSBURGH COKE and CHEMICAL CO.;805 Sqm/g
EXP.   : STATIC VOLUMETRIC
REF.   : LEWIS,W.K.,GILLILAND,E.R.,CHERTOW,B. and MILLIKEN,W.;J.Am.Chem.Soc.;
            72,1157(1950)

                    TEMP   (K)=   298.15

	TOTH		UNILAN	UNITS
B =	3.0142	C =	731.2870	kPa
M =	.3364	S =	4.6510	
NI =	9.2691	M =	7.1554	mmol/g
BIS=	.86448+000	BIS=	.27295+000	1/g

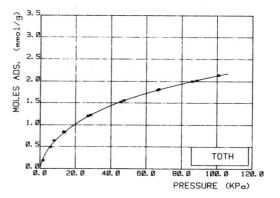

P(kPa)	N(mmol/g)	DN TOTH	DN UNILAN
1.7330	.1860	.0367	-.0155
6.0000	.4850	.0266	-.0081
8.0000	.6390	-.0291	-.0520
13.3300	.8430	-.0234	-.0219
13.8700	.8310	.0069	.0102
27.0700	1.2000	-.0085	.0183
28.5300	1.2200	.0030	.0306
45.7300	1.5200	.0096	.0342
47.4700	1.5600	-.0043	.0192
66.4000	1.8000	.0017	.0085
67.2000	1.8200	-.0091	-.0032
86.0000	2.0000	.0048	-.0096
88.8000	2.0200	.0108	-.0068
100.9000	2.1400	-.0042	-.0354
AVERAGE DN		.0128	.0195

                    DN=(NCAL-NEXP)

ETHYLENE

ACTIVATED CARBON

ADSORBENT : COLUMBIA GRADE L;MESH 20x40;1152 Sq m/g (N2)
EXP.   : STATIC VOLUMETRIC
REF.   : RAY,G.C.and BOX,E.O.;IND.ENG.CHEM.;42,1315(1950)

              TEMP   (K)=   310.92

	TOTH		UNILAN	UNITS
B =	11.9645	C =	9747.6801	kPa
M =	.5988	S =	6.8209	
NI =	7.3398	M =	16.4451	mmol/g
BIS=	.30072+000	BIS=	.29310+000	1/g
QIS=	26.4330	QIS=	25.0426	KJ/mol

P(kPa)	N(mmol/g)	DN TOTH	DN UNILAN
12.4000	.9115	-.0665	.0204
32.9400	1.5550	.0602	.1454
58.2700	2.1920	.0238	.0608
90.4000	2.7440	-.0094	-.0298
99.9700	2.8820	-.0239	-.0587
330.9000	4.3470	-.0118	-.1645
558.5000	4.8890	.0292	-.0909
837.7000	5.3450	-.0249	-.0658
1155.0000	5.6220	-.0178	.0402
1486.0000	5.7810	.0243	.1825
AVERAGE DN		.0292	.0859

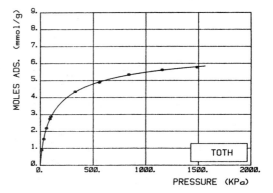

              TEMP   (K)=   338.70

	TOTH		UNILAN	UNITS
B =	6.8975	C =	9874.3800	kPa
M =	.4453	S =	6.1647	
NI =	8.8758	M =	15.0681	mmol/g
BIS=	.32680+000	BIS=	.16577+000	1/g
QIS=	26.4330	QIS=	25.0426	KJ/mol

P(kPa)	N(mmol/g)	DN TOTH	DN UNILAN
13.8700	.7020	-.0219	-.0767
41.5500	1.3600	-.0089	-.0168
77.8700	1.8860	.0022	.0185
99.9700	2.1200	.0110	.0316
100.0000	2.1100	.0213	.0419
296.5000	3.3280	.0037	.0043
520.5000	4.0200	-.0181	-.0349
803.2000	4.5210	-.0044	-.0224
1127.0000	4.9300	-.0216	-.0264
1479.0000	5.1860	.0268	.0444
AVERAGE DN		.0140	.0318

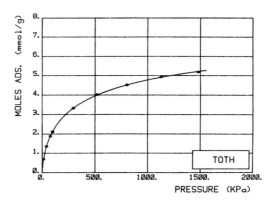

TEMP   (K)=   366.48

	TOTH		UNILAN		UNITS
B =	13.7543	C =	9939.9900		kPa
M =	.5057	S =	5.4322		
NI =	7.7092	M =	13.1737		mmol/g
BIS=	.13169+000	BIS=	.84984-001		1/g
QIS=	26.4330	QIS=	25.0426		KJ/mol

P(kPa)	N(mmol/g)	DN TOTH	DN UNILAN
12.6700	.3146	.0307	-.0045
36.9300	.7522	.0123	-.0066
68.1300	1.1400	.0017	.0032
94.9400	1.4070	-.0142	-.0028
99.9700	1.4390	-.0041	.0085
272.3000	2.3860	.0055	.0182
499.9000	3.0880	-.0214	-.0258
792.9000	3.5890	.0081	-.0040
1117.0000	3.9830	.0056	-.0010
1479.0000	4.3000	.0018	.0110
AVERAGE DN		.0105	.0086

TEMP   (K)=   394.20

	TOTH		UNILAN		UNITS
B =	19.5847	C =	82182.2002		kPa
M =	.5226	S =	7.0644		
NI =	7.1753	M =	16.4768		mmol/g
BIS=	.79305-001	BIS=	.54393-001		1/g
QIS=	26.4330	QIS=	25.0426		KJ/mol

P(kPa)	N(mmol/g)	DN TOTH	DN UNILAN
16.8000	.2950	-.0185	-.0450
43.2000	.5991	-.0231	-.0403
68.4000	.8247	-.0272	-.0320
99.9700	1.0500	-.0252	-.0181
100.1000	.9359	.0897	.0969
455.1000	2.3350	-.0034	.0111
748.1000	2.8770	-.0147	-.0140
1086.0000	3.2820	-.0102	-.0159
1458.0000	3.5790	.0159	.0124
AVERAGE DN		.0253	.0317

TEMP   (K)=   422.00

	TOTH		UNILAN		UNITS
B =	40.6011	C =	648734.0000		kPa
M =	.5933	S =	8.5800		
NI =	6.5859	M =	21.3281		mmol/g
BIS=	.44911-001	BIS=	.35787-001		1/g
QIS=	26.4330	QIS=	25.0426		KJ/mol

P(kPa)	N(mmol/g)	DN TOTH	DN UNILAN
14.9300	.1632	-.0059	-.0195
56.1400	.4778	.0034	-.0068
81.4700	.6314	.0093	.0049
96.8000	.7628	-.0350	-.0361
99.9700	.7834	-.0383	-.0388
275.8000	1.4490	.0076	.0211
555.0000	2.0730	.0525	.0581
875.6000	2.6010	.0147	.0121
1234.0000	2.9950	.0027	-.0004
1479.0000	3.2410	-.0410	-.0399
AVERAGE DN		.0210	.0238

TEMP   (K)=   449.80

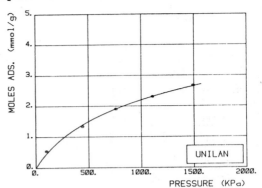

	TOTH		UNILAN	UNITS
B =	110.7510	C =	658685.0000	kPa
M =	.6847	S =	7.8891	
NI =	5.9652	M =	21.4402	mmol/g
BIS=	.23056-001	BIS=	.20583-001	1/g
QIS=	26.4330	QIS=	25.0426	KJ/mol

P(kPa)	N(mmol/g)	DN TOTH	DN UNILAN
99.9700	.5376	-.0718	-.0756
441.3000	1.3420	.0476	.0510
758.4000	1.9040	.0054	.0040
1107.0000	2.3090	.0060	.0035
1493.0000	2.6730	-.0232	-.0196
AVERAGE DN		.0308	.0307

TEMP   (K)=   477.59

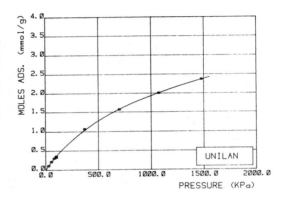

	TOTH		UNILAN	UNITS
B =	154.8920	C =	9963.1200	kPa
M =	.7033	S =	3.3648	
NI =	5.9573	M =	9.6037	mmol/g
BIS=	.18198-001	BIS=	.16433-001	1/g
QIS=	26.4330	QIS=	25.0426	KJ/mol

P(kPa)	N(mmol/g)	DN TOTH	DN UNILAN
26.9300	.1100	.0028	-.0027
54.1400	.2178	-.0030	-.0097
81.0700	.3042	.0034	-.0029
99.9700	.3277	.0412	.0355
100.3000	.3552	.0148	.0090
368.9000	1.0650	-.0302	-.0278
699.8000	1.5720	.0058	.0074
1072.0000	2.0120	.0007	.0006
1479.0000	2.3700	-.0017	.0011
AVERAGE DN		.0115	.0108

DN=(NCAL-NEXP)

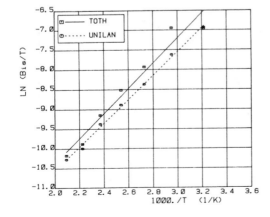

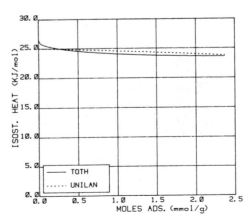

ETHYLENE
```
-------------------------------------------------------------------------
```
ACTIVATED CARBON
```
-------------------------------------------------------------------------
```
ADSORBENT : BPL;PITTSBURGH CHEMICAL Co.;988 Sqm/g
EXP.   : STATIC VOLUMETRIC
REF.   : REICH,R.,ZIEGLER,W.T.and ROGERS,K.A.;Ind.Eng.Chem.Process Des.Dev.
            19,336(1980)

                    TEMP (K)=   212.70
                    PSAT (kPa)=   742.00

            TOTH   UNILAN   UNITS
        ----------   -------------   -----
B =   .6060   C =   18.8780   kPa
M =   .2740   S =   4.8430
NI =   11.2930   M =   9.4343   mmol/g
DG =   -66.1626   DG = -64.2021   J/g
BIS= .12425+003   BIS= .11573+002   1/g
QIS=   38.0186   QIS=   26.4969   KJ/mol

    P(kPa)   N(mmol/g)   DN TOTH   DN UNILAN
    ------   ---------   -------   ---------
     .6900   1.8840   -.1495   -.2000
     .6900   1.8950   -.1605   -.2110
    1.3800   1.8620   .3936   .4064
    4.1400   3.0100   .2077   .2620
    5.5200   3.7210   -.2303   -.1778
    6.8900   3.7270   -.0218   .0264
   19.3000   4.7360   -.0062   .0023
   30.3400   5.3480   -.1655   -.1762
   35.1600   5.3270   .0020   -.0142
   73.0800   5.9960   .0452   .0122
   98.5900   6.4370   -.1142   -.1477
  106.2000   6.3460   .0456   .0127
  152.4000   6.6450   .0744   .0472
  237.2000   7.0080   .0960   .0829
  239.2000   7.1580   -.0469   -.0597
  277.2000   7.1020   .1328   .1259
  418.5000   7.4480   .1199   .1308
  422.6000   7.6400   -.0644   -.0532
  482.6000   7.6050   .0738   .0906
  599.8000   7.8840   -.0404   -.0156
  602.6000   8.0410   -.1939   -.1690
     AVERAGE DN   .1136   .1154
```

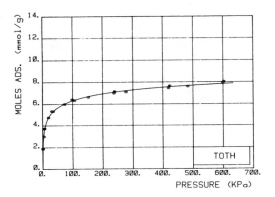

```
                    TEMP  (K)=    260.20
                    PSAT (kPa)=  3032.00

            TOTH              UNILAN         UNITS
        ----------        -------------      -----
B  =     2.3730        C  =   69.9790          kPa
M  =      .4130        S  =    3.2457
NI =     8.4070        M  =    7.3329        mmol/g
DG =   -61.6996        DG =  -60.8774          J/g
BIS= .22441+001        BIS= .89540+000        1/g
QIS=    38.0186        QIS=   26.4969        KJ/mol

    P(kPa)    N(mmol/g)   DN TOTH   DN UNILAN
    ------    ---------   -------   ---------
    10.3400     1.7800      -.0126     -.0158
    10.3400     1.7650       .0024     -.0008
    28.2700     2.7270      -.0203      .0019
    46.8800     3.2290      -.0016      .0196
    89.6300     3.9290      -.0105     -.0041
   115.1000     4.1700       .0122      .0149
   204.1000     4.7820      -.0131     -.0129
   273.0000     5.0460       .0072      .0097
   401.3000     5.4300      -.0186     -.0107
   552.3000     5.6960      -.0055      .0070
   709.5000     5.9300      -.0328     -.0185
   858.4000     6.0580      -.0110      .0028
  1062.0000     6.2280      -.0213     -.0109
  1216.0000     6.3060      -.0019      .0044
  1589.0000     6.4680       .0187      .0122
     AVERAGE  DN                      .0126      .0097
```

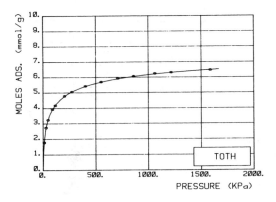
```

TEMP (K)=   301.40

	TOTH		UNILAN	UNITS
B =	7.4980	C =	222.7170	kPa
M =	.5030	S =	2.6867	
NI =	7.2510	M =	6.5455	mmol/g
BIS=	.33105+000	BIS=	.20031+000	1/g
QIS=	38.0186	QIS=	26.4969	KJ/mol

P(kPa)	N(mmol/g)	DN TOTH	DN UNILAN
17.2400	.9490	-.0063	-.0306
41.3700	1.5760	.0029	.0114
82.0500	2.2270	-.0148	.0058
116.5000	2.5700	.0003	.0198
157.2000	2.9050	-.0174	-.0016
277.9000	3.5140	-.0128	-.0062
435.7000	3.9880	-.0111	-.0085
510.2000	4.1380	.0011	.0035
670.2000	4.4220	-.0109	-.0071
751.5000	4.5300	-.0083	-.0034
929.4000	4.7320	-.0113	-.0040
1029.0000	4.8220	-.0090	-.0004
1188.0000	4.9450	-.0053	.0050
1366.0000	5.0700	-.0113	.0004
1419.0000	5.0960	-.0055	.0064
1696.0000	5.2470	-.0120	.0006
AVERAGE DN		.0088	.0072

DN=(NCAL-NEXP)

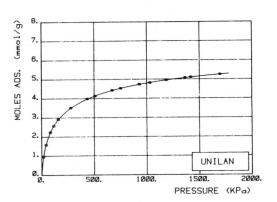

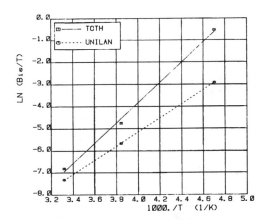

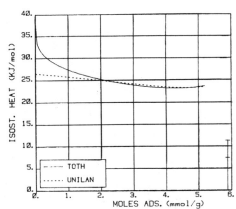

ETHYLENE

ACTIVATED CARBON

ADSORBENT : NUXIT-AL, (1.5 mm DIAMETER, 3-4 mm LONG)
EXP.   : STATIC VOLUMETRIC
REF.   : SZEPESY, L., ILLES, V., *Acta Chim. Hung.*,**35,** 37(1963)

TEMP   (K) =   293.15

TOTH		UNILAN		UNIT
B =	2.7786	C =	999.1430	kPa
M =	0.3378	S =	5.0563	
NI =	11.3111	M =	10.2711	mmol/g
BIS = 1.3383+00		BIS = .38900+00		1/g
QIS =	34.4932	QIS =	27.0710	KJ/mol

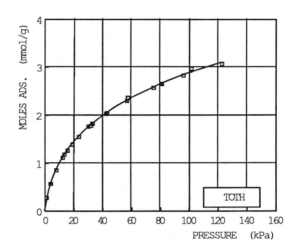

P(kPa)	N(mmol/g)	DN TOTH	DN UNILAN
1.1999	0.2726	-0.0202	-0.0972
3.8663	0.5649	-0.0046	-0.0827
7.7993	0.8504	0.0090	-0.0379
12.2656	1.0981	0.0107	-0.0072
13.4389	1.1704	-0.0055	-0.0174
15.3987	1.2547	-0.0027	-0.0059
18.7584	1.3836	0.0027	0.0109
23.4647	1.5402	0.0097	0.0287
30.1841	1.7562	-0.0078	0.0190
31.8640	1.7834	0.0096	0.0374
32.6639	1.8218	-0.0081	0.0201
42.2631	2.0351	0.0018	0.0301
42.7964	2.0422	0.0059	0.0340
56.7952	2.2997	0.0114	0.0310
56.8218	2.3108	0.0007	0.0203
57.5818	2.3603	-0.0361	-0.0171
75.0603	2.5767	0.0097	0.0118
80.9265	2.6588	0.0046	0.0003
96.1252	2.8306	0.0126	-0.0085
101.4580	2.9681	-0.0674	-0.0945
122.7629	3.0671	0.0403	-0.0105
AVERAGE DN		0.0134	0.0297

TEMP   (K) =   313.15

TOTH		UNILAN		UNIT
B =	3.8369	C =	1309.4000	kPa
M =	0.3543	S =	4.8050	
NI =	10.6430	M =	8.9671	mmol/g
BIS = .622849+00		BIS = .226566+00		1/g
QIS =	34.4932	QIS =	27.0710	KJ/mol

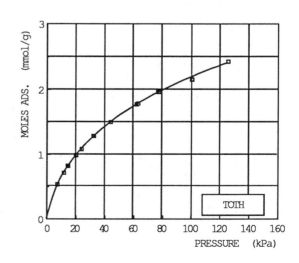

P(kPa)	N(mmol/g)	DN TOTH	DN UNILAN
7.3994	0.5332	0.0001	-0.0436
11.5990	0.7090	0.0007	-0.0249
14.7321	0.8179	0.0026	-0.0112
19.9983	0.9879	-0.0072	-0.0058
24.0646	1.0811	0.0071	0.0169
32.4639	1.2824	-0.0027	0.0175
44.1962	1.5014	-0.0018	0.0226
62.1281	1.7709	-0.0017	0.0166
63.1280	1.7807	0.0018	0.0195
77.1934	1.9681	-0.0127	-0.0053
77.5934	1.9637	-0.0037	0.0034
100.2581	2.1605	0.0333	0.0196
125.3893	2.4259	-0.0167	-0.0555
AVERAGE DN		0.0071	0.0202

TEMP   (K) =   333.15

	TOTH		UNILAN	UNIT
B =	5.8324	C =	1408.4900	kPa
M =	0.3913	S =	4.3226	
NI =	9.3994	M =	7.8507	mmol/g
BIS =	.287334+00	BIS =	.134603+00	1/g
QIS =	34.4932	QIS =	27.0710	KJ/mol

P (kPa)	N (mmol/g)	DN TOTH	DN UNILAN
7.9327	0.3641	-0.0065	-0.0428
19.1317	0.6528	0.0011	-0.0127
20.7316	0.7108	-0.0222	-0.0331
33.2638	0.9165	0.0061	0.0118
39.7300	1.0195	0.0046	0.0152
50.3957	1.1646	0.0079	0.0224
59.5949	1.2739	0.0117	0.0266
70.5273	1.4015	0.0047	0.0176
87.4592	1.5728	-0.0022	0.0039
94.5253	1.6299	0.0029	0.0053
118.1233	1.8325	-0.0133	-0.0256
125.0560	1.8722	-0.0035	-0.0205
AVERAGE DN		0.0072	0.0198

TEMP   (K) =   363.15

	TOTH		UNILAN	UNIT
B =	10.1388	C =	1500.0000	kPa
M =	0.4189	S =	3.2402	
NI =	9.1716	M =	7.1068	mmol/g
BIS =	.109867+00	BIS =	5.62903-02	1/g
QIS =	34.4932	QIS =	27.0710	KJ/mol

P (kPa)	N (mmol/g)	DN TOTH	DN UNILAN
10.5991	0.2102	0.0098	-0.0285
13.9321	0.2981	-0.0257	-0.0650
28.3976	0.4556	0.0067	-0.0238
36.2636	0.5564	-0.0074	-0.0301
43.3963	0.6171	0.0037	-0.0116
59.7949	0.7643	0.0031	0.0040
61.8614	0.7670	0.0174	0.0202
86.5260	0.9749	-0.0069	0.0156
119.8565	1.1797	-0.0051	0.0364
AVERAGE DN		0.0095	0.0261

DN= (NCAL-NEXP)

ETHYLENE

ACTIVATED CARBON

ADSORBENT : NUXIT-AL, (1.5 mm DIAMETER, 3-4 mm LONG)
EXP.   : STATIC VOLUMETRIC
REF.   : SZEPESY, L., ILLES, V., *Acta Chim. Hung.,***35,** 53(1963)

TEMP   (K) =   293.15

TOTH		UNILAN		UNIT
B =	8.2358	C =	146.7450	kPa
M =	0.5429	S =	2.3163	
NI =	7.3989	M =	6.6101	mmol/g
BIS =.371018+00		BIS =.237929+00		1/g
QIS =	23.5122	QIS =	22.3177	KJ/mol

P(kPa)	N(mmol/g)	DN TOTH	DN UNILAN
105.6820	2.9270	-0.0050	-0.0051
118.5502	3.0609	-0.0054	-0.0053
135.3702	3.2130	-0.0026	-0.0024
153.8113	3.3522	0.0078	0.0079
172.7591	3.4901	0.0059	0.0058
192.4162	3.6190	0.0027	0.0025
219.5713	3.7712	0.0037	0.0035
245.3078	3.9046	-0.0023	-0.0024
273.9828	4.0322	-0.0040	-0.0040
305.7988	4.1545	-0.0026	-0.0025
347.9500	4.2968	-0.0018	-0.0016
390.6079	4.4267	-0.0058	-0.0055
441.2704	4.5534	-0.0024	-0.0020
521.9250	4.7283	-0.0030	-0.0029
571.3716	4.8113	0.0056	0.0053
617.6772	4.8938	0.0004	-0.0004
AVERAGE DN		0.0038	0.0037

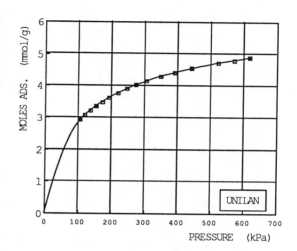

TEMP   (K) =   313.15

TOTH		UNILAN		UNIT
B =	14.1561	C =	821.3100	kPa
M =	0.5837	S =	3.7445	
NI =	6.8627	M =	9.0836	mmol/g
BIS =.190667+00		BIS =.162511+00		1/g
QIS =	23.5122	QIS =	22.3177	KJ/mol

P(kPa)	N(mmol/g)	DN TOTH	DN UNILAN
141.1457	2.5522	-0.0147	0.0049
167.6929	2.7195	0.0123	0.0223
203.3593	2.9457	0.0064	0.0064
234.2634	3.1055	0.0093	0.0032
264.6609	3.2483	0.0069	-0.0034
302.1512	3.3928	0.0142	0.0011
339.5401	3.5356	0.0043	-0.0095
379.9688	3.6753	-0.0083	-0.0212
411.7848	3.8029	-0.0460	-0.0569
468.3242	3.9126	-0.0136	-0.0192
530.2337	4.0188	0.0151	0.0174
583.0240	4.1317	0.0035	0.0140
648.3787	4.2312	0.0154	0.0372
AVERAGE DN		0.0131	0.0167

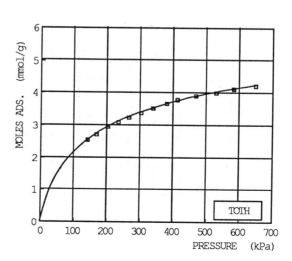

TEMP   (K) =   333.15

TOTH		UNILAN		UNIT
B =	15.0921	C =	775.0840	kPa
M =	0.5487	S =	3.0643	
NI =	6.9762	M =	7.6694	mmol/g
BIS =.137350+00		BIS =.095583+00		1/g
QIS =	23.5122	QIS =	22.3177	KJ/mol

P (kPa)	N (mmol/g)	DN TOTH	DN UNILAN
108.2151	1.7401	−0.0184	−0.0164
137.0927	1.9400	0.0069	0.0097
159.4855	2.0917	0.0060	0.0084
187.0459	2.2586	0.0030	0.0044
222.9150	2.4416	0.0055	0.0052
251.9953	2.5808	−0.0016	−0.0030
284.6219	2.7075	0.0046	0.0020
323.7334	2.8503	0.0037	0.0003
364.6687	2.9877	−0.0017	−0.0053
408.5424	3.1140	−0.0015	−0.0049
456.9758	3.2429	−0.0057	−0.0080
517.6694	3.3830	−0.0072	−0.0073
574.3101	3.5004	−0.0097	−0.0069
636.6249	3.5918	0.0121	0.0187
AVERAGE DN		0.0063	0.0072

TEMP   (K) =   363.15

TOTH		UNILAN		UNIT
B =	26.5305	C =	1214.2300	kPa
M =	0.5877	S =	2.8817	
NI =	5.9964	M =	6.7520	mmol/g
BIS =6.84284−02		BIS =5.18212−02		1/g
QIS =	23.5122	QIS =	22.3177	KJ/mol

P (kPa)	N (mmol/g)	DN TOTH	DN UNILAN
132.6344	1.2716	−0.0110	−0.0117
161.3094	1.4358	−0.0219	−0.0210
194.9493	1.5853	−0.0136	−0.0119
250.3741	1.7250	0.0680	0.0695
273.6788	1.8811	−0.0063	−0.0052
321.6055	2.0244	0.0025	0.0028
370.8495	2.1725	−0.0080	−0.0086
424.3491	2.3064	−0.0094	−0.0105
473.7957	2.4112	−0.0046	−0.0059
581.7068	2.6160	−0.0032	−0.0036
646.4535	2.7191	0.0004	0.0012
AVERAGE DN		0.0135	0.0138

DN = (NCAL−NEXP)

ETHYLENE
```
--------------------------------------------------------------------------
```
CARBON MOLECULAR SIEVE
```
--------------------------------------------------------------------------
```
ADSORBENT : MSC-5A;TAKEDA CHEMICAL Co.;0.56 cc/g;650 Sqm/g
EXP.   : STATIC GRAVIMETRIC (McBAIN QUARTZ SPRING BALANCE)
REF.   : NAKAHARA,T.,HIRATA,M.and OMORI,T.;J.Chem.Eng.Data;19,310(1974)

              TEMP   (K)=   278.65
              PSAT (kPa)= 4645.28
```

| | TOTH | | UNILAN | UNITS |
|---|---|---|---|---|
| B = | 1.0879 | C = | 7.3365 | kPa |
| M = | .4500 | S = | 3.0585 | |
| NI = | 3.4264 | M = | 3.0400 | mmol/g |
| DG = | -46.4644 | DG = | -44.5112 | J/g |
| BIS= | .65821+001 | BIS= | .33347+001 | 1/g |
| QIS= | 35.4943 | QIS= | 31.7527 | KJ/mol |

| P(kPa) | N(mmol/g) | DN TOTH | DN UNILAN |
|---|---|---|---|
| .1067 | .1568 | .0025 | -.0230 |
| .2933 | .3422 | -.0180 | -.0370 |
| 1.3330 | .7735 | -.0020 | .0090 |
| 3.7330 | 1.1910 | .0122 | .0254 |
| 6.8670 | 1.4690 | .0150 | .0211 |
| 10.8700 | 1.6860 | .0111 | .0115 |
| 16.2700 | 1.8890 | -.0091 | -.0120 |
| 21.8000 | 2.0280 | -.0201 | -.0239 |
| 32.3300 | 2.1850 | -.0130 | -.0161 |
| 46.6700 | 2.3170 | -.0023 | -.0037 |
| 71.7300 | 2.4630 | .0051 | .0048 |
| 90.8000 | 2.5340 | .0115 | .0105 |
| | AVERAGE DN | .0102 | .0165 |

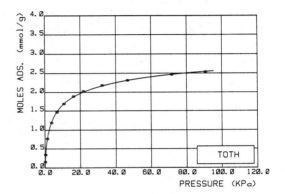

```
              TEMP   (K)=    303.15
```

	TOTH		UNILAN	UNITS
B =	2.0024	C =	17.7090	kPa
M =	.4966	S =	2.7613	
NI =	3.2412	M =	2.9414	mmol/g
BIS=	.20183+001	BIS=	.11944+001	1/g
QIS=	35.4943	QIS=	31.7527	KJ/mol

P(kPa)	N(mmol/g)	DN TOTH	DN UNILAN
1.6670	.4990	-.0082	-.0164
5.6930	.9268	.0184	.0244
9.6800	1.1940	-.0094	-.0048
13.4700	1.3400	.0005	.0025
20.0000	1.5330	-.0040	-.0052
29.2800	1.7220	-.0132	-.0164
40.9300	1.8430	.0189	.0157
69.9300	2.1000	-.0080	-.0080
86.4700	2.1740	.0028	.0049
	AVERAGE DN	.0093	.0109

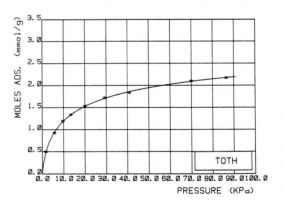
```

TEMP (K)=   323.15

	TOTH		UNILAN	UNITS
B =	3.0396	C =	70.3956	kPa
M =	.5000	S =	3.4204	
NI =	3.1948	M =	3.4382	mmol/g
BIS=	.92880+000	BIS=	.58597+000	1/g
QIS=	35.4943	QIS=	31.7527	KJ/mol

P(kPa)	N(mmol/g)	DN TOTH	DN UNILAN
.6400	.1283	.0104	-.0052
1.5730	.2852	-.0127	-.0238
2.8000	.4135	-.0109	-.0142
7.0670	.7058	-.0106	-.0020
12.6700	.9375	-.0082	.0006
19.4700	1.0800	.0398	.0448
27.0700	1.2620	.0107	.0115
39.8700	1.4540	.0015	-.0017
56.6700	1.6470	-.0261	-.0300
72.1300	1.7290	.0033	.0015
83.8700	1.8040	-.0034	-.0025
AVERAGE DN		.0125	.0125

DN=(NCAL-NEXP)

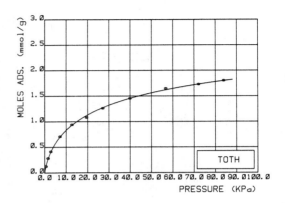

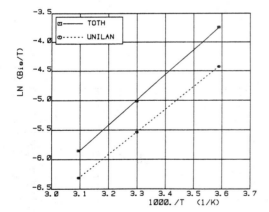

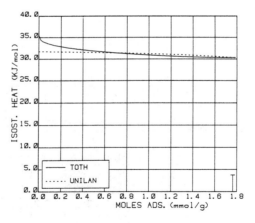

ETHYLENE

CARBON MOLECULAR SIEVE

ADSORBENT : MSC-5A;TAKEDA CHEMICAL Co.;0.56 cc/g;650 Sqm/g
EXP.   : STATIC GRAVIMETRIC (McBAIN QUARTZ SPRING BALANCE)
REF.   : NAKAHARA,T.,HIRATA,M.and MORI,H.;J.Chem.Eng.Data;27,317(1982)

```
          TEMP   (K)=    274.85
          PSAT (kPa)=   4267.71
```

	TOTH		UNILAN	UNITS
B =	.6323	C =	208.7710	kPa
M =	.2458	S =	6.8975	
NI =	6.4062	M =	6.6046	mmol/g
DG =	-57.1551	DG =	-54.2577	J/g
BIS=	.94492+002	BIS=	.51867+001	1/g
QIS=	62.5953	QIS=	31.5602	KJ/mol

P(kPa)	N(mmol/g)	DN TOTH	DN UNILAN
.9870	.8500	.0184	-.0185
1.0400	.9010	-.0149	-.0487
2.5600	1.2400	-.0151	-.0070
2.8130	1.3160	-.0523	-.0411
4.9200	1.5360	-.0297	-.0080
5.5070	1.4100	.1475	.1699
7.8260	1.7460	-.0245	-.0032
9.2930	1.7460	.0578	.0771
12.6100	1.9850	-.0320	-.0186
16.2700	2.1700	-.0899	-.0833
20.6700	2.2020	-.0009	-.0021
25.0400	2.3190	-.0200	-.0281
30.3100	2.3560	.0410	.0257
32.1100	2.4370	-.0103	-.0279
33.3500	2.4220	.0242	.0051
AVERAGE DN		.0386	.0376

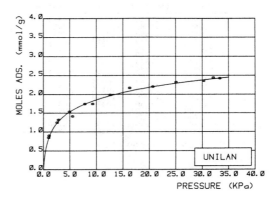

```
          TEMP   (K)=    303.15
```

	TOTH		UNILAN	UNITS
B =	1.1075	C =	335.3200	kPa
M =	.2879	S =	6.2321	
NI =	6.0473	M =	5.7278	mmol/g
BIS=	.10690+002	BIS=	.17575+001	1/g
QIS=	62.5953	QIS=	31.5602	KJ/mol

P(kPa)	N(mmol/g)	DN TOTH	DN UNILAN
.6670	.3700	-.0053	-.0487
1.0530	.4620	.0043	-.0233
1.3330	.5240	.0023	-.0157
2.2000	.6740	-.0023	.0004
2.8930	.7540	.0077	.0201
3.9070	.8520	.0170	.0375
5.6400	1.0140	-.0020	.0234
6.7870	1.1220	-.0330	-.0078
11.8400	1.3350	.0035	.0172
11.9200	1.3310	.0107	.0242
18.0100	1.5560	-.0133	-.0194
22.8700	1.6650	-.0006	-.0221
24.0100	1.6770	.0126	-.0124
AVERAGE DN		.0088	.0209

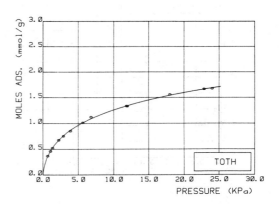

TEMP   (K)=   323.15

	TOTH		UNILAN	UNITS
B =	2.1941	C =	359.1110	kPa
M =	.3984	S =	5.2416	
NI =	4.7087	M =	5.6264	mmol/g
BIS=	.17601+001	BIS=	.75874+000	1/g
QIS=	62.5953	QIS=	31.5602	KJ/mol

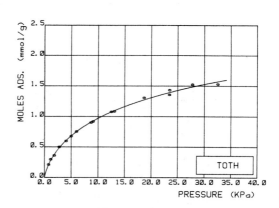

P(kPa)	N(mmol/g)	DN TOTH	DN UNILAN
.7600	.2150	-.0043	-.0345
1.1600	.2980	-.0156	-.0423
1.8130	.3640	.0142	-.0045
2.7200	.5020	-.0167	-.0252
4.0270	.5970	.0115	.0135
4.9730	.6780	.0050	.0119
5.9730	.7520	.0003	.0108
8.6930	.8980	.0095	.0241
9.0670	.9180	.0080	.0227
12.5200	1.0740	.0004	.0136
13.0800	1.0840	.0115	.0240
18.6500	1.3070	-.0340	-.0293
23.4100	1.3550	.0377	.0344
23.4300	1.4320	-.0389	-.0421
27.7700	1.5230	-.0378	-.0483
32.5500	1.5290	.0439	.0258
AVERAGE DN		.0181	.0254

DN=(NCAL-NEXP)

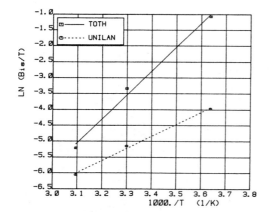

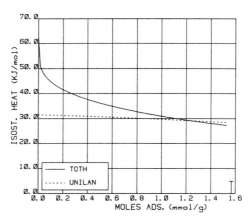

```
ETHYLENE
--------------------------------------------------------------------------------
SILICA GEL
--------------------------------------------------------------------------------
ADSORBENT : DAVISON CHEMICAL Co.;REF.GRADE;751 Sqm/g;mesh 14x20
EXP.      : STATIC VOLUMETRIC
REF.      : LEWIS,W.K.,GILLILAND,E.R.,CHERTOW,B. and MILLIKEN,W.;J.Am.Chem.Soc.;
            72,1157(1950)
```

TEMP   (K)=   298.15

	TOTH		UNILAN	UNITS
B =	63.4881	C =	7390.1000	kPa
M =	.8449	S =	5.1558	
NI =	2.5369	M =	8.1058	mmol/g
BIS=	.46233-001	BIS=	.45729-001	1/g

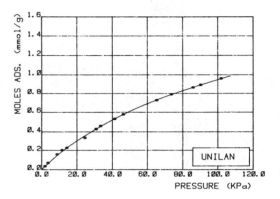

P(kPa)	N(mmol/g)	DN TOTH	DN UNILAN
1.8930	.0348	-.0006	-.0006
3.5330	.0687	-.0062	-.0061
8.7200	.1572	-.0116	-.0108
11.6300	.2030	-.0144	-.0133
14.2700	.2280	-.0022	-.0009
24.6000	.3353	.0218	.0229
30.8700	.4280	-.0003	.0003
33.4000	.4599	-.0054	-.0050
41.3900	.5320	.0017	.0015
46.2800	.5797	-.0011	-.0017
65.2900	.7270	.0047	.0033
73.6000	.7910	-.0011	-.0024
85.9100	.8660	.0022	.0018
90.2100	.8950	-.0014	-.0013
101.9000	.9620	-.0037	-.0020
AVERAGE DN		.0052	.0049

DN=(NCAL-NEXP)

ETHYLENE

SILICA GEL

ADSORBENT : DAVISON CHEMICAL Co.;REF.GRADE;751 Sqm/g;mesh 14x20
EXP.   : STATIC VOLUMETRIC
REF.   : LEWIS,W.K.,GILLILAND,E.R.,CHERTOW,B. and BAREIS,D.;J.Am.Chem.Soc.;
            72,1160(1950)

                    TEMP   (K)=   273.15
                    PSAT (kPa)= 4106.78

              TOTH   UNILAN   UNITS
        ------------   --------------   ------
B   =   5.7359   C   =   3013.2000   kPa
M   =   .4126   S   =   5.4288
NI =   7.0687   M   =   7.8471   mmol/g
DG =   -32.3854   DG =   -29.6902   J/g
BIS= .23268+000   BIS= .12412+000   1/g
QIS=   31.6480   QIS=   25.3664   KJ/mol

   P(kPa)   N(mmol/g)   DN TOTH   DN UNILAN

     .2533   .0120   .0086   .0017
    1.0130   .0597   .0105   -.0064
    1.1200   .0826   -.0062   -.0238
    3.1200   .1710   .0051   -.0179
    3.6400   .2061   -.0076   -.0304
    8.3200   .3695   -.0038   -.0167
    9.0530   .3839   .0041   -.0070
   13.3500   .5160   -.0105   -.0116
   17.7700   .6038   .0049   .0118
   30.1900   .8338   .0075   .0253
   39.4100   .9930   -.0134   .0053
   46.7200   1.0680   .0075   .0243
   63.4800   1.2560   .0071   .0145
   73.2700   1.3600   -.0028   -.0027
   84.2100   1.4490   .0033   .0057
   99.4000   1.5760   -.0057   -.0280
   AVERAGE DN   .0068   .0146

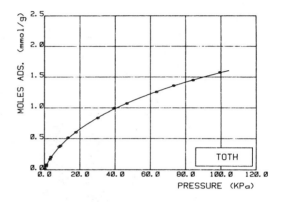

                    TEMP   (K)=   313.15

              TOTH   UNILAN   UNITS
        ------------   --------------   ------
B   =   18.0843   C   =   295.1580   kPa
M   =   .5092   S   =   1.7863
NI =   5.0893   M   =   2.3857   mmol/g

BIS= .44980-001   BIS= .34163-001   1/g
QIS=   31.6480   QIS=   25.3664   KJ/mol

   P(kPa)   N(mmol/g)   DN TOTH   DN UNILAN

    1.4130   .0277   -.0062   -.0094
    2.6670   .0386   .0002   -.0045
    5.0000   .0738   -.0053   -.0114
    9.8270   .1283   -.0050   -.0110
   13.4500   .1588   .0016   -.0033
   16.2300   .1912   -.0040   -.0078
   20.0400   .2186   .0032   .0010
   28.5500   .2894   .0032   .0041
   37.9500   .3656   -.0031   .0004
   45.6700   .4092   .0058   .0105
   59.2000   .4940   .0046   .0095
   69.9500   .5638   -.0049   -.0012
   82.7100   .6210   .0039   .0047
   92.8300   .6760   -.0024   -.0047
  101.8000   .7170   -.0027   -.0081
   AVERAGE DN   .0037   .0061

          DN=(NCAL-NEXP)

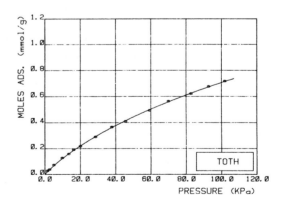

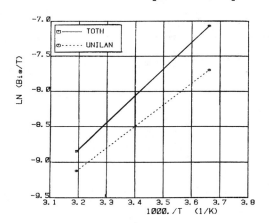

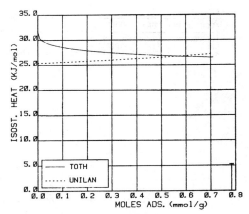

ETHYLENE

ZEOLITE MOLECULAR SIEVE

ADSORBENT : 13X;LINDE(UNION CARBIDE);.3 cc/g;525 Sqm/g (N2)
EXP.   : STATIC VOLUMETRIC
REF.   : HYUN,S.H.and DANNER,R.P.;J.Chem.Eng.Data;27,196(1982)

                    TEMP   (K)=   298.15

	TOTH		UNILAN	UNITS
B =	1.2615	C =	4.8016	kPa
M =	.5521	S =	2.4629	
NI =	3.3021	M =	3.0525	mmol/g
BIS=	.53738+001	BIS=	.37280+001	1/g
QIS=	38.7925	QIS=	37.3581	KJ/mol

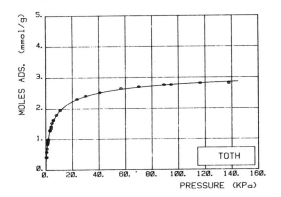

P(kPa)	N(mmol/g)	DN TOTH	DN UNILAN
.2400	.4100	-.1121	-.1265
.2700	.4100	-.0854	-.0989
.3400	.4100	-.0278	-.0388
.5700	.5800	-.0411	-.0454
.6100	.4100	.1523	.1490
.8100	.6900	-.0224	-.0221
1.1500	.8300	-.0169	-.0134
1.3100	.8900	-.0188	-.0145
1.5500	.8700	.0791	.0839
1.5900	.9400	.0211	.0260
1.6200	.9700	.0000	.0049
1.8500	1.0100	.0242	.0290
3.0700	1.2900	.0015	.0037
3.2000	1.3100	.0032	.0051
3.9100	1.3600	.0587	.0590
3.9800	1.4300	-.0019	-.0017
4.6500	1.5400	-.0295	-.0305
5.6000	1.6400	-.0312	-.0334
5.7000	1.6300	-.0119	-.0142
7.9900	1.8000	-.0060	-.0094
10.8600	1.9500	-.0016	-.0043
23.4000	2.3100	-.0111	-.0067
29.9800	2.4100	-.0114	-.0046
40.8000	2.5200	-.0074	.0010
56.6500	2.6500	-.0282	-.0210
70.1700	2.7000	-.0137	-.0093
88.9800	2.7600	-.0081	-.0087
94.4100	2.7600	.0073	.0051
116.2000	2.8100	.0085	-.0004
137.8000	2.8100	.0473	.0319
AVERAGE DN		.0297	.0301

TEMP   (K)=   323.15

	TOTH		UNILAN	UNITS
B =	2.5552	C =	15.0813	kPa
M =	.5581	S =	2.3108	
NI =	3.3856	M =	3.0561	mmol/g
BIS=	.16938+001	BIS=	.11761+001	1/g
QIS=	38.7925	QIS=	37.3581	KJ/mol

P(kPa)	N(mmol/g)	DN TOTH	DN UNILAN
.6700	.2500	.0093	-.0081
.7400	.2800	-.0004	-.0174
.9400	.3800	-.0464	-.0616
1.7900	.5600	-.0404	-.0473
1.8500	.5700	-.0392	-.0456
3.0000	.7000	.0139	.0148
4.3200	.9000	-.0258	-.0206
5.6000	.9700	.0284	.0356
5.6600	1.0100	-.0063	.0009
6.7100	1.0300	.0593	.0670
12.0400	1.3500	.0496	.0562
15.5800	1.4900	.0506	.0557
17.1000	1.5800	.0116	.0161
19.5600	1.6700	-.0050	-.0013
36.2100	2.0200	-.0284	-.0272
39.6500	2.0700	-.0325	-.0315
49.7700	2.1800	-.0305	-.0301
53.3800	2.2200	-.0371	-.0368
62.8800	2.2800	-.0208	-.0211
65.0400	2.2900	-.0155	-.0159
72.2300	2.3300	-.0086	-.0096
78.5300	2.3700	-.0122	-.0137
79.6100	2.3700	-.0063	-.0079
95.9300	2.4300	.0117	.0083
137.8000	2.5000	.0812	.0715
AVERAGE DN		.0268	.0289

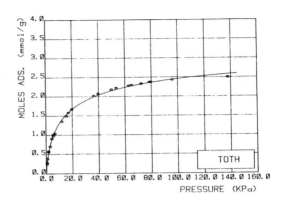

TEMP   (K)=   373.15

	TOTH		UNILAN	UNITS
B =	8.1584	C =	160.7310	kPa
M =	.5859	S =	2.9002	
NI =	3.3456	M =	3.7053	mmol/g
BIS=	.28852+000	BIS=	.22345+000	1/g
QIS=	38.7925	QIS=	37.3581	KJ/mol

P(kPa)	N(mmol/g)	DN TOTH	DN UNILAN
1.0500	.0700	.0097	.0014
1.3100	.1200	-.0231	-.0320
1.8500	.1300	.0005	-.0090
2.7000	.1800	-.0010	-.0104
3.7800	.2400	-.0053	-.0135
5.7700	.3200	.0047	-.0004
6.9500	.3800	-.0078	-.0110
12.7100	.5500	.0134	.0164
18.2800	.6900	.0166	.0220
19.1200	.7400	-.0143	-.0087
25.1500	.8700	-.0211	-.0153
32.4000	.9700	.0018	.0066
42.0800	1.1000	.0064	.0092
43.2300	1.1000	.0207	.0233
45.0500	1.1300	.0126	.0148
46.5300	1.1400	.0199	.0218
51.1200	1.2000	.0107	.0117
56.1200	1.2600	.0017	.0018
61.3700	1.3200	-.0090	-.0097
61.7700	1.3300	-.0154	-.0162
63.2900	1.3500	-.0219	-.0229
69.9300	1.3900	-.0063	-.0081
70.8800	1.4000	-.0088	-.0106
81.1900	1.4700	-.0026	-.0051
87.5300	1.5300	-.0204	-.0230
93.5700	1.5500	-.0028	-.0054
96.0700	1.5500	.0120	.0094
126.1000	1.7200	-.0058	-.0061
137.8000	1.7400	.0234	.0246
AVERAGE DN		.0110	.0128

DN=(NCAL-NEXP)

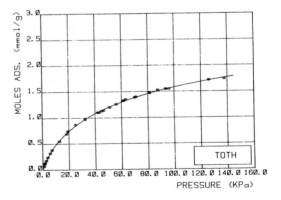

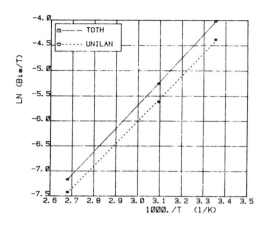

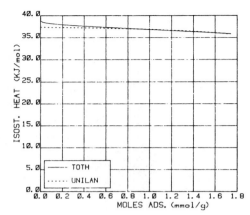

ETHYLENE

ZEOLITE

ADSORBENT : 13X; UNION CARBIDE: 1/16 in PELLETS
EXP.   : STATIC VOLUMETRIC
REF.   : KAUL, B. K., *IND. ENG. CHEM. RES.* **26**, 928 (1987)

TEMP (K)   =   323.15

TOTH		UNILAN		UNIT
B =	4.5079	C =	16.6877	kPa
M =	.6875	S =	1.9025	
NI =	3.1854	M =	3.0624	mmol/g
BIS =.95755+000		BIS =.84916+000		1/g
QIS =	33.3082	QIS =	33.1801	KJ/mol

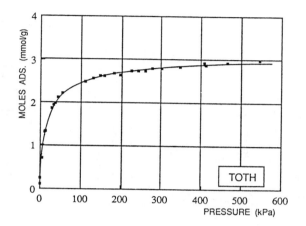

P (kPa)	N (mmol/g)	DN TOTH	DN UNILAN
0.2758	0.1138	-0.0272	-0.0312
0.2758	0.1147	-0.0281	-0.0321
1.0342	0.2458	0.0279	0.0264
4.4126	0.7009	0.0819	0.0887
10.2731	1.3116	-0.0676	-0.0670
12.1347	1.3375	0.0060	0.0046
28.0614	1.8631	-0.0175	-0.0253
32.1293	1.9394	-0.0166	-0.0240
36.3351	1.9795	0.0118	0.0049
43.3677	2.1223	-0.0357	-0.0412
55.0197	2.2124	-0.0046	-0.0073
110.3841	2.4944	0.0170	0.0228
130.7924	2.5590	0.0142	0.0212
146.7192	2.6269	-0.0143	-0.0069
158.4402	2.6295	0.0081	0.0156
181.6754	2.6786	0.0016	0.0087
196.8437	2.6425	0.0613	0.0679
225.5946	2.7295	0.0122	0.0177
242.8313	2.7379	0.0232	0.0278
259.9302	2.7455	0.0329	0.0365
276.2017	2.7964	-0.0031	-0.0004
300.6779	2.7906	0.0227	0.0239
346.1140	2.8285	0.0160	0.0143
407.4768	2.9316	-0.0539	-0.0594
410.9241	2.8696	0.0098	0.0040
466.7712	2.9284	-0.0254	-0.0346
546.7498	2.9940	-0.0639	-0.0777
AVERAGE DN		0.0261	0.0293

TEMP (K)   =   373.15

TOTH		UNILAN		UNIT
B =	12.0273	C =	97.6009	kPa
M =	.6583	S =	2.0892	
NI =	3.2128	M =	3.1755	mmol/g
BIS =.22790+000		BIS =.19216+000		1/g
QIS =	33.3083	QIS =	33.1801	KJ/mol

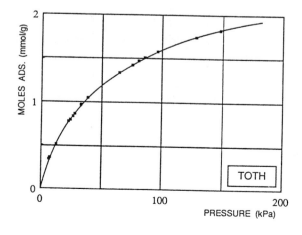

P (kPa)	N (mmol/g)	DN TOTH	DN UNILAN
6.6189	0.3400	-0.0090	-0.0143
7.3773	0.3587	0.0010	-0.0035
12.1347	0.5113	0.0066	0.0056
22.4078	0.7767	-0.0029	-0.0008
23.9936	0.8008	0.0055	0.0078
25.9930	0.8356	0.0098	0.0121
27.3030	0.8660	0.0040	0.0062
32.1982	0.9744	-0.0197	-0.0177
38.6103	1.0515	0.0006	0.0020
63.9828	1.3362	0.0039	0.0028
75.2901	1.4312	0.0047	0.0031
80.3232	1.4731	0.0011	-0.0007
85.2185	1.5151	-0.0059	-0.0077
96.1121	1.5820	-0.0017	-0.0034
127.8277	1.7489	-0.0019	-0.0016
148.1671	1.8287	0.0027	0.0050
AVERAGE DN		0.0051	0.0059

TEMP (K)　=　423.15

TOTH		UNILAN		UNIT
B =	33.9877	C =	433.0554	kPa
M =	.7035	S =	2.3662	
NI =	2.8377	M =	3.3043	mmol/g
BIS =	.66473−001	BIS =	.59916−001	1/g
QIS =	33.3083	QIS =	33.1801	KJ/mol

P (kPa)	N (mmol/g)	DN TOTH	DN UNILAN
5.1710	0.0870	−0.0010	−0.0041
6.1363	0.0995	0.0011	−0.0022
15.5820	0.2244	0.0020	−0.0001
16.4094	0.2351	0.0012	−0.0008
18.5467	0.2623	−0.0011	−0.0026
19.7878	0.2739	0.0014	0.0001
32.1982	0.3988	0.0031	0.0037
40.1272	0.4684	0.0038	0.0051
42.8850	0.4934	0.0017	0.0032
48.5387	0.5412	−0.0013	0.0003
53.9166	0.5831	−0.0032	−0.0015
60.3976	0.6259	−0.0006	0.0009
69.4296	0.6862	−0.0022	−0.0008
78.5306	0.7393	−0.0007	0.0002
82.0469	0.7575	0.0009	0.0017
95.4916	0.8320	−0.0023	−0.0022
108.9363	0.9003	−0.0062	−0.0067
123.8978	0.9592	−0.0001	−0.0012
138.0319	1.0185	−0.0035	−0.0051
155.1997	1.0797	−0.0028	−0.0047
192.5690	1.1805	0.0127	0.0110
214.3562	1.2412	0.0106	0.0095
242.7624	1.3197	0.0003	0.0006
277.4427	1.4040	−0.0106	−0.0080
AVERAGE DN		0.0031	0.0032

DN= (NCAL−NEXP)

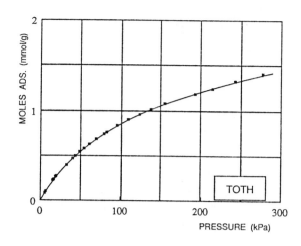

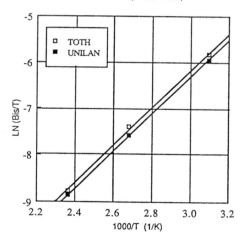

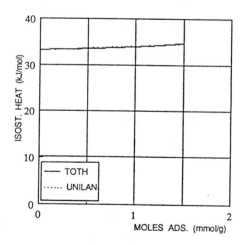

N-HEPTANE

SILICA GEL

ADSORBENT : DAVISON CHEM.Co.
EXP.   : STATIC GRAVIMETRIC
REF.   : SIRCAR,S.and MYERS,A.L.;AIChE J.;19,159(1973)

TEMP (K)= 303.15
PSAT (kPa)= 7.79

	TOTH		UNILAN	UNITS
B =	.3454	C =	9976.9800	kPa
M =	.9396	S =	12.8958	
NI =	2.5531	M =	11.7593	mmol/g
DG =	-20.1545	DG =	-20.7970	J/g
BIS=	.19948+002	BIS=	.45914+002	1/g

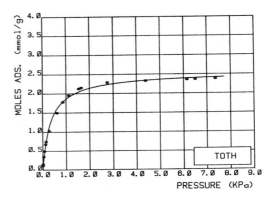

P(kPa)	N(mmol/g)	DN TOTH	DN UNILAN
.0055	.1110	-.0687	-.0209
.0092	.1500	-.0798	-.0073
.0387	.3460	-.0789	.0798
.0680	.4950	-.0639	.1036
.1187	.6800	-.0187	.1169
.1440	.7510	.0061	.1199
.2800	1.0280	.1068	.1120
.6133	1.4980	.1069	-.0213
.8666	1.7810	.0102	-.1519
1.1200	1.9560	-.0414	-.2129
1.5470	2.1300	-.0804	-.2424
1.6530	2.1500	-.0756	-.2326
2.7600	2.2880	-.0528	-.1396
4.3860	2.3300	.0082	.0280
6.1460	2.3600	.0330	.1511
6.5060	2.3660	.0349	.1710
7.3590	2.3790	.0377	.2139
AVERAGE DN		.0532	.1250

DN=(NCAL-NEXP)

HYDROGEN

ACTIVATED CARBON

ADSORBENT : COLUMBIA GRADE L;MESH 20x40;1152 Sq m/g (N2)
EXP.   : STATIC VOLUMETRIC
REF.   : RAY,G.C.and BOX,E.O.;IND.ENG.CHEM.;42,1315(1950)

TEMP   (K)=   310.92

	TOTH		UNILAN	UNITS
B =	39150.2998	C =	139725.0000	kPa
M =	.8915	S =	.4107	
NI =	28.3010	M =	26.8939	mmol/g
BIS=	.51577-003	BIS=	.51166-003	1/g
QIS=	1.2223	QIS=	1.1597	KJ/mol

P(kPa)	N(mmol/g)	DN TOTH	DN UNILAN
24.9400	.0083	-.0033	-.0034
64.3800	.0218	-.0090	-.0091
99.3400	.0327	-.0129	-.0131
99.9600	.0307	-.0108	-.0110
1014.0000	.1741	-.0255	-.0250
1772.0000	.3584	-.0126	-.0124
AVERAGE DN		.0124	.0123

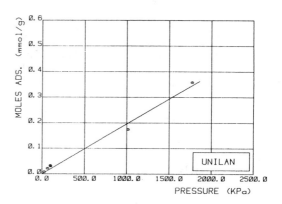

TEMP   (K)=   338.70

	TOTH		UNILAN	UNITS
B =	2694.5000	C =	79437.5996	kPa
M =	.6965	S =	.7177	
NI =	22.1147	M =	18.2203	mmol/g
BIS=	.73925-003	BIS=	.70278-003	1/g
QIS=	1.2223	QIS=	1.1597	KJ/mol

P(kPa)	N(mmol/g)	DN TOTH	DN UNILAN
99.9600	.0241	.0018	.0008
327.5000	.0768	.0067	.0045
599.9000	.1484	.0021	-.0001
1028.0000	.2713	-.0185	-.0189
1314.0000	.3073	.0121	.0139
AVERAGE DN		.0082	.0076

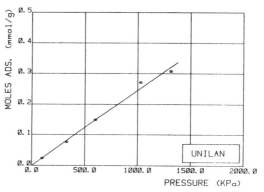

TEMP   (K)=   366.48

	TOTH		UNILAN	UNITS
B =	40602.7002	C =	87502.0996	kPa
M =	.9178	S =	.6074	
NI =	19.1765	M =	14.9991	mmol/g
BIS=	.55600-003	BIS=	.55500-003	1/g
QIS=	1.2223	QIS=	1.1597	KJ/mol

P(kPa)	N(mmol/g)	DN TOTH	DN UNILAN
56.7800	.0105	-.0001	-.0001
99.7600	.0170	.0012	.0011
99.9600	.0174	.0008	.0008
448.2000	.0768	.0044	.0043
903.4002	.1638	-.0012	-.0012
1368.0000	.2456	-.0009	-.0010
AVERAGE DN		.0014	.0014

DN=(NCAL-NEXP)

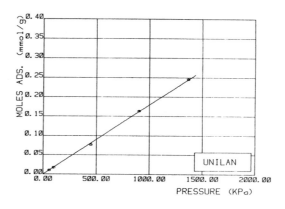

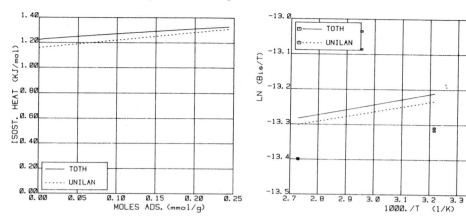

HYDROGEN

ACTIVATED CARBON

ADSORBENT : PCB, Calgon Corporation
EXP.   : Static
REF.   : Ritter, J. A. and R. T. Yang, *Ind. Eng. Chem.*
            Res. **26**, 1679 (1987).

TEMP (K)   =   296.

TOTH		UNILAN		UNIT
B =	147.3007	C =	69991.9400	kPa
M =	.432146	S =	2.3220	
NI =	47.6334	M =	11.3767	mmol/g
BIS =	.11265-002	BIS =	.86978-003	1/g
QIS =	13.8964	QIS =	12.4748	KJ/mol

P (kPa)	N (mmol/g)	DN TOTH	DN UNILAN
586.0495	0.1874	0.0248	0.0113
689.4700	0.3748	-0.1291	-0.1427
2068.4099	0.6157	0.0249	0.0225
3206.0356	0.9101	0.0122	0.0178
4467.7656	1.1689	0.0367	0.0439
5777.7588	1.4633	0.0125	0.0129
6377.5977	1.6329	-0.0398	-0.0451
	AVERAGE DN	0.0400	0.0423

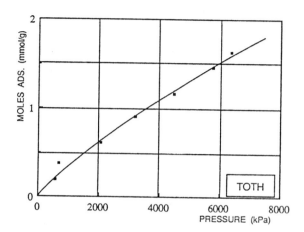

TEMP (K)   =   373.

TOTH		UNILAN		UNIT
B =	854.3569	C =	52494.4000	kPa
M =	.5108	S =	.3472	
NI =	79.8687	M =	6.9652	mmol/g
BIS =	.45144-003	BIS =	.41979-003	1/g
QIS =	13.8964	QIS =	12.4748	KJ/mol

P (kPa)	N (mmol/g)	DN TOTH	DN UNILAN
551.5760	0.0758	-0.0001	-0.0020
1241.0460	0.1651	0.0007	-0.0012
2137.3569	0.2811	-0.0031	-0.0037
3309.4561	0.4149	0.0040	0.0051
4688.3960	0.5800	-0.0017	-0.0001
5901.8633	0.7138	-0.0000	0.0003
6515.4917	0.7807	0.0001	-0.0010
	AVERAGE DN	0.0014	0.0019

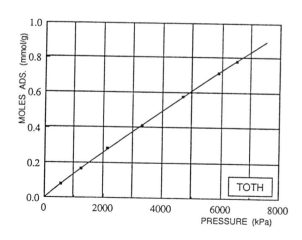

TEMP (K)   =   480.

TOTH		UNILAN		UNIT
B	= 1122.6680	C	=39982.4200	kPa
M	= .5522	S	= .3812	
NI	= 17.5369	M	= 1.9671	mmol/g
BIS	=.209512-03	BIS	=.201129-03	1/g
QIS	= 13.8964	QIS	= 12.4748	KJ/mol

P (kPa)	N (mmol/g)	DN TOTH	DN UNILAN
482.6290	0.0223	0.0018	0.0017
1309.9930	0.0580	0.0052	0.0058
2206.3040	0.1026	0.0011	0.0024
3426.6660	0.1785	-0.0221	-0.0203
4943.5000	0.2008	0.0182	0.0193
6150.0723	0.2677	-0.0008	-0.0014
6667.1748	0.2900	-0.0030	-0.0047
AVERAGE DN		0.0074	0.0079

DN= (NCAL-NEXP)

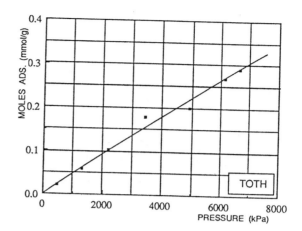

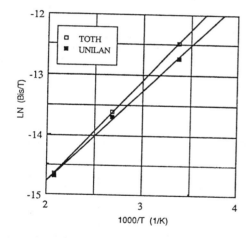

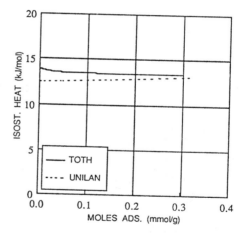

HYDROGEN SULFIDE

ACTIVATED CARBON

ADSORBENT : PCB, Calgon Corp.
EXP.   : Static
REF.   : Ritter, J. A. and R. T. Yang, *Ind. Eng. Chem.*
            Res. **26,** 1679 (1987).

              TEMP (K)   =   296.00
              PSAT (kPa) =   1913.19

TOTH		UNILAN		UNIT
B =	447.3047	C =	180.2157	kPa
M =	1.1562	S =	.27931−02	
NI =	10.4034	M =	10.9016	mmol/g
DG =	−65.1265	DG =	−65.7811	J/g
BIS=.130553+00		BIS=.148867+00		1/g
QIS=	16.2626	QIS=	18.0364	KJ/mol

P (kPa)	N (mmol/g)	DN TOTH	DN UNILAN
55.8471	2.6322	−0.1627	−0.0531
159.2676	4.9521	0.1630	0.1622
324.7404	7.0936	−0.0052	−0.0827
679.1279	8.7889	−0.1391	−0.1735
1151.4149	9.2797	0.0858	0.1462
AVERAGE DN		0.1112	0.1235

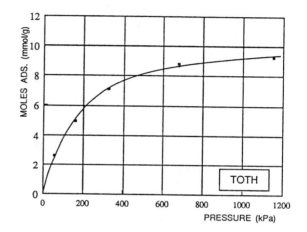

              TEMP (K)   =   373.00
              PSAT (kPa) =   8916.33

TOTH		UNILAN		UNIT
B =	134.5472	C =	982.3661	kPa
M =	.7573	S =	1.5242	
NI =	11.2283	M =	10.8951	mmol/g
DG =	−79.2399	DG =	−79.2875	J/g
BIS=.537918−01		BIS=.493459−01		1/g
QIS=	16.2626	QIS=	18.0364	KJ/mol

P (kPa)	N (mmol/g)	DN TOTH	DN UNILAN
90.3206	1.2135	−0.0149	−0.0257
238.5566	2.4716	0.0179	0.0213
441.9503	3.6673	−0.0024	0.0025
730.8382	4.7737	−0.0025	−0.0022
992.8368	5.4875	−0.0140	−0.0156
1220.3619	5.9336	0.0118	0.0108
AVERAGE DN		0.0106	0.0130

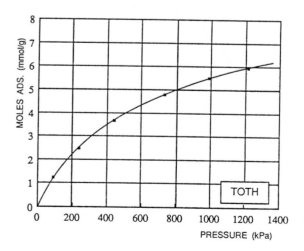

TEMP (K)   =   480.00

TOTH		UNILAN		UNIT
B	= 163.9034	C	= 8998.1150	kPa
M	= .6386	S	= 2.5969	
NI	= 12.2129	M	= 12.6394	mmol/g
BIS	=.165963−01	BIS	=.144059−01	1/g
QIS	= 16.2626	QIS	= 18.0364	KJ/mol

P (kPa)	N (mmol/g)	DN TOTH	DN UNILAN
172.3675	0.5666	−0.0013	−0.0132
403.3400	1.1377	−0.0004	0.0005
710.1541	1.7310	0.0047	0.0127
1034.2050	2.2530	−0.0059	−0.0027
1296.2036	2.5965	0.0025	−0.0035
AVERAGE DN		0.0030	0.0065

DN= (NCAL−NEXP)

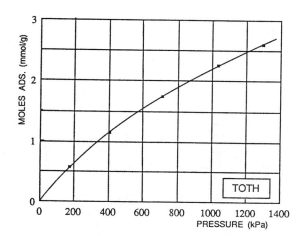

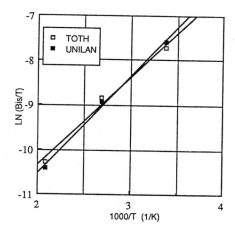

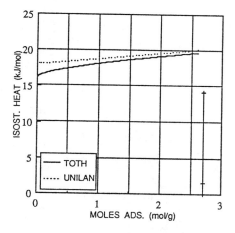

HYDROGEN SULFIDE

H-MORDENITE

ADSORBENT : Norton Co.; TYPE Z-900H; 1/16 in. Pellets
EXP.   : Static Volumetric
REF.   : Talu, O., Zwiebel, I.; *AIChE J.* **32**, 1263 (1986)

TEMP (K) =		283.15
PSAT (kPa) =		1369.38

TOTH		UNILAN		UNIT
B =	.3252	C =	982.0691	kPa
M =	.2425	S =	10.8325	
NI =	4.5675	M =	7.8357	mmol/g
DG =	-54.2715	DG =	-55.9098	J/g
BIS=	.11047+04	BIS=	.43906+02	1/g
QIS=	44.5314	QIS=	37.4425	KJ/mol

P (kPa)	N (mmol/g)	DN TOTH	DN UNILAN
0.41	1.13	-0.0017	-0.0097
0.54	1.21	0.0090	0.0059
0.66	1.28	0.0067	0.0062
0.84	1.35	0.0196	0.0212
0.94	1.42	-0.0112	-0.0089
1.10	1.48	-0.0160	-0.0132
1.30	1.54	-0.0168	-0.0137
1.52	1.59	-0.0109	-0.0079
1.87	1.65	0.0037	0.0062
2.28	1.72	0.0054	0.0072
2.67	1.78	0.0028	0.0039
3.07	1.84	-0.0064	-0.0060
4.83	1.98	0.0185	0.0171
5.87	2.06	0.0091	0.0074
7.09	2.13	0.0072	0.0054
8.61	2.21	-0.0032	-0.0045
9.83	2.27	-0.0160	-0.0167
	AVERAGE DN	0.0097	0.0095

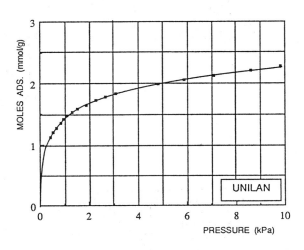

TEMP (K) =		303.15
PSAT (kPa) =		2275.94

TOTH		UNILAN		UNIT
B =	.3681	C =	820.1335	kPa
M =	.1932	S =	9.9139	
NI =	5.7074	M =	6.8722	mmol/g
DG =	-55.9188	DG =	-54.4018	J/g
BIS=	.25379+004	BIS=	.21527+002	1/g
QIS=	44.5314	QIS=	37.4425	KJ/mol

P (kPa)	N (mmol/g)	DN TOTH	DN UNILAN
0.49	0.89	0.0306	0.0010
0.58	0.96	0.0073	-0.0147
0.76	1.03	0.0148	0.0036
0.86	1.10	-0.0188	-0.0256
1.10	1.16	-0.0045	-0.0038
1.36	1.23	-0.0086	-0.0026
1.70	1.30	-0.0075	0.0027
1.89	1.38	-0.0531	-0.0414
2.57	1.45	-0.0215	-0.0068
3.32	1.51	0.0053	0.0208
4.03	1.57	0.0121	0.0272
4.87	1.63	0.0183	0.0322
5.73	1.69	0.0157	0.0281
6.78	1.75	0.0157	0.0261
8.09	1.82	0.0092	0.0170
9.60	1.88	0.0112	0.0160
11.67	1.96	0.0024	0.0034
14.32	2.04	-0.0026	-0.0059
17.10	2.11	-0.0072	-0.0145
20.26	2.18	-0.0146	-0.0259
25.95	2.26	-0.0030	-0.0202
27.62	2.28	0.0001	-0.0187
	AVERAG DN	0.0129	0.0163

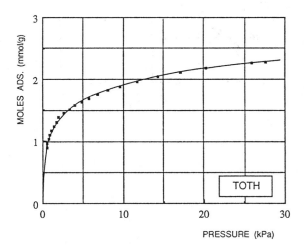

TEMP (K)   =   338.15
PSAT (kPa) =   4801.53

TOTH		UNILAN		UNIT
B =	.6684	C =	265.0502	kPa
M =	.2550	S =	7.0752	
NI =	4.7882	M =	5.2415	mmol/g
DG =	−54.3484	DG =	−53.6063	J/g
BIS =	.65349+002	BIS =	.46453+001	1/g
QIS =	44.5314	QIS =	37.4425	KJ/mol

P (kPa)	N (mmol/g)	DN TOTH	DN UNILAN
0.48	0.39	0.0813	0.0340
0.55	0.48	0.0205	−0.0209
0.68	0.51	0.0384	0.0066
0.85	0.60	0.0020	−0.0196
1.16	0.72	−0.0381	−0.0457
1.62	0.81	−0.0354	−0.0294
2.21	0.90	−0.0331	−0.0166
2.62	0.97	−0.0502	−0.0289
4.11	1.10	−0.0326	−0.0029
5.92	1.21	−0.0156	0.0164
7.83	1.30	−0.0044	0.0266
11.02	1.42	0.0031	0.0302
14.82	1.51	0.0265	0.0481
17.94	1.59	0.0208	0.0379
22.44	1.67	0.0286	0.0398
25.66	1.72	0.0317	0.0390
31.70	1.81	0.0256	0.0267
38.63	1.91	0.0044	−0.0005
46.49	1.98	0.0084	−0.0023
54.16	2.05	−0.0007	−0.0160
62.07	2.11	−0.0064	−0.0257
70.31	2.17	−0.0169	−0.0397
79.18	2.24	−0.0398	−0.0658
AVERAGE DN		0.0245	0.0269

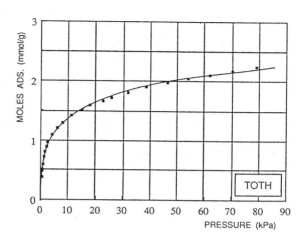

TEMP (K)   =   368.15
PSAT (kPa) =   8224.12

TOTH		UNILAN		UNIT
B =	.8459	C =	401.3063	kPa
M =	.2225	S =	6.2261	
NI =	6.3612	M =	4.9047	mmol/g
DG =	−57.0008	DG =	−53.4962	J/g
BIS =	.41309+002	BIS =	.15195+001	1/g
QIS =	44.5314	QIS =	37.4425	KJ/mol

P (kPa)	N (mmol/g)	DN TOTH	DN UNILAN
3.19	0.64	0.0233	−0.0045
5.27	0.83	−0.0265	−0.0290
17.81	1.22	−0.0070	0.0225
33.56	1.45	0.0110	0.0341
54.32	1.66	0.0027	0.0102
77.47	1.81	0.0074	−0.0016
101.87	1.95	−0.0104	−0.0348
AVERAGE DN		0.0126	0.0195

DN = (NCAL−NEXP)

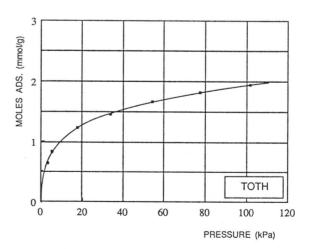

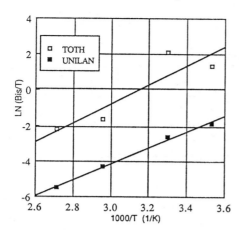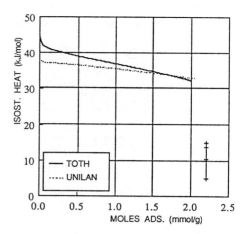

METHANE

ACTIVATED CARBON

ADSORBENT : Fiber Carbon, KF-1500, Toyobo Co., Ltd.
EXP.   : Gravimetric
REF.   : Kuro-Oka, M., Suzuki, T., Nitta, T. and
            Katayama, T., J. Chem. Eng. Japan, 17, 588
            (1984)

TEMP (K)   =   273.15

TOTH	UNILAN	UNIT
B = 5.7038	C = 1596.9990	kPa
M = .2678	S = 3.8716	
NI = 65.8450	M = 9.3707	mmol/g
BIS =.224460+00	BIS =.826008-01	1/g
QIS = 23.5638	QIS = 16.9834	KJ/mol

P(kPa)	N(mmol/g)	DN TOTH	DN UNILAN
0.4010	0.0390	-0.0145	-0.0245
0.4560	0.0450	-0.0176	-0.0285
1.3210	0.0850	-0.0166	-0.0379
2.0170	0.1330	-0.0356	-0.0618
3.0160	0.1420	-0.0067	-0.0370
5.1570	0.2250	-0.0174	-0.0506
6.6530	0.2570	-0.0038	-0.0364
9.8880	0.3550	-0.0122	-0.0401
13.7200	0.4390	-0.0011	-0.0212
17.2600	0.5190	-0.0009	-0.0134
21.1400	0.5780	0.0217	0.0171
26.5200	0.7010	0.0034	0.0082
31.8200	0.7820	0.0180	0.0299
35.5600	0.8620	0.0017	0.0175
40.3800	0.9300	0.0121	0.0313
53.0600	1.1250	0.0066	0.0284
63.5000	1.2740	-0.0005	0.0174
72.6700	1.3980	-0.0081	0.0028
86.2400	1.5640	-0.0134	-0.0175
99.3400	1.7090	-0.0142	-0.0368
AVERAGE DN		0.0113	0.0279

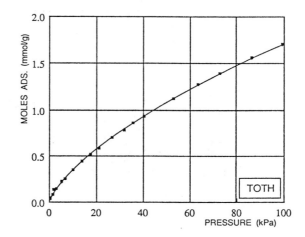

TEMP (K)   =   298.15

TOTH	UNILAN	UNIT
B = 16.3726	C = 2176.4760	kPa
M = .4390	S = 3.6007	
NI = 15.2740	M = 8.0149	mmol/g
BIS =.649494-01	BIS =.463887-01	1/g
QIS = 23.5638	QIS = 16.9834	KJ/mol

P(kPa)	N(mmol/g)	DN TOTH	DN UNILAN
0.9500	0.0430	-0.0212	-0.0254
2.1390	0.0670	-0.0205	-0.0277
2.6670	0.0770	-0.0200	-0.0282
4.4960	0.1060	-0.0146	-0.0249
6.6640	0.1480	-0.0185	-0.0298
9.4130	0.1820	-0.0072	-0.0185
13.3800	0.2370	-0.0013	-0.0112
15.8800	0.2760	-0.0040	-0.0126
20.0800	0.3280	0.0018	-0.0043
26.5200	0.4040	0.0086	0.0063
31.6400	0.4700	0.0042	0.0046
37.3300	0.5290	0.0102	0.0130
42.9800	0.5970	0.0035	0.0081
49.2600	0.6560	0.0094	0.0152
56.1800	0.7170	0.0166	0.0228
63.4500	0.8030	-0.0012	0.0046
68.2500	0.8470	-0.0018	0.0032
80.0700	0.9520	-0.0049	-0.0034
91.3600	1.0460	-0.0075	-0.0110
98.9200	1.1050	-0.0080	-0.0156
AVERAGE DN		0.0092	0.0145

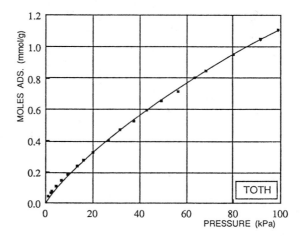

TEMP (K)   =   323.15

TOTH		UNILAN		UNIT
B	= 10.7588	C	= 2876.5030	kPa
M	= .3000	S	= 3.7367	
NI	= 56.0199	M	= 5.8747	mmol/g
BIS	=.547476-01	BIS	=.307891-01	1/g
QIS	= 23.5638	QIS	= 16.9834	KJ/mol

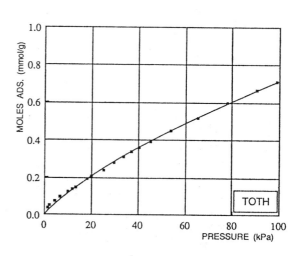

P (kPa)	N (mmol/g)	DN TOTH	DN UNILAN
1.5110	0.0340	-0.0119	-0.0169
2.1810	0.0480	-0.0173	-0.0234
4.4470	0.0740	-0.0163	-0.0246
6.5730	0.0940	-0.0131	-0.0221
10.0500	0.1220	-0.0059	-0.0145
11.7100	0.1380	-0.0060	-0.0141
13.4000	0.1440	0.0038	-0.0037
18.2600	0.1900	0.0007	-0.0045
20.0300	0.2030	0.0027	-0.0016
25.1800	0.2340	0.0135	0.0118
29.5500	0.2770	0.0043	0.0045
33.6300	0.3090	0.0027	0.0045
37.0200	0.3370	-0.0008	0.0022
40.3000	0.3590	0.0002	0.0042
45.1300	0.3870	0.0053	0.0103
53.6500	0.4460	0.0021	0.0081
65.0700	0.5170	0.0019	0.0072
77.8800	0.5970	-0.0033	-0.0009
90.4600	0.6670	-0.0037	-0.0062
99.0700	0.7110	-0.0020	-0.0088
AVERAGE DN		0.0059	0.0097

DN= (NCAL-NEXP)

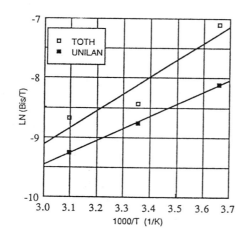

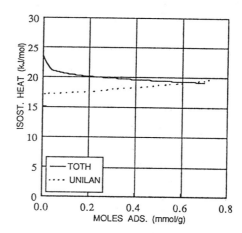

METHANE

ACTIVATED CARBON

ADSORBENT : Atrition Resistant Beads; Taiyo Kaken Co, Japan;
            800-1200 g/m2
EXP.   : STATIC VOLUMETRIC
REF.   : KAUL, B. K., *IND. ENG. CHEM. RES.* **26**, 928 (1987)

TEMP (K)   =   310.95

TOTH		UNILAN		UNIT
B =	21.6354	C =	3999.9650	kPa
M =	.4368	S =	3.2305	
NI =	16.1377	M =	9.5003	mmol/g
BIS=	.36616-001	BIS=	.23998-001	1/g
QIS=	22.5175	QIS=	18.2361	KJ/mol

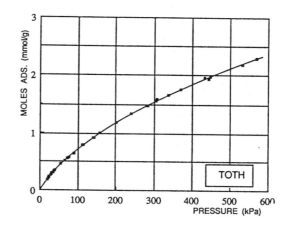

P (kPa)	N (mmol/g)	DN TOTH	DN UNILAN
18.5467	0.1816	0.0034	-0.0188
22.1320	0.2128	0.0023	-0.0205
23.9246	0.2262	0.0035	-0.0194
23.9936	0.2342	-0.0039	-0.0269
27.8546	0.2614	-0.0004	-0.0233
28.3372	0.2655	-0.0007	-0.0235
29.7162	0.2775	-0.0020	-0.0247
31.3709	0.2895	-0.0014	-0.0239
34.6803	0.3096	0.0034	-0.0186
36.9556	0.3315	-0.0018	-0.0232
39.2998	0.3426	0.0040	-0.0169
55.2265	0.4609	-0.0055	-0.0212
65.4996	0.5148	0.0056	-0.0062
70.8775	0.5603	-0.0071	-0.0169
75.7038	0.5737	0.0082	0.0003
88.1832	0.6438	0.0097	0.0066
111.9010	0.7914	-0.0113	-0.0063
113.9005	0.7950	-0.0047	0.0010
140.1003	0.9199	-0.0020	0.0109
140.5829	0.9137	0.0065	0.0195
157.3371	0.9985	-0.0018	0.0147
199.0500	1.1760	-0.0030	0.0191
239.1082	1.3290	-0.0020	0.0218
282.5448	1.4789	0.0017	0.0239
304.2631	1.5517	0.0013	0.0216
306.8141	1.5802	-0.0189	0.0011
337.1508	1.6543	0.0033	0.0195
370.2454	1.7431	0.0144	0.0254
433.6766	1.9652	-0.0291	-0.0310
444.0187	1.9251	0.0388	0.0344
446.7766	1.9679	0.0033	-0.0017
533.6498	2.1843	0.0053	-0.0220
570.1917	2.2914	-0.0162	-0.0539
AVERAGE DN		0.0069	0.0187

TEMP (K)   =   422.05

TOTH		UNILAN		UNIT
B =	5619.4730	C =	2842.4650	kPa
M =	1.1731	S =	.7841	
NI =	2.2482	M =	3.7268	mmol/g
BIS=	.50192-002	BIS=	.50867-002	1/g
QIS=	22.5175	QIS=	18.2361	KJ/mol

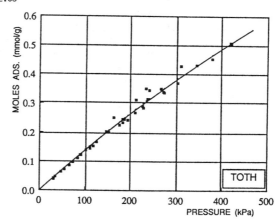

P (kPa)	N (mmol/g)	DN TOTH	DN UNILAN
30.2677	0.0424	0.0005	0.0009
30.6125	0.0397	0.0037	0.0040
34.6803	0.0468	0.0023	0.0026
46.8150	0.0634	0.0027	0.0031
47.5734	0.0625	0.0046	0.0050
54.3302	0.0723	0.0042	0.0045
66.0512	0.0870	0.0056	0.0059
66.1202	0.0861	0.0065	0.0069
71.7738	0.0977	0.0027	0.0030
83.7017	0.1124	0.0041	0.0044
89.0795	0.1213	0.0024	0.0027
90.8032	0.1236	0.0025	0.0027
102.3174	0.1379	0.0036	0.0038
110.5910	0.1495	0.0029	0.0031

111.3494	0.1454	0.0080	0.0081
117.7615	0.1530	0.0088	0.0089
125.4146	0.1673	0.0045	0.0045
147.0639	0.2021	−0.0023	−0.0025
152.3729	0.1999	0.0067	0.0065
162.3012	0.2498	−0.0307	−0.0309
175.6080	0.2262	0.0097	0.0093
182.1580	0.2449	−0.0009	−0.0013
182.6406	0.2338	0.0108	0.0105
186.8464	0.2431	0.0067	0.0063
193.6032	0.2423	0.0159	0.0154
210.2883	0.2663	0.0121	0.0116
211.5294	0.3114	−0.0314	−0.0319
227.1804	0.2895	0.0092	0.0087
227.4562	0.2837	0.0153	0.0148
233.7993	0.3507	−0.0440	−0.0446
236.4193	0.3145	−0.0048	−0.0054
241.3145	0.3444	−0.0289	−0.0295
266.8249	0.3502	−0.0051	−0.0056
272.2717	0.3400	0.0115	0.0110
273.2370	0.3355	0.0170	0.0165
302.4705	0.3703	0.0152	0.0149
309.5721	0.4287	−0.0354	−0.0356
344.7350	0.4310	0.0007	0.0008
377.8296	0.4515	0.0151	0.0158
419.1978	0.5077	0.0011	0.0027
419.8872	0.5068	0.0027	0.0043
AVERAGE DN		0.0098	0.0100

DN= (NCAL−NEXP)

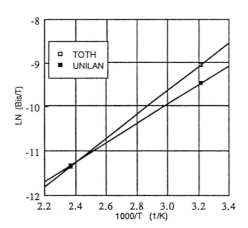

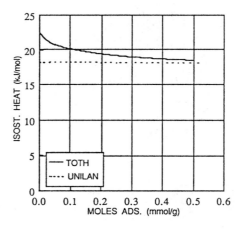

```
METHANE
-----------------------------------------------------------------------
ACTIVATED CARBON
-----------------------------------------------------------------------
ADSORBENT : COLUMBIA G GRADE;1157 Sq m/g;MESH 8x14
EXP.      : STATIC
REF.      : PAYNE,H.K.,STUDERVANT,G.A. and LELAND,T,W,;IND.ENG.CHEM.FUNDAM.;7,363(1968)
```

TEMP   (K)=   283.15

	TOTH		UNILAN	UNITS
B =	1000.0000	C =	517.5630	kPa
M =	1.0830	S =	.0915	
NI =	7.5332	M =	7.6471	mmol/g
BIS=	.30105-001	BIS=	.34831-001	1/g
QIS=	13.1990	QIS=	12.5742	KJ/mol

P(kPa)	N(mmol/g)	DN TOTH	DN UNILAN
517.1000	3.6680	.0450	.1538
1710.0000	5.7480	.1008	.1213
2759.0000	6.4790	-.0529	-.0408
4124.0000	6.9600	-.1839	-.1665
5509.0000	7.1480	-.1841	-.1584
6891.0000	7.1870	-.1082	-.0748
8622.0000	7.1110	.0608	.1025
10340.0000	6.9030	.3307	.3791
AVERAGE DN		.1333	.1496

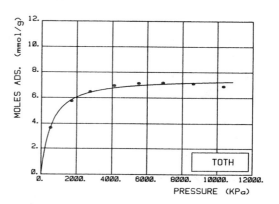

TEMP   (K)=   293.15

	TOTH		UNILAN	UNITS
B =	999.9990	C =	988.4610	kPa
M =	1.0767	S =	5.0112-005	
NI =	6.9544	M =	8.0706	mmol/g
BIS=	.27728-001	BIS=	.19905-001	1/g
QIS=	13.1990	QIS=	12.5742	KJ/mol

P(kPa)	N(mmol/g)	DN TOTH	DN UNILAN
182.7000	1.8600	-.1982	-.6116
496.4000	3.1220	.1510	-.4588
1034.0000	4.3210	.2591	-.2153
1744.0000	5.2400	.1208	-.1078
2758.0000	5.9750	-.0923	-.0661
3785.0000	6.3880	-.2326	-.0352
4844.0000	6.6400	-.3157	.0180
6181.0000	6.7890	-.3300	.1186
7932.0000	6.7730	-.2040	.3566
9656.0000	6.6900	-.0509	.5783
11740.0000	6.4520	.2446	.9550
13740.0000	6.1540	.5814	1.3362
AVERAGE DN		.2317	.4048

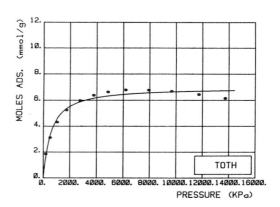

TEMP   (K)=   303.15

	TOTH		UNILAN	UNITS
B =	1000.0000	C =	680.2300	kPa
M =	1.0516	S =	.0314	
NI =	6.7621	M =	6.8345	mmol/g
BIS=	.23917-001	BIS=	.25327-001	1/g
QIS=	13.1990	QIS=	12.5742	KJ/mol

P(kPa)	N(mmol/g)	DN TOTH	DN UNILAN
162.0000	1.4350	-.1532	-.1203
461.9000	2.6780	.0701	.0861
744.6000	3.4160	.1588	.1556
1396.0000	4.4780	.1410	.1172
2399.0000	5.3880	-.0368	-.0634
3527.0000	5.9290	-.1797	-.1996
4854.0000	6.2740	-.2689	-.2796
6553.0000	6.4850	-.2927	-.2933
8284.0000	6.5050	-.1967	-.1892
10030.0000	6.4000	-.0138	.0004
11770.0000	6.2440	.1976	.2170
13750.0000	5.9740	.5140	.5383
AVERAGE DN		.1853	.1883

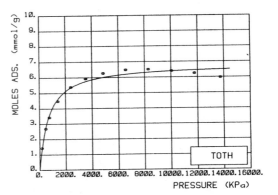

TEMP   (K)=   313.15

	TOTH		UNILAN	UNITS
B =	999.9990	C =	864.0690	kPa
M =	1.0202	S =	.4982	
NI =	6.6690	M =	6.7173	mmol/g
BIS=	.19914-001	BIS=	.21088-001	1/g
QIS=	13.1990	QIS=	12.5742	KJ/mol

P(kPa)	N(mmol/g)	DN TOTH	DN UNILAN
162.0000	1.1100	-.0561	-.0244
506.8000	2.4170	.0667	.0828
1375.0000	3.9830	.1526	.1274
2751.0000	5.1340	-.0144	-.0483
4158.0000	5.6940	-.1307	-.1582
5543.0000	5.9820	-.1742	-.1942
6915.0000	6.1550	-.1919	-.2051
8625.0000	6.1800	-.0868	-.0932
10350.0000	6.1050	.0787	.0780
12480.0000	5.9230	.3399	.3446
AVERAGE DN		.1292	.1356

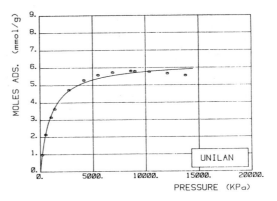

TEMP   (K)=   323.15

	TOTH		UNILAN	UNITS
B =	999.9990	C =	967.8940	kPa
M =	1.0050	S =	.0008	
NI =	6.3117	M =	6.3267	mmol/g
BIS=	.17550-001	BIS=	.17561-001	1/g
QIS=	13.1990	QIS=	12.5742	KJ/mol

P(kPa)	N(mmol/g)	DN TOTH	DN UNILAN
168.9000	.9780	-.0369	-.0380
517.1000	2.1530	.0544	.0501
1031.0000	3.1520	.1175	.1112
1380.0000	3.6230	.1020	.0955
2751.0000	4.7220	-.0376	-.0420
4175.0000	5.2850	-.1472	-.1490
5514.0000	5.5760	-.1942	-.1941
6888.0000	5.7300	-.1845	-.1828
8611.0000	5.8090	-.1249	-.1216
8991.0000	5.7910	-.0828	-.0793
10340.0000	5.7610	.0197	.0241
12070.0000	5.6650	.1865	.1919
13780.0000	5.5450	.3602	.3664
AVERAGE DN		.1268	.1266

DN=(NCAL-NEXP)

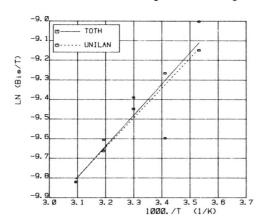

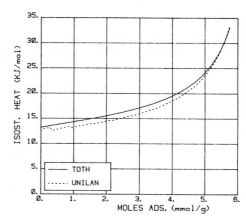

METHANE
```
----------------------------------------------------------------
```
ACTIVATED CARBON
```
----------------------------------------------------------------
```
ADSORBENT : COLUMBIA GRADE L;MESH 20x40;1152 Sq m/g (N2)
EXP.   : STATIC VOLUMETRIC
REF.   : RAY,G.C.and BOX,E.O.;IND.ENG.CHEM.;42,1315(1950)

TEMP   (K)=   310.92

	TOTH		UNILAN	UNITS
B =	43.3427	C =	541649.0000	kPa
M =	.5941	S =	8.3282	
NI =	8.0668	M =	25.0642	mmol/g
BIS=	.36636-001	BIS=	.29724-001	l/g
QIS=	20.5544	QIS=	18.4309	KJ/mol

P(kPa)	N(mmol/g)	DN TOTH	DN UNILAN
25.8700	.2535	.0324	.0179
40.2700	.3970	.0186	.0067
61.6000	.5868	−.0005	−.0063
81.3400	.7327	−.0054	−.0053
92.9400	.8155	−.0114	−.0080
99.9700	.8703	−.0215	−.0163
101.3000	.8901	−.0330	−.0274
337.8000	1.8790	.0176	.0407
575.7000	2.5190	.0073	.0184
848.0000	3.0360	−.0058	−.0081
1148.0000	3.4560	−.0157	−.0258
1479.0000	3.7680	.0194	.0086
	AVERAGE DN	.0157	.0158

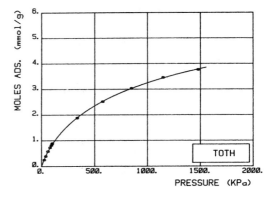

TEMP   (K)=   338.70

	TOTH		UNILAN	UNITS
B =	40.1021	C =	560988.0000	kPa
M =	.5271	S =	7.7932	
NI =	9.9448	M =	23.7995	mmol/g
BIS=	.25465-001	BIS=	.18580-001	l/g
QIS=	20.5544	QIS=	18.4309	KJ/mol

P(kPa)	N(mmol/g)	DN TOTH	DN UNILAN
26.9300	.1854	.0041	−.0173
50.6700	.3203	.0052	−.0180
74.5400	.4446	.0022	−.0183
97.6000	.5533	.0001	−.0160
99.9700	.5632	.0007	−.0149
310.3000	1.2700	.0088	.0287
561.9000	1.8990	−.0460	−.0179
844.6000	2.3140	.0153	.0326
1162.0000	2.6930	.0511	.0483
1479.0000	3.1130	−.0344	−.0588
	AVERAGE DN	.0168	.0271

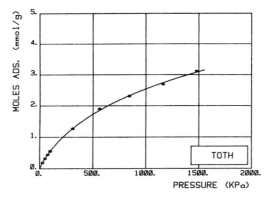

TEMP   (K)=   366.48

	TOTH		UNILAN	UNITS
B =	921.9860	C =	9990.4200	kPa
M =	.9608	S =	3.2278	
NI =	4.7213	M =	10.5583	mmol/g
BIS=	.11807-001	BIS=	.12562-001	1/g
QIS=	20.5544	QIS=	18.4309	KJ/mol

P(kPa)	N(mmol/g)	DN TOTH	DN UNILAN
37.7300	.1484	-.0074	.0002
66.6700	.2500	-.0072	.0040
99.5600	.3539	-.0013	.0122
99.9700	.3533	.0007	.0141
406.8000	1.1670	-.0123	-.0140
803.2000	1.8180	.0070	-.0113
1134.0000	2.1910	.0215	.0120
1479.0000	2.5340	-.0171	-.0007
AVERAGE DN		.0093	.0086

TEMP   (K)=   394.20

	TOTH		UNILAN	UNITS
B =	157.4790	C =	=656780.0000	kPa
M =	.6453	S =	7.0214	
NI =	7.1759	M =	20.3075	mmol/g
BIS=	.92575-002	BIS=	.80849-002	1/g
QIS=	20.5544	QIS=	18.4309	KJ/mol

P(kPa)	N(mmol/g)	DN TOTH	DN UNILAN
51.6000	.1257	.0035	-.0037
98.8000	.2352	-.0021	-.0100
99.9700	.2355	.0001	-.0078
413.7000	.7680	.0010	.0042
741.2000	1.1780	-.0029	.0039
1093.0000	1.5160	.0032	.0059
1472.0000	1.8230	-.0012	-.0068
AVERAGE DN		.0020	.0060

TEMP   (K)=   422.00

	TOTH		UNILAN	UNITS
B =	693.4730	C =	9919.1801	kPa
M =	.8248	S =	2.4015	
NI =	5.2087	M =	7.8824	mmol/g
BIS=	.65673-002	BIS=	.63557-002	1/g
QIS=	20.5544	QIS=	18.4309	KJ/mol

P(kPa)	N(mmol/g)	DN TOTH	DN UNILAN
50.8000	.0873	.0037	.0022
99.2000	.1606	.0116	.0098
99.9700	.1690	.0045	.0026
320.6000	.4813	.0157	.0147
555.0000	.7884	-.0068	-.0073
937.7000	1.1720	-.0125	-.0130
1203.0000	1.3830	-.0065	-.0065
1479.0000	1.5620	.0113	.0129
AVERAGE DN		.0091	.0086

DN=(NCAL-NEXP)

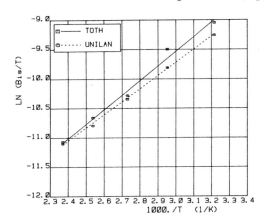
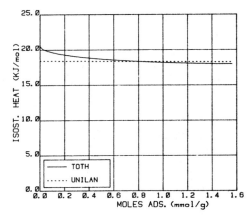

METHANE
```
-------------------------------------------------------------------------
ACTIVATED CARBON
-------------------------------------------------------------------------
ADSORBENT : BPL;PITTSBURGH CHEMICAL Co.;988 Sqm/g
EXP.      : STATIC VOLUMETRIC
REF.      : REICH,R.,ZIEGLER,W.T.and ROGERS,K.A.;Ind.Eng.Chem.Process Des.Dev.
            19,336(1980)
```

TEMP   (K)=   212.70

	TOTH		UNILAN	UNITS
B =	5.8050	C =	184.1600	kPa
M =	.4810	S =	2.8244	
NI =	9.2790	M =	8.3229	mmol/g
BIS=	.42377+000	BIS=	.23757+000	1/g
QIS=	20.8773	QIS=	17.8542	KJ/mol

P(kPa)	N(mmol/g)	DN TOTH	DN UNILAN
11.7200	1.1380	-.0279	-.0699
22.7500	1.6690	-.0139	-.0215
53.0900	2.5610	-.0048	.0186
72.3900	2.9240	.0092	.0349
113.1000	3.5180	-.0126	.0083
134.4000	3.7360	-.0032	.0141
235.1000	4.4840	-.0078	-.0034
266.8000	4.6320	.0111	.0133
444.7000	5.2840	.0166	.0148
474.4000	5.3590	.0221	.0206
654.3000	5.7500	.0204	.0219
695.7000	5.9440	-.1017	-.0993
708.8000	5.8480	.0160	-.0187
967.3000	6.2090	.0052	.0132
985.9000	6.2220	.0129	.0212
1219.0000	6.4610	-.0013	.0101
1274.0000	6.4930	.0120	.0239
1463.0000	6.7380	-.0941	-.0812
1615.0000	6.7150	.0251	.0382
1964.0000	7.0030	-.0799	-.0684
2370.0000	7.1410	-.0516	-.0445
2725.0000	7.2560	-.0489	-.0473
3451.0000	7.2610	.1338	.1215
AVERAGE DN		.0318	.0360

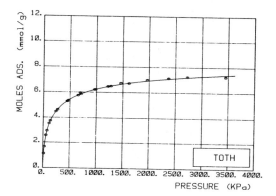

TEMP   (K)=   260.20

	TOTH		UNILAN	UNITS
B =	31.1610	C =	641.0250	kPa
M =	.6100	S =	2.0965	
NI =	7.6890	M =	7.1056	mmol/g
BIS=	.59217-001	BIS=	.45836-001	1/g
QIS=	20.8773	QIS=	17.8542	KJ/mol

P(kPa)	N(mmol/g)	DN TOTH	DN UNILAN
44.1300	.7930	-.0299	-.0535
82.0500	1.2070	-.0154	-.0239
130.3000	1.6180	-.0103	-.0057
162.0000	1.8300	.0018	.0116
243.4000	2.2970	-.0030	.0126
302.7000	2.5610	.0001	.0166
409.5000	2.9480	.0001	.0155
504.7000	3.2160	.0074	.0209
618.5000	3.4930	.0011	.0125
728.8000	3.7050	.0080	.0176
862.5000	3.9330	.0034	.0115
948.7000	4.1220	-.0604	-.0529
1002.0000	4.1250	.0081	.0153
1168.0000	4.3250	.0061	.0127
1337.0000	4.4870	.0153	.0219
1513.0000	4.6490	.0068	.0135
1589.0000	4.7460	-.0304	-.0234
1665.0000	4.7550	.0172	.0242
2023.0000	5.0570	-.0550	-.0473
2408.0000	5.2460	-.0473	-.0390
2766.0000	5.4110	-.0621	-.0539
3121.0000	5.5070	-.0321	-.0243
3465.0000	5.5700	.0102	.0172
3826.0000	5.5940	.0829	.0886
AVERAGE DN		.0214	.0265

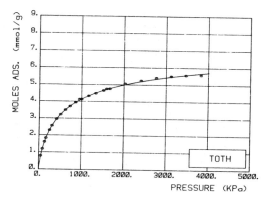

TEMP   (K)=   301.40

	TOTH		UNILAN		UNITS
B =	177.1240	C =	1162.7900		kPa
M =	.7730	S =	1.4303		
NI =	6.0290	M =	5.8140		mmol/g
BIS=	.18651-001	BIS=	.17260-001		1/g
QIS=	20.8773	QIS=	17.8542		KJ/mol

P(kPa)	N(mmol/g)	DN TOTH	DN UNILAN
72.3900	.4770	-.0297	-.0370
131.7000	.7640	-.0262	-.0306
181.3000	.9660	-.0186	-.0205
274.4000	1.2980	-.0158	-.0147
398.5000	1.6440	.0001	.0025
467.5000	1.8110	.0040	.0064
606.0000	2.0970	.0129	.0145
677.7000	2.2300	.0126	.0135
823.2000	2.4630	.0161	.0157
872.9000	2.5600	-.0084	-.0093
894.9000	2.5640	.0184	.0174
1037.0000	2.7480	.0183	.0162
1162.0000	2.8940	.0148	.0122
1261.0000	3.0650	-.0538	-.0569
1329.0000	3.0640	.0128	.0095
1444.0000	3.1740	.0061	.0026
1588.0000	3.2840	.0135	.0099
1644.0000	3.3850	-.0450	-.0486
1719.0000	3.3950	-.0006	-.0041
2002.0000	3.6170	-.0392	-.0423
2367.0000	3.8060	-.0326	-.0349
2732.0000	3.9740	-.0391	-.0403
3096.0000	4.0490	.0218	.0215
3447.0000	4.1950	-.0117	-.0111
3829.0000	4.2500	.0396	.0410
AVERAGE DN		.0205	.0213

DN=(NCAL-NEXP)

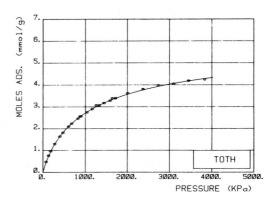

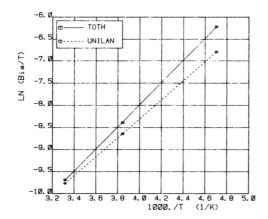

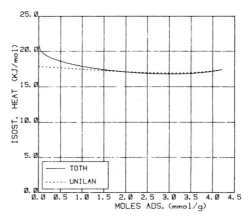

METHANE

ACTIVATED CARBON

ADSORBENT : PCB, Calgon Corp.
EXP.   : STATIC
REF.   : Ritter, J. A. and R. T. Yang, *Ind. Eng. Chem.*
              Res. **26**, 1679 (1987).

TEMP (K)   =   296.

TOTH		UNILAN		UNIT
B =	758.4196	C =	558.5903	kPa
M =	1.0418	S =	3.8488−04	
NI =	6.0913	M =	6.1476	mmol/g
BIS =.257913−01		BIS =.270831−01		1/g
QIS =	17.9044	QIS =	17.4202	KJ/mol

P (kPa)	N (mmol/g)	DN TOTH	DN UNILAN
275.7880	2.0299	−0.0198	0.0024
1137.6255	4.0822	0.0540	0.0401
2413.1450	5.0414	−0.0357	−0.0494
3757.6116	5.3983	−0.0406	−0.0467
5239.9722	5.5767	−0.0236	−0.0216
6274.1772	5.6213	0.0169	0.0232
6687.8589	5.6213	0.0439	0.0518
AVERAGE DN		0.0335	0.0336

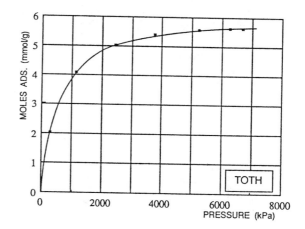

TEMP (K)   =   373.

TOTH		UNILAN		UNIT
B =	400.5754	C =	2011.8540	kPa
M =	.8148	S =	1.1869	
NI =	5.2403	M =	5.0467	mmol/g
BIS =.103904−01		BIS =.973860−02		1/g
QIS =	17.9044	QIS =	17.4202	KJ/mol

P (kPa)	N (mmol/g)	DN TOTH	DN UNILAN
482.6290	1.1153	−0.0296	−0.0324
1123.8361	1.8515	0.0251	0.0259
1620.2545	2.2307	0.0474	0.0484
1999.4630	2.5430	−0.0275	−0.0266
3447.3501	3.1676	−0.0458	−0.0450
4929.7104	3.4799	0.0110	0.0114
6156.9673	3.6806	0.0212	0.0206
6584.4385	3.7699	−0.0075	−0.0086
AVERAGE DN		0.0269	0.0274

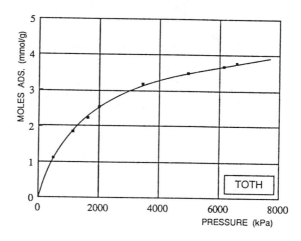

TEMP (K)   =   480.

TOTH		UNILAN		UNIT
B	=167256.400	C	= 5354.8700	kPa
M	= 1.4226	S	= 9.6299-04	
NI	= 2.9692	M	= 3.8650	mmol/g
BIS	=.252346-02	BIS	=.288040-02	1/g
QIS	= 17.9044	QIS	= 17.4202	KJ/mol

P (kPa)	N (mmol/g)	DN TOTH	DN UNILAN
637.7598	0.3569	0.0306	0.0543
1296.2036	0.7584	-0.0201	-0.0053
2378.6716	1.2046	-0.0053	-0.0157
3709.3486	1.6061	-0.0008	-0.0242
5329.6030	1.9184	0.0193	0.0093
6246.5981	2.0745	-0.0008	0.0064
6687.8589	2.1415	-0.0120	0.0049
	AVERAGE DN	0.0127	0.0172

DN= (NCAL-NEXP)

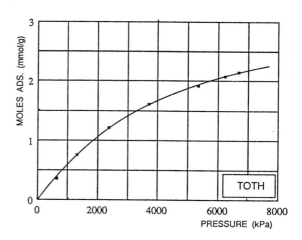

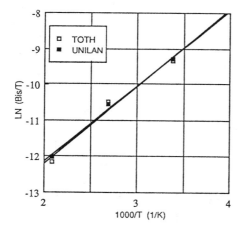

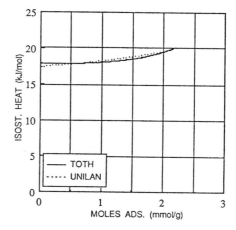

METHANE

ACTIVATED CARBON

ADSORBENT : NUXIT-AL, (1.5 mm DIAMETER, 3-4 mm LONG)
EXP.   : STATIC VOLUMETRIC
REF.   : SZEPESY, L., ILLES, V., *Acta Chim. Hung.*,**35**, 37(1963)

TEMP   (K) = 293.15

TOTH		UNILAN		UNIT
B =	23.8331	C =	41793.6992	kPa
M =	.4771	S =	6.1749	
NI =	12.2353	M =	13.1050	mmol/g
BIS =	3.8723-02	BIS =	2.9738-02	1/g
QIS =	19.4510	QIS =	18.8501	KJ/mol

P(kPa)	N(mmol/g)	DN TOTH	DN UNILAN
16.0253	0.1919	-0.0045	-0.0124
27.3577	0.2972	-0.0023	-0.0069
31.4640	0.3061	0.0248	0.0216
34.6371	0.3654	-0.0074	-0.0097
46.5294	0.4556	-0.0017	-0.0008
48.0626	0.4640	0.0016	0.0028
54.5687	0.5216	-0.0072	-0.0048
62.6080	0.5725	-0.0004	0.0029
75.6336	0.6662	-0.0057	-0.0022
83.4462	0.7103	0.0005	0.0035
83.6329	0.6876	0.0244	0.0274
94.0587	0.8089	-0.0326	-0.0310
99.8715	0.8143	-0.0032	-0.0029
112.8304	0.8785	0.0070	0.0040
124.8694	0.9424	0.0092	0.0023
AVERAGE DN		0.0088	0.0090

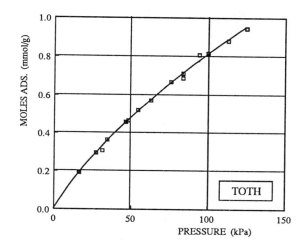

TEMP   (K) = 313.15

TOTH		UNILAN		UNIT
B =	44.3338	C =	35270.0000	kPa
M =	.5506	S =	5.6636	
NI =	8.7141	M =	10.7120	mmol/g
BIS =	2.31742-02	BIS =	2.01183-02	1/g
QIS =	19.4510	QIS =	18.8501	KJ/mol

P(kPa)	N(mmol/g)	DN TOTH	DN UNILAN
13.8388	0.1115	-0.0072	-0.0102
25.4778	0.1870	-0.0065	-0.0081
41.2632	0.2735	0.0005	0.0012
46.5694	0.3083	-0.0047	-0.0034
64.0879	0.3864	0.0094	0.0118
79.6199	0.4730	-0.0012	0.0009
82.9929	0.4881	-0.0005	0.0014
101.3647	0.5711	-0.0005	-0.0006
128.9890	0.6858	0.0004	-0.0051
AVERAGE DN		0.0035	0.0048

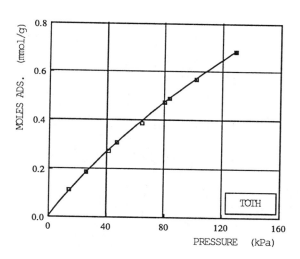

TEMP   (K) =   333.15

TOTH	UNILAN	UNIT
B = 53.4762	C =42143.1016	kPa
M = .5607	S = 5.5327	
NI = 7.0700	M = 9.3178	mmol/g
BIS =1.62085-02	BIS =1.39928-02	1/g
QIS = 19.4510	QIS = 18.8501	KJ/mol

P(kPa)	N(mmol/g)	DN TOTH	DN UNILAN
15.0920	0.0817	-0.0054	-0.0087
25.6112	0.1187	0.0047	0.0017
32.5572	0.1539	-0.0012	-0.0037
51.8223	0.2240	0.0048	0.0040
64.7145	0.2766	-0.0005	-0.0004
79.1533	0.3230	0.0033	0.0041
102.4313	0.4034	-0.0011	0.0000
122.4563	0.4667	-0.0035	-0.0029
AVERAGE DN		0.0031	0.0032

TEMP   (K) =   363.15

TOTH	UNILAN	UNIT
B = 66.0408	C =34000.8008	kPa
M = .5638	S = 4.3373	
NI = 5.7028	M = 10.5860	mmol/g
BIS =1.01916-02	BIS =8.28903-03	1/g
QIS = 19.4510	QIS = 18.8501	KJ/mol

P(kPa)	N(mmol/g)	DN TOTH	DN UNILAN
10.3325	0.0317	-0.0000	-0.0036
26.1044	0.0727	0.0023	-0.0031
27.9976	0.0723	0.0076	0.0023
55.7019	0.1455	0.0022	-0.0014
62.9547	0.1593	0.0049	0.0023
85.6594	0.2133	0.0003	0.0018
108.7774	0.2691	-0.0084	-0.0019
AVERAGE DN		0.0037	0.0023

DN= (NCAL-NEXP)

METHANE

ACTIVATED CARBON

ADSORBENT : NUXIT-AL, (1.5 mm DIAMETER, 3-4 mm LONG)
EXP.   : STATIC VOLUMETRIC
REF.   : SZEPESY, L., ILLES, V., *Acta Chim. Hung.*, **35**, 53 (1963)

TEMP   (K) =   293.15

TOTH		UNILAN		UNIT
B =	49.5007	C =	1226.8199	kPa
M =	0.6459	S =	2.3938	
NI =	5.5493	M =	5.9467	mmol/g
BIS =	3.21749-02	BIS =	2.68091-02	1/g
QIS =	16.3530	QIS =	17.3218	KJ/mol

P(kPa)	N(mmol/g)	DN TOTH	DN UNILAN
98.6905	0.8031	-0.0226	-0.0273
126.1496	0.8933	0.0340	0.0322
157.5604	1.0865	-0.0104	-0.0099
185.0194	1.1967	-0.0037	-0.0021
217.7474	1.3158	0.0033	0.0056
257.6695	1.4604	-0.0032	-0.0006
297.9968	1.5849	-0.0025	-0.0002
344.2010	1.7111	0.0003	0.0020
397.1940	1.8446	-0.0006	0.0004
450.7949	1.9659	-0.0018	-0.0015
513.9204	2.0913	-0.0001	-0.0005
581.0989	2.2149	-0.0028	-0.0034
681.1066	2.3688	0.0016	0.0013
AVERAGE DN		0.0067	0.0067

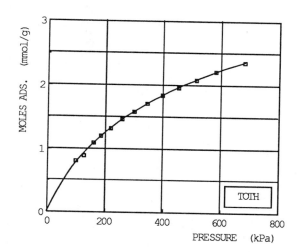

TEMP   (K) =   313.15

TOTH		UNILAN		UNIT
B =	71.6075	C =	2821.4900	kPa
M =	0.6450	S =	2.7183	
NI =	6.1072	M =	7.0520	mmol/g
BIS =	2.11602-02	BIS =	1.80609-02	1/g
QIS =	16.3530	QIS =	17.3218	KJ/mol

P(kPa)	N(mmol/g)	DN TOTH	DN UNILAN
101.2237	0.5711	-0.0063	-0.0110
117.9423	0.6470	-0.0110	-0.0140
150.7716	0.7750	-0.0098	-0.0101
179.3452	0.8665	0.0016	0.0031
218.5580	0.9946	0.0029	0.0061
265.9781	1.1302	0.0088	0.0127
311.2704	1.2480	0.0136	0.0174
368.7216	1.3979	0.0047	0.0075
421.4106	1.5322	-0.0121	-0.0107
474.5050	1.6259	0.0031	0.0027
535.4013	1.7442	-0.0003	-0.0030
601.6678	1.8620	-0.0036	-0.0088
AVERAGE DN		0.0065	0.0089

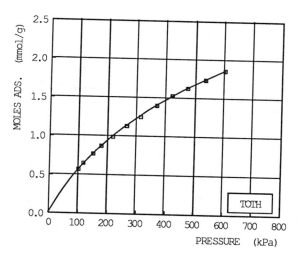

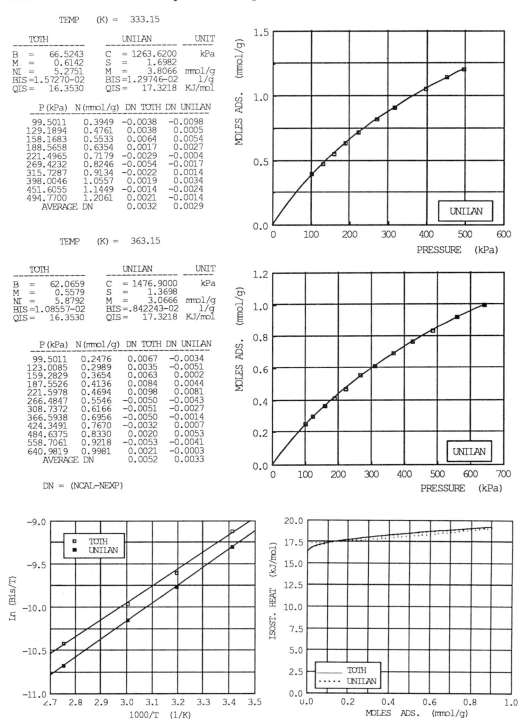

TEMP   (K) =   333.15

TOTH		UNILAN		UNIT
B =	66.5243	C =	1263.6200	kPa
M =	0.6142	S =	1.6982	
NI =	5.2751	M =	3.8066	mmol/g
BIS=	1.57270-02	BIS=	1.29746-02	1/g
QIS=	16.3530	QIS=	17.3218	KJ/mol

P(kPa)	N(mmol/g)	DN TOTH	DN UNILAN
99.5011	0.3949	-0.0038	-0.0098
129.1894	0.4761	0.0038	0.0005
158.1683	0.5533	0.0064	0.0054
188.5658	0.6354	0.0017	0.0027
221.4965	0.7179	-0.0029	-0.0004
269.4232	0.8246	-0.0054	-0.0017
315.7287	0.9134	-0.0022	0.0014
398.0046	1.0557	0.0019	0.0034
451.6055	1.1449	-0.0014	-0.0024
494.7700	1.2061	0.0021	-0.0014
AVERAGE DN		0.0032	0.0029

TEMP   (K) =   363.15

TOTH		UNILAN		UNIT
B =	62.0659	C =	1476.9000	kPa
M =	0.5579	S =	1.3698	
NI =	5.8792	M =	3.0666	mmol/g
BIS=	1.08557-02	BIS=	.842243-02	1/g
QIS=	16.3530	QIS=	17.3218	KJ/mol

P(kPa)	N(mmol/g)	DN TOTH	DN UNILAN
99.5011	0.2476	0.0067	-0.0034
123.0085	0.2989	0.0035	-0.0051
159.2829	0.3654	0.0063	0.0002
187.5526	0.4136	0.0084	0.0044
221.5978	0.4694	0.0098	0.0081
266.4847	0.5546	-0.0050	-0.0043
308.7372	0.6166	-0.0051	-0.0027
366.5938	0.6956	-0.0050	-0.0014
424.3491	0.7670	-0.0032	0.0007
484.6375	0.8330	0.0020	0.0053
558.7061	0.9218	-0.0053	-0.0041
640.9819	0.9981	0.0021	-0.0003
AVERAGE DN		0.0052	0.0033

DN = (NCAL-NEXP)

```
METHANE
--------------------------------------------------------------------------------
CARBON MOLECULAR SIEVE
--------------------------------------------------------------------------------
ADSORBENT : MSC-5A;TAKEDA CHEMICAL Co.;0.56 cc/g;650 Sqm/g
EXP.      : STATIC GRAVIMETRIC (McBAIN QUARTZ SPRING BALANCE)
REF.      : NAKAHARA,T.,HIRATA,M.and OMORI,T.;J.Chem.Eng.Data;19,310(1974)
```

TEMP   (K)=   278.55

	TOTH		UNILAN	UNITS
B =	17.9138	C =	483.4160	kPa
M =	.7150	S =	3.4894	
NI =	2.6595	M =	4.4687	mmol/g
BIS=	.10887+000	BIS=	.10042+000	1/g
QIS=	27.4210	QIS=	24.9114	KJ/mol

P(kPa)	N(mmol/g)	DN TOTH	DN UNILAN
.6667	.0125	.0171	.0158
3.6000	.1434	-.0024	-.0037
5.7330	.2182	-.0080	-.0082
11.7300	.3740	-.0018	.0001
20.4000	.5423	.0108	.0128
32.4000	.7542	-.0115	-.0115
43.6000	.8851	-.0048	-.0064
57.3300	.9973	.0182	.0162
70.6700	1.1220	.0005	-.0003
87.2000	1.2400	-.0083	-.0057
AVERAGE DN		.0084	.0081

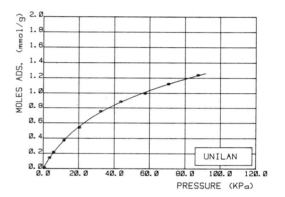

TEMP   (K)=   303.15

	TOTH		UNILAN	UNITS
B =	80.2946	C =	525.0840	kPa
M =	.9107	S =	2.4880	
NI =	2.2205	M =	3.9591	mmol/g
BIS=	.45329-001	BIS=	.45654-001	1/g
QIS=	27.4210	QIS=	24.9114	KJ/mol

P(kPa)	N(mmol/g)	DN TOTH	DN UNILAN
2.8000	.0374	.0113	.0117
6.6670	.1060	.0053	.0063
15.2700	.2306	.0052	.0063
26.1300	.3802	-.0102	-.0100
39.0700	.5049	.0002	-.0009
54.0000	.6420	-.0063	-.0078
69.4700	.7355	.0141	.0138
82.9300	.8415	-.0065	-.0044
AVERAGE DN		.0074	.0077

DN=(NCAL-NEXP)

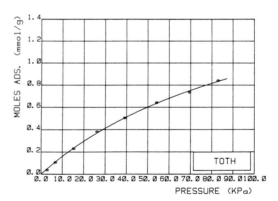

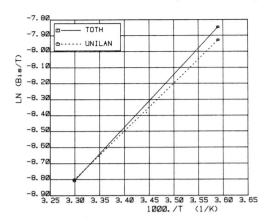

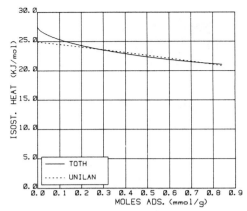

NITROGEN

ACTIVATED CARBON

ADSORBENT : CHAR 1
EXP.   : STATIC VOLUMETRIC
REF.   : LOPEZ-GONZALEZ,J. de D.,CARPENTER,F.G.and DEITZ,V.R.;J. of RESEARCH
OF THE NATIONAL BUREAU OF STANDARDS;55,11(1955).

TEMP   (K)=   90.00
PSAT (kPa)=   359.69

	TOTH		UNILAN	UNITS
B =	.1691	C =	61.3927	kPa
M =	.1506	S =	11.4574	
NI =	27.4502	M =	30.7401	mmol/g
DG =	−92.6076	DG =	−80.3695	J/g
BIS=	.27457+007	BIS=	.15467+004	1/g
QIS=	20.4922	QIS=	13.0862	KJ/mol

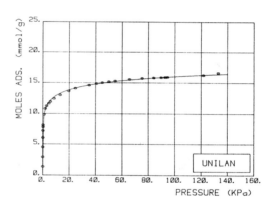

P(kPa)	N(mmol/g)	DN TOTH	DN UNILAN
.0003	1.4400	−.1165	−.9479
.0054	2.9660	.4078	.0274
.0241	4.5330	.3664	.3516
.0626	6.0490	−.0267	.0941
.1262	7.1860	−.2773	−.1091
.1917	7.7810	−.3234	−.1457
.2446	8.1040	−.3205	−.1427
.9712	9.8510	−.1636	−.0426
1.8700	10.8100	−.1977	−.1231
2.8420	11.2600	−.0553	−.0118
4.5320	11.6500	.2134	.2241
5.5750	11.9600	.1946	.1920
8.4530	12.4500	.2862	.2603
12.9800	13.0500	.2793	.2356
19.5300	13.6900	.1967	.1436
24.7100	14.1300	.0738	.0192
35.1100	14.6200	.0515	.0004
40.6100	14.8800	−.0169	−.0643
45.0700	15.0300	−.0306	−.0746
49.9600	15.1700	−.0366	−.0764
54.8500	15.3000	−.0457	−.0811
65.5700	15.5500	−.0662	−.0916
75.2500	15.7300	−.0707	−.0869
84.0900	15.8300	−.0302	−.0379
89.6700	15.8800	.0007	−.0017
92.4800	15.9100	.0094	.0097
93.2300	15.9200	.0095	.0105
94.4900	15.9400	.0063	.0085
121.8000	16.2500	.0118	.0391
133.0000	16.6500	−.2800	−.2429
AVERAGE DN		.1388	.1299

TEMP (K)= 77.70
PSAT (kPa)= 106.19

TOTH		UNILAN		UNITS
B =	.1153	C =	34.7855	kPa
M =	.1341	S =	13.7518	
NI =	28.2370	M =	33.5865	mmol/g
DG =	-91.8038	DG =	-75.2340	J/g
BIS=	.18098+009	BIS=	.21278+005	1/g
QIS=	20.4922	QIS=	13.0862	KJ/mol

P(kPa)	N(mmol/g)	DN TOTH	DN UNILAN
.0000	1.4000	.3371	-.5591
.0031	4.7450	.6084	.6871
.0098	6.3300	.3079	.4894
.0244	7.9430	-.2102	-.0155
.0445	8.9770	-.4951	-.3178
.0921	9.9520	-.5413	-.4053
.1593	10.5200	-.3989	-.3039
.2548	11.0000	-.2665	-.2104
.2963	11.2000	-.2695	-.2262
.7560	12.0700	.0820	.0476
1.8370	13.1000	.1967	.1017
2.6970	13.6000	.1847	.0707
4.3750	14.2800	.1113	-.0186
5.1180	14.5100	.0758	-.0570
6.2120	14.7900	.0345	-.1005
7.9000	15.1300	-.0120	-.1469
8.8670	15.2900	-.0321	-.1659
14.1400	15.8100	.0060	-.1160
18.8300	16.0800	.0724	-.0362
25.8100	16.2800	.2368	.1488
34.1800	16.5400	.2962	.2318
37.7200	16.6100	.3370	.2822
43.5000	16.7300	.3762	.3363
49.3900	16.8200	.4269	.4013
53.6000	16.9000	.4369	.4212
56.4500	16.9300	.4637	.4545
67.0800	17.1200	.4614	.4752
73.2400	17.2400	.4362	.4625
75.7700	17.2700	.4427	.4739
81.3600	17.2900	.4990	.5409
81.4700	17.4300	.3604	.4025
87.5100	17.5500	.3167	.3698
88.0300	17.6000	.2730	.3271
88.9500	17.6500	.2340	.2898
92.8400	17.8400	.0895	.1520
98.5600	17.9600	.0327	.1050
99.0500	18.0900	-.0921	-.0189
102.9000	18.3100	-.2719	-.1923
104.2000	18.7800	-.7287	-.6470
104.9000	19.2600	-1.2017	-1.1188
106.1000	21.0400	-2.9698	-2.8849
AVERAGE DN		.3714	.3613

DN=(NCAL-NEXP)

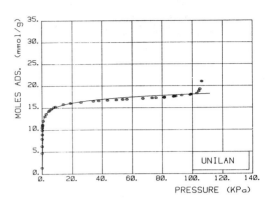

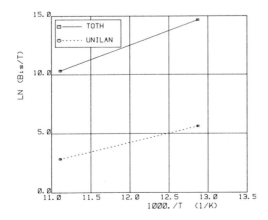

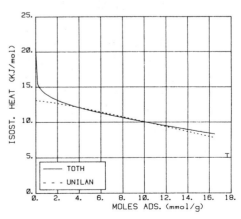

NITROGEN
```
--------------------------------------------------------------------------------
```
ACTIVATED CARBON
```
--------------------------------------------------------------------------------
```
ADSORBENT : BPL;PITTSBURGH COKE & CHEM.Co;1050-1150 Sq m/g;MESH 12x30;.5 g/cc
EXP.   : STATIC GRAVIMETRIC
REF.   : MEREDITH,J.M.and PLANK,C.A.;J.CHEM.ENG.DATA; 12,259(1967)

TEMP   K)=   303.15

TOTH		UNILAN		UNITS
B =	10.9107	C =	714.6020	kPa
M =	.3524	S =	2.4611	
NI =	9.0079	M =	1.7706	mmol/g
BIS=	.25763-001	BIS=	.14759-001	1/g
QIS=	-57.8552	QIS=	8.6677	KJ/mol

P(kPa)	N(mmol/g)	DN TOTH	DN UNILAN
8.1330	.0643	-.0137	-.0196
15.5300	.0999	-.0139	-.0190
26.7300	.1321	.0000	-.0025
36.8000	.1642	.0045	.0041
48.4000	.1999	.0070	.0082
58.2000	.2320	.0047	.0065
74.2600	.2642	.0174	.0191
81.6600	.2963	.0047	.0059
95.0600	.3320	.0024	.0019
96.7300	.3641	-.0256	-.0264
AVERAGE DN		.0094	.0113

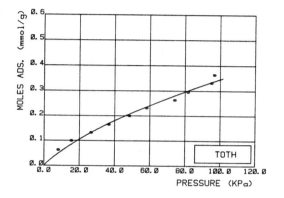

TEMP   (K)=   313.15

TOTH		UNILAN		UNITS
B =	4.1391	C =	99.9998	kPa
M =	.2612	S =	1.0346	
NI =	4.8902	M =	.4416	mmol/g
BIS=	.55392-001	BIS=	.13660-001	1/g
QIS=	-57.8552	QIS=	8.6677	KJ/mol

P(kPa)	N(mmol/g)	DN TOTH	DN UNILAN
6.1330	.0321	.0051	-.0028
13.4700	.0643	.0002	-.0057
23.4600	.0964	-.0033	-.0053
36.6000	.1285	-.0053	-.0035
58.4000	.1606	.0027	.0066
82.4600	.1928	.0065	.0085
98.6600	.2249	-.0045	-.0055
AVERAGE DN		.0039	.0054

DN=(NCAL-NEXP)

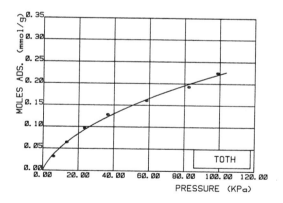

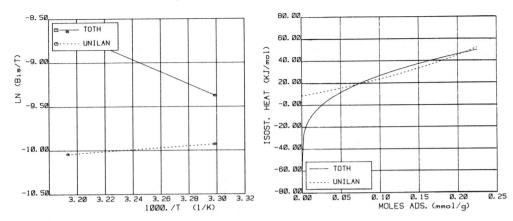

NITROGEN

ACTIVATED CARBON

ADSORBENT : ASC(METAL INPREG.BPL);PITTSBURGH COKE & CHEM.Co;800-850 Sq m/g;MESH 12x30
EXP.   : STATIC GRAVIMETRIC
REF.   : MEREDITH,J.M.and PLANK,C.A.;J.CHEM.ENG.DATA; 12,259(1967)

TEMP   (K)=   303.15

	TOTH		UNILAN		UNITS
B =	13.9249	C =	898.4780		kPa
M =	.3539	S =	2.4278		
NI =	9.9819	M =	1.4502		mmol/g
BIS=	.14735-001	BIS=	.94219-002		1/g
QIS=	-78.7103	QIS=	9.0878		KJ/mol

P(kPa)	N(mmol/g)	DN TOTH	DN UNILAN
14.3300	.0607	-.0087	-.0115
22.6000	.0714	.0045	.0029
26.8000	.0892	-.0019	-.0030
39.8000	.1214	-.0015	-.0011
54.8000	.1499	.0040	.0054
69.4600	.1785	.0060	.0074
83.2600	.2106	.0008	.0014
96.4600	.2392	-.0035	-.0043
97.1900	.2392	-.0022	-.0030
AVERAGE DN		.0037	.0044

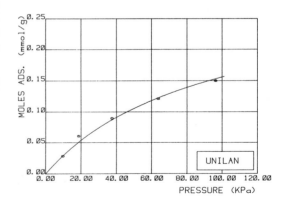

TEMP   (K)=   313.15

	TOTH		UNILAN		UNITS
B =	4.0089	C =	99.1470		kPa
M =	.2524	S =	.6099		
NI =	3.8842	M =	.3107		mmol/g
BIS=	.41263-001	BIS=	.86741-002		1/g
QIS=	-78.7103	QIS=	9.0878		KJ/mol

P(kPa)	N(mmol/g)	DN TOTH	DN UNILAN
9.7330	.0286	.0075	.0005
18.6000	.0607	-.0048	-.0099
37.5300	.0892	-.0019	-.0022
63.8000	.1214	-.0013	.0012
96.2600	.1499	.0020	.0032
AVERAGE DN		.0035	.0034

DN=(NCAL-NEXP)

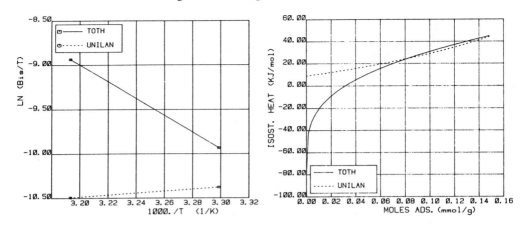

NITROGEN

ACTIVATED CARBON
==
ADSORBENT : COLUMBIA GRADE L;MESH 20x40;1152 Sq m/g (N2)
EXP.   : STATIC VOLUMETRIC
REF.   : RAY,G.C.and BOX,E.O.;IND.ENG.CHEM.;42,1315(1950)

TEMP   (K)=   310.92

TOTH		UNILAN		UNITS
B =	109.0120	C =	757155.0000	kPa
M =	.6260	S =	7.5243	
NI =	6.8156	M =	19.9362	mmol/g
BIS=	.97938-002	BIS=	.83789-002	1/g
QIS=	15.9218	QIS=	14.5675	KJ/mol

P(kPa)	N(mmol/g)	DN TOTH	DN UNILAN
28.5300	.0912	.0051	-.0018
54.0100	.1741	-.0012	-.0097
78.5400	.2417	-.0006	-.0088
98.2500	.2967	-.0039	-.0113
99.9700	.2919	.0053	-.0020
510.2000	1.0550	.0075	.0183
817.0000	1.4490	-.0028	.0060
1134.0000	1.7760	-.0174	-.0163
1486.0000	2.0280	.0130	.0040
AVERAGE DN		.0063	.0087

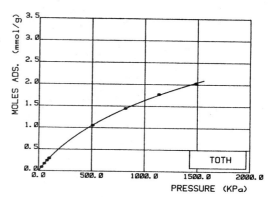

TEMP   (K)=   338.70

TOTH		UNILAN		UNITS
B =	105.2490	C =	623924.0000	kPa
M =	.5498	S =	6.6154	
NI =	11.2012	M =	20.8695	mmol/g
BIS=	.66149-002	BIS=	.53143-002	1/g
QIS=	15.9218	QIS=	14.5675	KJ/mol

P(kPa)	N(mmol/g)	DN TOTH	DN UNILAN
99.9700	.1997	-.0084	-.0215
451.6000	.6707	.0125	.0107
779.1000	1.0390	-.0059	.0001
1131.0000	1.3520	-.0061	-.0022
1486.0000	1.6120	.0047	-.0004
AVERAGE DN		.0075	.0070

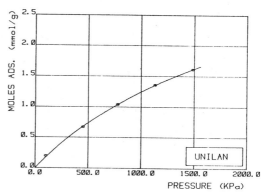

TEMP   (K)=   366.48

TOTH		UNILAN		UNITS
B =	380.2750	C =	702095.0000	kPa
M =	.6910	S =	6.3561	
NI =	7.3403	M =	19.5010	mmol/g
BIS=	.41297-002	BIS=	.38346-002	1/g
QIS=	15.9218	QIS=	14.5675	KJ/mol

P(kPa)	N(mmol/g)	DN TOTH	DN UNILAN
27.6000	.0354	.0007	-.0010
53.4700	.0672	.0012	-.0013
82.0100	.1021	.0007	-.0022
99.2000	.1265	-.0034	-.0065
99.9700	.1229	.0011	-.0020
379.2000	.4045	.0105	.0110
727.4000	.7116	.0024	.0063
1096.0000	1.0040	-.0224	-.0201
1486.0000	1.2130	.0131	.0098
AVERAGE DN		.0062	.0067

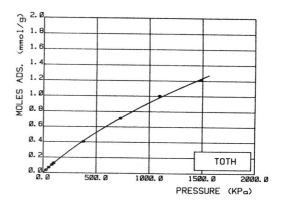

TEMP   (K)=   394.20

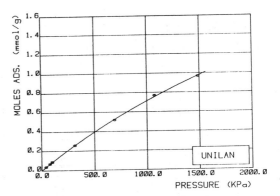

	TOTH		UNILAN		UNITS
B =	679.2120	C =	798561.0000		kPa
M =	.7258	S =	6.0671		
NI =	7.4799	M =	20.0425		mmol/g
BIS=	.30713-002	BIS=	.29246-002		1/g
QIS=	15.9218	QIS=	14.5675		KJ/mol

P(kPa)	N(mmol/g)	DN TOTH	DN UNILAN
39.2000	.0362	-.0005	-.0016
77.0700	.0681	.0009	-.0007
98.9400	.0886	-.0009	-.0026
99.9700	.0870	.0015	-.0001
313.7000	.2560	.0033	.0026
692.9000	.5171	.0061	.0081
1072.0000	.7680	-.0152	-.0133
1486.0000	.9677	.0074	.0057
AVERAGE DN		.0045	.0043

TEMP   (K)=   422.00

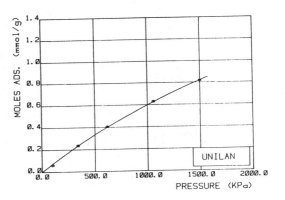

	TOTH		UNILAN		UNITS
B =	473.1860	C =	962387.0000		kPa
M =	.6772	S =	6.2902		
NI =	7.1974	M =	16.9428		mmol/g
BIS=	.28344-002	BIS=	.26476-002		1/g
QIS=	15.9218	QIS=	14.5675		KJ/mol

P(kPa)	N(mmol/g)	DN TOTH	DN UNILAN
99.9700	.0614	.0139	.0120
341.3000	.2355	.0009	.0002
620.5000	.4045	-.0042	-.0026
1058.0000	.6297	-.0048	-.0028
1500.0000	.8192	.0041	.0024
AVERAGE DN		.0056	.0040

DN=(NCAL-NEXP)

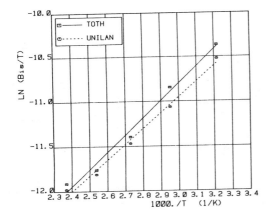

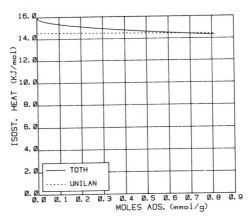

NITROGEN

ZEOLITE

ADSORBENT : 10X Linde; 1/16 in. pellets; 672 m²/gm
EXP.   : Static Volumetric
REF.   : Nolan, J. T., T. W. McKeehan and R. P. Danner,
            J. Chem. Eng. Data, **26**, 112 (1981)

TEMP (K)	=	78.32
PSAT (kPa)	=	113.35

TOTH		UNILAN		UNIT
B = 9.3359E-5		C = .0872		kPa
M = 9.7684		S = .0066		
NI = 7.3249		M = 7.3903		mmol/g
DG = -31.7882		DG = -34.5045		J/g
BIS =.123323+02		BIS =.551863+02		1/g
QIS = 4.5874		QIS = 8.0061		KJ/mol

P (kPa)	N (mmol/g)	DN TOTH	DN UNILAN
0.2400	4.1223	0.4185	1.2991
0.3067	6.2415	-0.4921	-0.4867
0.3733	6.6385	0.0548	-0.6470
0.3733	6.5761	0.1173	-0.5845
3.0400	6.9910	0.3339	0.1934
3.5067	7.0133	0.3116	0.1978
5.1200	7.0713	0.2536	0.1953
7.8266	7.1382	0.1867	0.1707
10.7200	7.1784	0.1466	0.1524
11.9066	7.1962	0.1287	0.1404
14.9733	7.2319	0.0930	0.1156
16.3200	7.2453	0.0796	0.1058
16.5200	7.2274	0.0975	0.1241
20.3066	7.2810	0.0440	0.0778
22.9599	7.3033	0.0216	0.0590
27.1466	7.3256	-0.0007	0.0411
33.0133	7.3657	-0.0408	0.0051
33.7866	7.3657	-0.0408	0.0056
39.8532	7.3970	-0.0720	-0.0227
39.9332	7.3970	-0.0720	-0.0227
53.0132	7.4683	-0.1434	-0.0902
53.4132	7.4639	-0.1390	-0.0856
63.2932	7.5576	-0.2327	-0.1774
66.2932	7.5353	-0.2103	-0.1546
66.3598	7.5219	-0.1970	-0.1412
78.8398	7.6379	-0.3130	-0.2557
79.7198	7.5933	-0.2683	-0.2110
AVERAGE DN		0.1670	0.2134

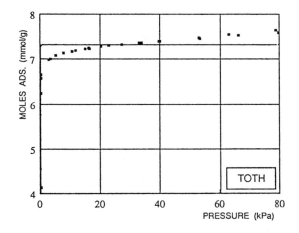

TEMP (K)	=	172.04

TOTH		UNILAN		UNIT
B = 1.1240		C = 1125.3750		kPa
M = .2594		S = 6.5943		
NI = 9.7951		M = 10.0364		mmol/g
BIS =.892818+01		BIS =.706949+00		1/g
QIS = 4.5874		QIS = 8.0061		KJ/mol

P (kPa)	N (mmol/g)	DN TOTH	DN UNILAN
1.3200	0.6871	-0.0666	-0.2159
7.0800	1.2849	0.0519	0.0259
16.6266	1.8069	0.0379	0.0712
30.5866	2.2485	0.0125	0.0634
48.0265	2.6054	-0.0122	0.0365
70.6132	2.9133	-0.0229	0.0143
95.9998	3.1631	-0.0289	-0.0061
128.9463	3.3996	-0.0267	-0.0211
164.2262	3.5825	-0.0116	-0.0220
198.0795	3.7208	0.0046	-0.0189
226.3728	3.8234	0.0125	-0.0207
252.8127	3.8948	0.0326	-0.0085
AVERAGE DN		0.0267	0.0437

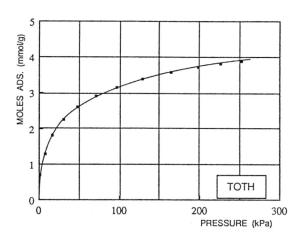

TEMP (K)   =   227.59

TOTH		UNILAN		UNIT
B =	1.9875	C =	2792.9590	kPa
M =	.1480	S =	4.9384	
NI =	123.1357	M =	6.3291	mmol/g
BIS=.224791+01		BIS=.605789-01		1/g
QIS=	4.5874	QIS=	8.0061	KJ/mol

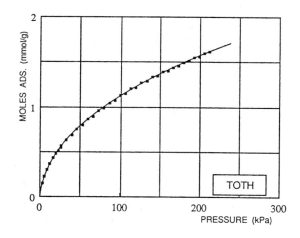

P(kPa)	N(mmol/g)	DN TOTH	DN UNILAN
2.8667	0.1517	-0.0021	-0.0659
5.4667	0.2275	-0.0045	-0.0728
8.5600	0.2989	-0.0071	-0.0708
12.1733	0.3658	-0.0073	-0.0614
16.2266	0.4328	-0.0099	-0.0523
19.9333	0.4774	-0.0025	-0.0345
22.5599	0.5086	0.0003	-0.0250
26.4399	0.5487	0.0068	-0.0092
26.4666	0.5621	-0.0063	-0.0222
32.9066	0.6291	-0.0030	-0.0060
39.8932	0.6826	0.0120	0.0199
46.1332	0.7495	0.0009	0.0165
53.0132	0.7941	0.0134	0.0352
59.7865	0.8610	-0.0012	0.0251
66.2665	0.8967	0.0103	0.0395
73.2398	0.9592	-0.0043	0.0268
79.5465	0.9815	0.0146	0.0464
86.4931	1.0440	-0.0045	0.0272
93.1731	1.0707	0.0085	0.0394
99.7197	1.1243	-0.0075	0.0219
106.3197	1.1466	0.0065	0.0339
113.0664	1.2001	-0.0111	0.0137
119.7464	1.2135	0.0098	0.0317
126.4530	1.2670	-0.0104	0.0082
133.2663	1.2849	0.0046	0.0195
139.6796	1.3340	-0.0144	-0.0033
146.3196	1.3473	0.0024	0.0094
153.0263	1.3875	-0.0080	-0.0055
159.3996	1.4009	0.0062	0.0043
166.1063	1.4410	-0.0056	-0.0123
172.8262	1.4589	0.0042	-0.0075
179.2662	1.4946	-0.0055	-0.0221
186.0795	1.5079	0.0079	-0.0140
192.8928	1.5481	-0.0059	-0.0333
199.3328	1.5570	0.0095	-0.0231
205.8395	1.5927	-0.0021	-0.0401
212.4128	1.6195	-0.0050	-0.0485
AVERAGE DN		0.0067	0.0283

TEMP (K)   =   273.15

TOTH		UNILAN		UNIT
B =	10.3238	C =	2841.9710	kPa
M =	.3515	S =	3.6671	
NI =	10.8951	M =	3.9409	mmol/g
BIS=.323137-01		BIS=.167939-01		1/g
QIS=	4.5874	QIS=	8.0061	KJ/mol

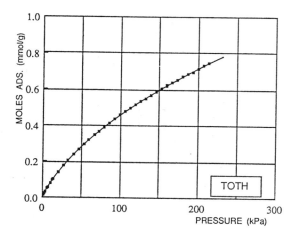

P(kPa)	N(mmol/g)	DN TOTH	DN UNILAN
1.3467	0.0134	0.0009	-0.0035
2.4267	0.0232	0.0010	-0.0055
3.6133	0.0335	0.0009	-0.0074
5.4000	0.0495	-0.0010	-0.0110
6.6400	0.0580	-0.0002	-0.0110
9.9866	0.0812	-0.0000	-0.0120
12.1066	0.0973	-0.0023	-0.0145
13.4933	0.1022	0.0015	-0.0106
19.8933	0.1405	0.0006	-0.0105
26.5333	0.1793	-0.0029	-0.0121
32.7466	0.2066	0.0004	-0.0067
39.9599	0.2396	0.0004	-0.0042
46.5732	0.2681	0.0004	-0.0021
53.1065	0.2945	0.0008	0.0002
59.5465	0.3199	0.0006	0.0016
66.3998	0.3462	-0.0001	0.0024
73.0131	0.3681	0.0018	0.0055

79.6531	0.3908	0.0020	0.0067
86.0531	0.4154	-0.0011	0.0042
92.9331	0.4372	-0.0008	0.0052
99.5464	0.4586	-0.0015	0.0048
106.2264	0.4796	-0.0021	0.0042
112.8264	0.4997	-0.0027	0.0036
119.4797	0.5180	-0.0018	0.0042
126.1864	0.5371	-0.0022	0.0034
132.8263	0.5483	0.0048	0.0099
139.5063	0.5661	0.0048	0.0092
146.2530	0.5885	0.0002	0.0037
152.7863	0.6050	0.0003	0.0029
159.3063	0.6210	0.0006	0.0022
159.5063	0.6233	-0.0011	0.0004
165.9063	0.6384	-0.0006	-0.0002
172.6529	0.6545	-0.0005	-0.0013
179.6396	0.6714	-0.0010	-0.0032
185.9062	0.6853	-0.0003	-0.0039
192.4128	0.6969	0.0029	-0.0021
199.1595	0.7129	0.0018	-0.0047
205.7728	0.7321	-0.0028	-0.0110
212.4795	0.7446	-0.0009	-0.0107
AVERAGE DN		0.0014	0.0057

DN= (NCAL−NEXP)

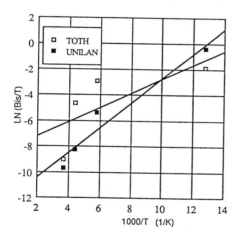

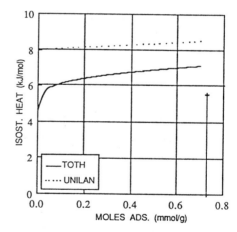

OXYGEN
_____

ZEOLITE
_____

ADSORBENT : 10X Linde; 1/16 in. pellets; 672 m²/gm
EXP.   : Static Volumetric
REF.   : Nolan, J. T., T. W. McKeehan and R. P. Danner,
            J. Chem. Eng. Data, **26**, 112 (1981)

TEMP (K)   =   172.04

TOTH		UNILAN		UNIT
B =	9.1457	C =	12909.1900	kPa
M =	.33474	S =	5.3604	
NI =	56.5868	M =	24.6580	mmol/g
BIS =.108803+00		BIS =.542317-01		1/g
QIS =	15.0343	QIS =	12.2252	KJ/mol

P (kPa)	N (mmol/g)	DN TOTH	DN UNILAN
7.0133	0.3346	-0.0327	-0.0830
16.7200	0.6157	-0.0083	-0.0558
27.6133	0.8834	0.0086	-0.0205
38.9199	1.1332	0.0165	0.0068
52.3332	1.4053	0.0171	0.0253
64.9865	1.6462	0.0093	0.0285
78.0665	1.8782	-0.0005	0.0243
91.4931	2.0968	-0.0069	0.0181
106.6664	2.3378	-0.0241	-0.0042
125.0797	2.5921	-0.0257	-0.0184
143.4530	2.7750	0.0265	0.0157
163.1729	3.0337	0.0043	-0.0310
AVERAGED DN		0.0150	0.0276

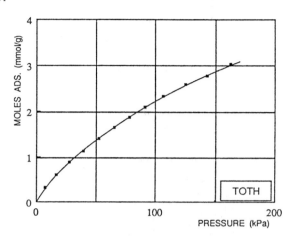

TEMP (K)   =   227.59

TOTH		UNILAN		UNIT
B =	84.4141	C =	2253.8140	kPa
M =	.5048	S =	1.6451	
NI =	34.0984	M =	6.9064	mmol/g
BIS =.985200-02		BIS =.879119-02		1/g
QIS =	15.0343	QIS =	12.2252	KJ/mol

P (kPa)	N (mmol/g)	DN TOTH	DN UNILAN
2.2267	0.0089	0.0023	0.0014
3.8133	0.0178	0.0011	-0.0002
5.7333	0.0268	0.0015	-0.0003
7.2800	0.0357	-0.0001	-0.0022
10.8933	0.0491	0.0034	0.0009
11.0933	0.0535	-0.0001	-0.0027
14.4666	0.0669	0.0020	-0.0008
14.5333	0.0669	0.0023	-0.0005
18.2400	0.0848	0.0012	-0.0018
20.2400	0.0937	0.0012	-0.0019
24.8933	0.1160	-0.0006	-0.0036
33.2799	0.1517	-0.0001	-0.0029
39.9999	0.1829	-0.0029	-0.0054
46.7065	0.2097	-0.0019	-0.0040
53.1732	0.2365	-0.0023	-0.0039
60.1198	0.2632	-0.0012	-0.0023
66.4665	0.2855	0.0017	0.0010
73.2131	0.3123	0.0012	0.0011
79.8531	0.3391	0.0001	0.0004
86.3598	0.3524	0.0115	0.0122
92.7731	0.3926	-0.0045	-0.0034
99.5064	0.4149	-0.0017	-0.0003
106.1864	0.4417	-0.0039	-0.0022
112.9064	0.4595	0.0028	0.0047
119.6397	0.4908	-0.0041	-0.0021

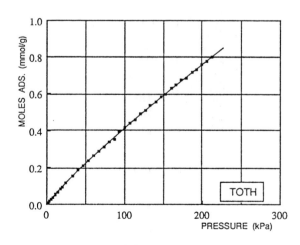

126.1197	0.5086	0.0012	0.0033
132.7597	0.5354	-0.0021	0.0000
139.5730	0.5577	-0.0005	0.0015
146.1863	0.5844	-0.0043	-0.0024
152.7596	0.5978	0.0050	0.0067
159.5063	0.6291	-0.0032	-0.0018
165.9063	0.6469	0.0007	0.0017
172.7196	0.6737	-0.0032	-0.0026
179.1996	0.6871	0.0050	0.0051
186.2395	0.7183	-0.0029	-0.0034
192.4529	0.7317	0.0041	0.0030
199.2928	0.7584	-0.0004	-0.0022
206.1062	0.7763	0.0038	0.0012
212.4128	0.8030	-0.0027	-0.0062
AVERAGE DN		0.0024	0.0027

TEMP (K)   =   273.15

TOTH		UNILAN		UNIT
B =	832.9233	C =	2495.0710	kPa
M =	.6873	S =	.0014	
NI =	28.3739	M =	4.0108	mmol/g
BIS =	.362879-02	BIS =	.365054-02	1/g
QIS =	15.0343	QIS =	12.2252	KJ/mol

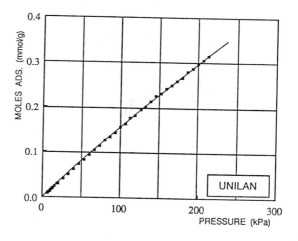

P(kPa)	N(mmol/g)	DN TOTH	DN UNILAN
6.6133	0.0098	0.0007	0.0007
8.0933	0.0120	0.0008	0.0009
9.5066	0.0147	0.0003	0.0005
9.5466	0.0138	0.0013	0.0014
10.6533	0.0165	0.0004	0.0004
12.0000	0.0183	0.0007	0.0009
13.4533	0.0210	0.0003	0.0005
15.4133	0.0232	0.0011	0.0014
18.7466	0.0290	0.0005	0.0009
19.9333	0.0299	0.0015	0.0019
26.5733	0.0415	0.0002	0.0007
33.1466	0.0509	0.0011	0.0017
39.8266	0.0620	0.0002	0.0009
46.2666	0.0718	0.0003	0.0011
53.0799	0.0816	0.0009	0.0019
59.7598	0.0919	0.0008	0.0020
66.2932	0.1031	-0.0004	0.0007
73.0131	0.1124	0.0004	0.0017
79.6531	0.1254	-0.0025	-0.0012
86.2264	0.1325	0.0002	0.0013
92.9331	0.1432	-0.0005	0.0008
99.5064	0.1548	-0.0022	-0.0009
106.4931	0.1637	-0.0008	0.0004
112.7597	0.1758	-0.0035	-0.0023
119.4130	0.1820	0.0001	0.0011
126.1197	0.1941	-0.0021	-0.0011
132.7330	0.2008	0.0010	0.0019
139.2663	0.2128	-0.0015	-0.0008
145.9863	0.2253	-0.0041	-0.0036
152.6530	0.2324	-0.0016	-0.0013
159.3329	0.2427	-0.0021	-0.0018
166.0796	0.2498	0.0006	0.0005
172.5862	0.2588	0.0010	0.0006
179.1329	0.2672	0.0020	0.0013
186.0395	0.2788	0.0004	-0.0005
192.3862	0.2864	0.0019	0.0008
199.3328	0.2976	0.0007	-0.0009
205.7062	0.3052	0.0022	0.0004
212.4128	0.3145	0.0024	-0.0000
AVERAGE DN		0.0012	0.0011

DN= (NCAL-NEXP)

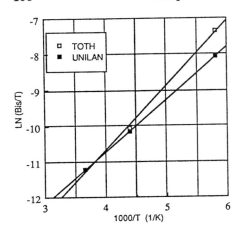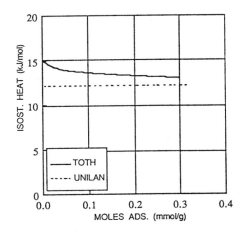

PENTANE

ACTIVATED CARBON

ADSORBENT : Fiber Carbon, KF-1500. Toyobo Co., Ltd.
EXP.   : Gravimetric
REF.   : Kuro-Oka, M., Suzuki, T., Nitta, T. and
            Katayama, T., J. Chem. Eng. Japan, **17**, 588
            (1984)

TEMP (K)   =   273.15
PSAT (kPa) =   24.49

TOTH		UNILAN		UNIT
B =	.0610	C =	.0065	kPa
M =	.2766	S =	5.3485	
NI =	5.4083	M =	4.8392	mmol/g
DG =	-91.1305	DG =	-89.5461	J/g
BIS =	.302500+06	BIS =	.332371+05	1/g
QIS =	60.4258	QIS =	58.2097	KJ/mol

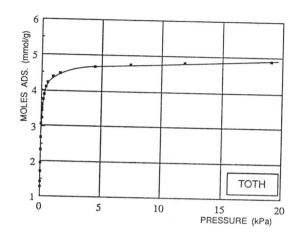

P (kPa)	N (mmol/g)	DN TOTH	DN UNILAN
0.0004	1.2950	-0.1356	-0.1016
0.0007	1.4220	-0.0277	0.0104
0.0015	1.7080	0.0320	0.0586
0.0028	1.9400	0.0973	0.1043
0.0057	2.3300	0.0514	0.0324
0.0112	2.6800	0.0247	-0.0150
0.0253	3.0700	0.0088	-0.0417
0.0374	3.2300	0.0192	-0.0290
0.0586	3.4300	0.0062	-0.0328
0.0877	3.6100	-0.0147	-0.0396
0.1300	3.7600	-0.0178	-0.0246
0.1880	3.8900	-0.0179	-0.0050
0.3270	4.1000	-0.0472	-0.0035
0.4900	4.2200	-0.0460	0.0177
0.9060	4.3800	-0.0387	0.0441
1.5450	4.4900	-0.0192	0.0633
4.4020	4.6700	0.0157	0.0472
7.3600	4.7500	0.0241	0.0124
11.8200	4.8200	0.0267	-0.0302
19.0300	4.8800	0.0318	-0.0721
AVERAGE DN		0.0351	0.0393

TEMP (K)   =   298.15
PSAT (kPa) =   68.51

TOTH		UNILAN		UNIT
B =	.1068	C =	.0412	kPa
M =	.2967	S =	4.8562	
NI =	5.3260	M =	4.6870	mmol/g
DG =	-86.9381	DG =	-86.1345	J/g
BIS =	.248186+05	BIS =	.373173+04	1/g
QIS =	60.4257	QIS =	58.2097	KJ/mol

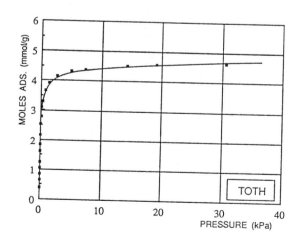

P (kPa)	N (mmol/g)	DN TOTH	DN UNILAN
0.0003	0.4090	-0.0268	-0.0905
0.0008	0.6670	-0.0392	-0.0635
0.0013	0.8440	-0.0622	-0.0625
0.0026	1.0470	-0.0089	0.0185
0.0044	1.2620	-0.0033	0.0351
0.0103	1.6110	0.0419	0.0768
0.0164	1.8380	0.0442	0.0683
0.0324	2.1700	0.0562	0.0588
0.0638	2.5400	0.0283	0.0106
0.1020	2.7900	0.0099	-0.0173
0.1860	3.0900	-0.0050	-0.0355
0.2820	3.3000	-0.0273	-0.0533
0.6810	3.6700	-0.0312	-0.0315
1.3510	3.9300	-0.0394	-0.0122
2.5620	4.1500	-0.0506	-0.0042
4.8780	4.3100	-0.0258	0.0220
7.1990	4.3900	-0.0058	0.0307
14.1200	4.5200	0.0178	0.0132
19.0800	4.5700	0.0289	-0.0013
30.6500	4.6400	0.0467	-0.0300
AVERAGE DN		0.0300	0.0368

	TEMP (K)	=	323.15
	PSAT(kPa)	=	159.58

TOTH			UNILAN			UNIT
B	=	.1763	C	=	.1886	kPa
M	=	.2889	S	=	4.6734	
NI	=	5.4514	M	=	4.5999	mmol/g
DG	=	−84.5823	DG	=	−83.4454	J/g
BIS	=	.595346+04	BIS	=	.750502+03	1/g
QIS	=	60.4258	QIS	=	58.2097	KJ/mol

P (kPa)	N (mmol/g)	DN TOTH	DN UNILAN
0.0003	0.1970	−0.0494	−0.1196
0.0007	0.3200	−0.0707	−0.1553
0.0015	0.4400	−0.0571	−0.1368
0.0041	0.6940	−0.0613	−0.1024
0.0061	0.7830	−0.0267	−0.0470
0.0120	0.9980	0.0008	0.0134
0.0173	1.1230	0.0227	0.0487
0.0319	1.3530	0.0600	0.0982
0.0540	1.5990	0.0616	0.1000
0.0806	1.7990	0.0578	0.0914
0.1640	2.1800	0.0345	0.0526
0.3200	2.5500	0.0028	0.0052
0.5890	2.8800	−0.0250	−0.0322
1.2230	3.2300	−0.0298	−0.0381
1.8830	3.4400	−0.0475	−0.0509
3.6890	3.7200	−0.0480	−0.0389
8.1120	4.0100	−0.0445	−0.0249
13.9100	4.1700	−0.0256	−0.0113
27.9400	4.3200	0.0301	0.0124
44.1000	4.3900	0.0796	0.0245
AVERAGE DN		0.0418	0.0602

DN= (NCAL−NEXP)

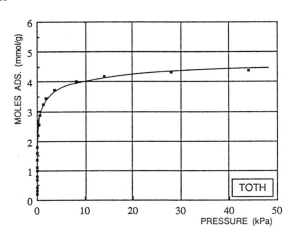

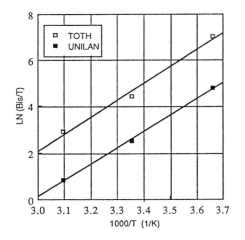

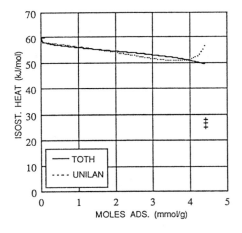

PENTANE
```
--------------------------------------------------------------------------------
```
CARBON MOLECULAR SIEVE
```
--------------------------------------------------------------------------------
```
ADSORBENT : MSC-5A;TAKEDA CHEMICAL Co.;0.56 cc/g;650 Sqm/g
EXP.   : STATIC GRAVIMETRIC (McBAIN QUARTZ SPRING BALANCE)
REF.   : NAKAHARA,T.,HIRATA,M.and OMORI,T.;J.Chem.Eng.Data;19,310(1974)

                    TEMP (K)=   303.15
                    PSAT (kPa)=   82.19

            TOTH   UNILAN   UNITS
        ----------   ---------------- ------
B   =   .0893   C = 12836.0000   kPa
M   =   .1797   S =   22.3249
NI =   2.0196   M =   4.2836   mmol/g
DG =   -34.5040   DG =   -28.4849   J/g
BIS= .35257+007   BIS= .93456+005   1/g
```

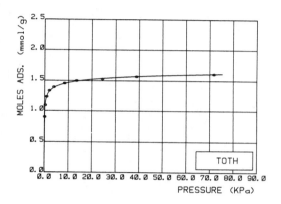

| P(kPa) | N(mmol/g) | DN TOTH | DN UNILAN |
|--------|-----------|---------|-----------|
| .0400  | .8981     | -.0105  | .0273     |
| .2800  | 1.0920    | .0253   | .0201     |
| .8267  | 1.2340    | .0010   | -.0180    |
| 2.0000 | 1.3360    | -.0121  | -.0353    |
| 3.8930 | 1.3930    | -.0067  | -.0284    |
| 8.3330 | 1.4550    | -.0024  | -.0174    |
| 13.5300| 1.5010    | -.0090  | -.0169    |
| 24.4000| 1.5270    | .0100   | .0137     |
| 38.8000| 1.5680    | .0021   | .0172     |
| 71.8000| 1.6090    | .0022   | .0353     |
| AVERAGE DN |        | .0081   | .0230     |

DN=(NCAL-NEXP)
```

PROPANE

ACTIVATED CARBON

ADSORBENT : ATRITION RESISTANT BEADS; TAIYO KAKEN Co.,
            JAPAN; 800-1200 g/m2
EXP.   : STATIC VOLUMETRIC
REF.   : KAUL, B. K., *IND. ENG. CHEM. RES.* **26**, 928 (1987)

                    TEMP (K)   =   310.95
                    PSAT (kPa) =   1295.74

TOTH		UNILAN		UNIT
B =	2.4498	C =	27.4231	kPa
M =	.4940	S =	2.8180	
NI =	5.9375	M =	5.3917	mmol/g
DG =	-54.8570	DG =	-54.5496	J/g
BIS =.25027+001		BIS =.15046+001		1/g
QIS =	19.8966	QIS =	27.0072	KJ/mol

P (kPa)	N (mmol/g)	DN TOTH	DN UNILAN
0.3447	0.2223	-0.0068	-0.0403
0.4826	0.2220	0.0593	0.0241
0.8963	0.6476	-0.2000	-0.2319
1.3789	0.5715	0.0333	0.0099
1.4479	0.5736	0.0515	0.0293
3.9989	1.1561	0.0091	0.0180
5.0331	1.1664	0.1524	0.1667
6.2052	1.6564	-0.1885	-0.1709
8.9631	1.6854	0.0632	0.0827
9.4457	1.6986	0.0917	0.1111
20.5462	2.4918	-0.0480	-0.0407
23.2351	2.6531	-0.1024	-0.0978
25.7862	2.7364	-0.0950	-0.0928
26.7514	2.5165	0.1569	0.1583
44.8845	3.1648	-0.0447	-0.0525
50.5382	3.2300	-0.0097	-0.0187
57.2260	3.3459	-0.0220	-0.0317
73.3596	3.4846	0.0417	0.0319
97.6979	3.7273	0.0229	0.0153
111.0047	3.8038	0.0424	0.0365
129.6204	3.9333	0.0262	0.0226
133.0677	4.0523	-0.0741	-0.0772
148.2361	4.0526	0.0018	0.0004
207.8752	4.2679	0.0122	0.0159
211.3915	4.2262	0.0647	0.0686
259.1028	4.4262	-0.0095	-0.0038
296.5410	4.5469	-0.0508	-0.0449
AVERAGE DN		0.0623	0.0628

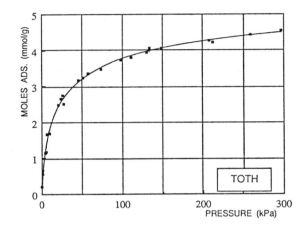

                    TEMP (K)   =   422.05

TOTH		UNILAN		UNIT
B =	4.4037	C =	1236.3320	kPa
M =	.3163	S =	3.8809	
NI =	13.8530	M =	7.3699	mmol/g
BIS =.44799+000		BIS =.13056+000		1/g
QIS =	19.8966	QIS =	27.0072	KJ/mol

P (kPa)	N (mmol/g)	DN TOTH	DN UNILAN
2.1374	0.1254	-0.0030	-0.0490
9.7215	0.3761	-0.0061	-0.0697
19.7878	0.5929	-0.0027	-0.0480
30.3367	0.7723	-0.0044	-0.0287
49.6418	1.0181	0.0039	0.0071
61.7076	1.1488	0.0038	0.0175
66.6718	1.1948	0.0072	0.0241
81.5643	1.3317	0.0059	0.0291
95.8363	1.4348	0.0190	0.0446
125.3456	1.6458	0.0164	0.0398
141.0656	1.8109	-0.0508	-0.0314
148.9255	1.7854	0.0207	0.0377
169.6786	1.9675	-0.0471	-0.0379
180.5033	1.8831	0.0929	0.0975
189.3285	2.0674	-0.0477	-0.0471

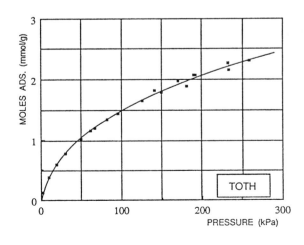

```
191.3279   2.0763  -0.0469  -0.0473
232.7651   2.2668  -0.0513  -0.0724
233.6614   2.1513   0.0680   0.0465
258.9649   2.3039   0.0171  -0.0181
      AVERAGE DN      0.0271   0.0417

     DN= (NCAL-NEXP)
```

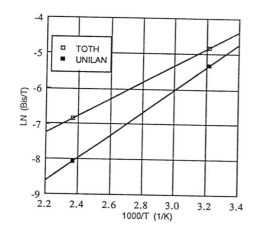

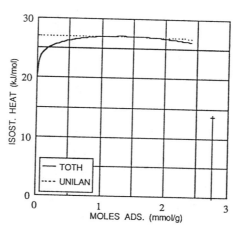

PROPANE

ACTIVATED CARBON

ADSORBENT : BLACK PEARLS I;GODFREY L. CABOT Co. OF BOSTON;705 Sqm/g
EXP.   : STATIC VOLUMETRIC
REF.   : LEWIS,W.K.,GILLILAND,E.R.,CHERTOW,B. and HOFFMAN,W.H.;J.Am.Chem.Soc.;
            72,1153(1950)

TEMP (K)= 298.15
PSAT (kPa)= 947.94

	TOTH		UNILAN	UNITS
B =	.6778	C =	311.3060	kPa
M =	.1651	S =	6.0550	
NI =	13.7027	M =	6.2313	mmol/g
DG =	-38.1376	DG =	-34.8042	J/g
BIS=	.35819+003	BIS=	.17464+001	1/g

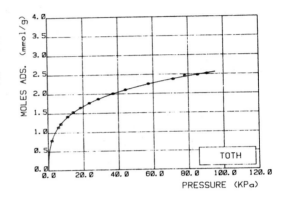

P(kPa)	N(mmol/g)	DN TOTH	DN UNILAN
2.0930	.7910	.0037	-.0953
5.8670	1.1410	.0052	-.0085
7.2270	1.2230	.0054	.0059
11.2000	1.4160	-.0024	.0213
14.2700	1.5320	-.0087	.0231
18.5300	1.6460	.0013	.0377
23.6000	1.7690	-.0016	.0349
28.7300	1.8720	-.0032	.0304
37.0700	2.0060	-.0010	.0246
44.2000	2.1050	-.0029	.0145
57.2700	2.2540	-.0042	-.0032
71.4700	2.3730	.0072	-.0096
78.0900	2.4540	-.0206	-.0454
85.4300	2.4790	.0089	-.0247
90.6900	2.5100	.0146	-.0252
AVERAGE DN		.0061	.0269

DN=(NCAL-NEXP)

PROPANE

ACTIVATED CARBON

ADSORBENT : COLUMBIA G GRADE;1157 Sq m/g;MESH 8x14
EXP.   : STATIC
REF.   : PAYNE,H.K.,STUDERVANT,G.A. and LELAND,T,W,;IND.ENG.CHEM.FUNDAM.;7,363(1968)

TEMP (K)=   293.15
PSAT (kPa)=   832.87

	TOTH		UNILAN	UNITS
B =	1.2626	C =	17.9169	kPa
M =	.4124	S =	3.5343	
NI =	8.4514	M =	7.6408	mmol/g
DG =	-73.7439	DG =	-72.8800	J/g
BIS=	.11702+002	BIS=	.50351+001	1/g
QIS=	37.9920	QIS=	29.9927	KJ/mol

P(kPa)	N(mmol/g)	DN TOTH	DN UNILAN
.8480	.8920	.1708	.1485
2.3720	1.8100	.0085	.0358
5.6190	2.7400	-.1151	-.0867
13.8200	3.6730	-.1079	-.1172
35.1100	4.5560	-.0073	-.0526
74.8100	5.2230	.0678	.0255
133.9000	5.6940	.1104	.0915
272.3000	6.2910	.0598	.0761
448.2000	6.6830	-.0026	.0263
606.7000	6.9820	-.1225	-.0961
AVERAGE DN		.0773	.0756

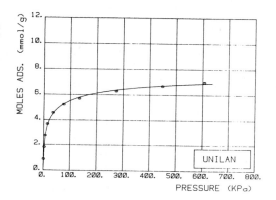

TEMP (K)=   303.15
PSAT (kPa)=   1074.30

	TOTH		UNILAN	UNITS
B =	.8342	C =	32.2851	kPa
M =	.3036	S =	4.6224	
NI =	9.4275	M =	8.1555	mmol/g
DG =	-77.3807	DG =	-75.9111	J/g
BIS=	.43171+002	BIS=	.70059+001	1/g
QIS=	37.9920	QIS=	29.9927	KJ/mol

P(kPa)	N(mmol/g)	DN TOTH	DN UNILAN
.3378	.6870	.0593	-.0476
1.6620	1.6140	-.0190	.0004
6.1290	2.6020	-.0167	.0528
18.6900	3.6000	-.0307	.0054
45.9200	4.4020	-.0106	-.0196
75.4800	4.8280	.0103	-.0174
134.5000	5.3050	.0346	-.0018
248.2000	5.8370	.0045	-.0231
424.0000	6.2150	.0346	.0280
641.2000	6.5300	.0133	.0274
903.2000	6.8280	-.0562	-.0254
AVERAGE DN		.0263	.0226

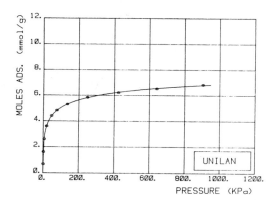

```
           TEMP   (K)=    313.15
           PSAT (kPa)=   1363.80
```

TOTH		UNILAN		UNITS
B =	1.3026	C =	31.8807	kPa
M =	.3715	S =	3.7499	
NI =	8.4626	M =	7.3963	mmol/g
DG =	-75.2400	DG =	-74.1152	J/g
BIS=	.10815+002	BIS=	.34224+001	1/g
QIS=	37.9920	QIS=	29.9927	KJ/mol

P(kPa)	N(mmol/g)	DN TOTH	DN UNILAN
1.6620	1.1320	.0486	.0190
6.8740	2.2630	-.0158	.0192
21.9800	3.3850	-.0503	-.0364
67.6600	4.4330	-.0067	-.0299
137.5000	5.0560	.0192	-.0063
279.2000	5.6190	.0460	.0371
499.9000	6.0810	.0164	.0235
699.8000	6.3220	.0020	.0129
886.1000	6.4780	-.0051	.0038
1093.0000	6.6380	-.0396	-.0364
	AVERAGE DN	.0250	.0224

```
           TEMP   (K)=    323.15
           PSAT (kPa)=   1707.30
```

TOTH		UNILAN		UNITS
B =	1.8802	C =	39.5194	kPa
M =	.4189	S =	3.3509	
NI =	7.8710	M =	7.0093	mmol/g
DG =	-73.1341	DG =	-72.2998	J/g
BIS=	.46849+001	BIS=	.20260+001	1/g
QIS=	37.9920	QIS=	29.9927	KJ/mol

P(kPa)	N(mmol/g)	DN TOTH	DN UNILAN
3.2820	1.2340	.0423	.0334
12.5300	2.3960	-.0211	.0054
39.6900	3.5480	-.0358	-.0391
68.5700	4.0740	-.0160	-.0338
132.7000	4.6750	.0118	-.0090
275.8000	5.2650	.0527	.0481
496.4000	5.7470	.0151	.0257
734.3000	6.0340	-.0088	.0043
944.6000	6.1860	-.0055	.0033
1151.0000	6.2930	.0019	.0034
1369.0000	6.4150	-.0250	-.0330
	AVERAGE DN	.0215	.0217

```
           TEMP   (K)=    333.15
           PSAT (kPa)=   2111.30
```

TOTH		UNILAN		UNITS
B =	2.0446	C =	44.6454	kPa
M =	.4232	S =	3.3027	
NI =	7.5268	M =	6.6924	mmol/g
DG =	-73.4624	DG =	-72.6662	J/g
BIS=	.38475+001	BIS=	.17065+001	1/g
QIS=	37.9920	QIS=	29.9927	KJ/mol

P(kPa)	N(mmol/g)	DN TOTH	DN UNILAN
2.4060	.9460	-.0041	-.0339
10.5400	1.9830	.0109	.0386
31.4300	3.0150	-.0112	.0013
70.4700	3.7910	-.0056	-.0162
142.1000	4.4310	.0039	-.0123
282.7000	4.9870	.0276	.0229
527.4000	5.4940	-.0172	-.0085
799.8000	5.7670	-.0176	-.0080
1120.0000	5.9410	.0076	.0091
1355.0000	6.0480	.0050	-.0024
	AVERAGE DN	.0111	.0153

```
           DN=(NCAL-NEXP)
```

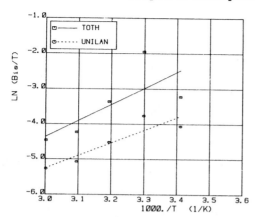
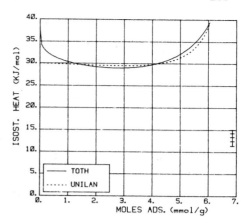

PROPANE

ACTIVATED CARBON

ADSORBENT : COLUMBIA GRADE L;MESH 20x40;1152 Sq m/g (N2)
EXP.   : STATIC VOLUMETRIC
REF.   : RAY,G.C.and BOX,E.O.;IND.ENG.CHEM.;42,1315(1950)

TEMP (K)=   310.92
PSAT (kPa)= 1294.80

	TOTH		UNILAN	UNITS
B =	2.1295	C =	16.4689	kPa
M =	.5193	S =	2.5755	
NI =	6.1207	M =	5.6076	mmol/g
DG =	-64.2888	DG =	-63.5229	J/g
BIS=	.36903+001	BIS=	.22320+001	1/g
QIS=	26.4622	QIS=	28.4511	KJ/mol

P(kPa)	N(mmol/g)	DN TOTH	DN UNILAN
2.2660	1.0440	.0968	.0685
15.6000	2.8190	-.0561	-.0659
31.7400	3.4800	-.0637	-.0688
59.6000	3.9680	-.0155	-.0063
89.7400	4.2070	.0593	.0799
99.9700	4.3420	.0019	.0251
293.0000	4.9400	.0532	.0702
479.2000	5.2940	-.0759	-.0892
679.1000	5.3040	.0490	.0045
AVERAGE DN		.0524	.0531

TEMP (K)=   338.70
PSAT (kPa)= 2364.30

	TOTH		UNILAN	UNITS
B =	3.3678	C =	44.9072	kPa
M =	.5144	S =	2.8766	
NI =	6.0938	M =	5.6959	mmol/g
DG =	-64.5060	DG =	-64.2675	J/g
BIS=	.16194+001	BIS=	.10986+001	1/g
QIS=	26.4622	QIS=	28.4511	KJ/mol

P(kPa)	N(mmol/g)	DN TOTH	DN UNILAN
6.0000	1.0690	.0983	.1270
27.2000	2.4690	-.0712	-.0626
53.2000	3.0780	-.0625	-.0803
92.6700	3.5120	.0000	-.0290
99.9700	3.6350	-.0576	-.0870
244.8000	4.1880	.0952	.0840
424.0000	4.4750	.1697	.1794
617.1000	4.7100	.1486	.1688
803.2000	5.2890	-.2960	-.2726
AVERAGE DN		.1110	.1212

TEMP (K)= 366.48
PSAT (kPa)= 4002.90

	TOTH		UNILAN	UNITS
B =	4.1392	C =	47.8317	kPa
M =	.5381	S =	2.5831	
NI =	5.1868	M =	4.7819	mmol/g
DG =	-65.3295	DG =	-64.7214	J/g
BIS=	.11281+001	BIS=	.77608+000	1/g
QIS=	26.4622	QIS=	28.4511	KJ/mol

P(kPa)	N(mmol/g)	DN TOTH	DN UNILAN
4.0000	.6942	-.0051	-.0102
21.0700	1.6770	.0578	.0713
45.3300	2.3860	-.0372	-.0378
93.3400	2.9540	-.0264	-.0363
99.9700	3.0410	-.0605	-.0705
279.2000	3.5840	.1130	.1137
461.9000	3.9220	.0628	.0681
634.3000	4.2440	-.1009	-.0976
AVERAGE DN		.0580	.0632

TEMP (K)= 394.20

	TOTH		UNILAN	UNITS
B =	1.9797	C =	85510.5996	kPa
M =	.3026	S =	9.6161	
NI =	8.8543	M =	15.7850	mmol/g
BIS=	.30369+001	BIS=	.47199+000	1/g
QIS=	26.4622	QIS=	28.4511	KJ/mol

P(kPa)	N(mmol/g)	DN TOTH	DN UNILAN
7.0670	.7099	.0579	-.0480
38.6700	1.6770	-.0023	.0074
83.2100	2.2380	-.0167	.0169
98.5400	2.3700	-.0188	.0154
99.9700	2.4060	-.0436	-.0094
303.4000	3.2510	.0230	.0266
482.6000	3.5990	.0722	.0539
651.5000	3.9780	-.0493	-.0813
AVERAGE DN		.0355	.0324

TEMP (K)= 422.00

	TOTH		UNILAN	UNITS
B =	2.4061	C =	814091.0000	kPa
M =	.2662	S =	10.8641	
NI =	13.7390	M =	20.5374	mmol/g
BIS=	.17817+001	BIS=	.21291+000	1/g
QIS=	26.4622	QIS=	28.4511	KJ/mol

P(kPa)	N(mmol/g)	DN TOTH	DN UNILAN
13.0700	.6798	.0146	-.1039
41.7300	1.2640	-.0090	-.0327
71.0700	1.6110	-.0132	.0111
96.2700	1.8310	-.0126	.0324
99.9700	1.8940	-.0469	.0001
313.7000	2.7850	.0600	.0990
493.0000	3.1440	.1573	.1508
692.9000	3.8190	-.1550	-.2108
AVERAGE DN		.0586	.0801

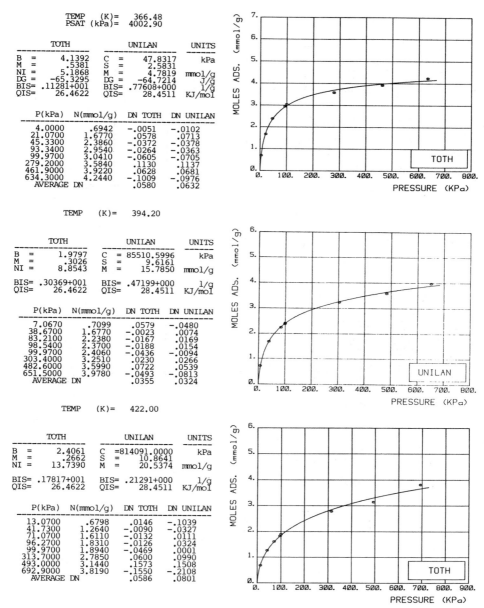

TEMP   (K)=   449.80

	TOTH		UNILAN	UNITS
B =	9.6843	C =	759775.0000	kPa
M =	.4901	S =	10.2699	
NI =	5.9373	M =	18.7644	mmol/g
BIS=	.21601+000	BIS=	.12973+000	1/g
QIS=	26.4622	QIS=	28.4511	KJ/mol

P(kPa)	N(mmol/g)	DN TOTH	DN UNILAN
9.0670	.3295	-.0249	-.0592
44.5300	.9154	-.0050	-.0111
73.0700	1.2020	.0063	.0115
97.8700	1.3710	.0372	.0460
99.9700	1.4440	-.0206	-.0117
286.1000	2.2640	-.0042	-.0043
461.9000	2.6780	-.0055	-.0105
668.8000	2.9900	.0012	.0003
	AVERAGE DN	.0131	.0193

TEMP   (K)=   477.59

	TOTH		UNILAN	UNITS
B =	8.9170	C =	720535.0000	kPa
M =	.3853	S =	9.3861	
NI =	10.8622	M =	19.5342	mmol/g
BIS=	.14750+000	BIS=	.68362-001	1/g
QIS=	26.4622	QIS=	28.4511	KJ/mol

P(kPa)	N(mmol/g)	DN TOTH	DN UNILAN
13.6000	.2876	-.0352	-.0764
38.0000	.5847	-.0518	-.0771
94.8100	.9604	.0030	.0213
99.9700	1.0040	-.0092	.0117
310.3000	1.8130	.0385	.0745
486.1000	2.2520	.0371	.0393
665.3000	2.6670	-.0420	-.0806
	AVERAGE DN	.0310	.0544

TEMP   (K)=   505.37

	TOTH		UNILAN	UNITS
B =	13.8086	C =	592457.0000	kPa
M =	.4691	S =	9.5905	
NI =	5.8033	M =	11.7345	mmol/g
BIS=	.90437-001	BIS=	.63455-001	1/g
QIS=	26.4622	QIS=	28.4511	KJ/mol

P(kPa)	N(mmol/g)	DN TOTH	DN UNILAN
25.2000	.2958	.0000	.0001
96.8100	.7466	-.0001	.0003
	AVERAGE DN	.0001	.0002

DN=(NCAL-NEXP)

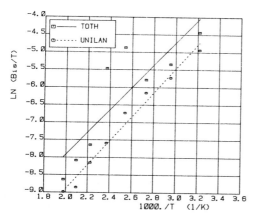

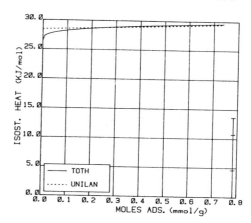

PROPANE

ACTIVATED CARBON

ADSORBENT : NUXIT-AL,(1.5 mm DIAMETER, 3-4 mm LONG)
EXP.   : STATIC VOLUMETRIC
REF.   : SZEPESY, L., ILLES, V., *Acta Chim. Hung.*,**35**, 37(1963)

TEMP   (K) =   293.15
PSAT (kPa) =   832.82

TOTH		UNILAN		UNIT
B =	0.7664	C =	995.2710	kPa
M =	0.3537	S =	8.6799	
NI =	6.2866	M =	11.5871	mmol/g
DG =	-60.7973	DG =	-61.3361	J/g
BIS=3.25097+01		BIS=.961706+01		1/g
QIS =	42.4385	QIS =	38.4125	KJ/mol

P(kPa)	N(mmol/g)	DN TOTH	DN UNILAN
0.0667	0.3592	-0.0770	-0.1374
0.2400	0.6501	-0.0306	-0.0606
0.5333	0.9745	-0.0329	-0.0242
0.8399	1.1833	-0.0181	0.0088
1.3732	1.4193	0.0185	0.0559
2.4665	1.7049	0.0933	0.1280
3.9330	2.0623	0.0442	0.0659
4.9329	2.2167	0.0441	0.0571
15.3320	3.0631	-0.0150	-0.0476
18.3318	3.2068	-0.0362	-0.0732
22.8247	3.3759	-0.0570	-0.0972
29.8641	3.5419	-0.0446	-0.0849
40.7299	3.7288	-0.0317	-0.0658
59.5283	3.9416	-0.0101	-0.0262
85.6594	4.1277	0.0166	0.0301
101.0581	4.2098	0.0271	0.0581
125.9893	4.3178	0.0384	0.0971
AVERAGE DN		0.0374	0.0658

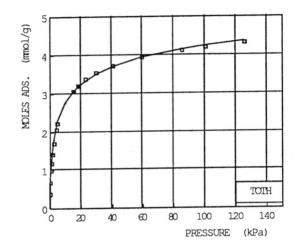

TEMP   (K) =   313.15
PSAT (kPa) = 1363.76

TOTH		UNILAN		UNIT
B =	1.7139	C =	996.9510	kPa
M =	0.4588	S =	7.1798	
NI =	5.5735	M =	10.9126	mmol/g
DG =	-56.7081	DG =	-58.7398	J/g
BIS=4.48457+00		BIS=2.60522+00		1/g
QIS =	42.4385	QIS =	38.4125	KJ/mol

P(kPa)	N(mmol/g)	DN TOTH	DN UNILAN
0.4666	0.3922	-0.0129	-0.0282
2.1998	0.9963	0.0094	0.0372
4.6663	1.4702	-0.0041	0.0241
7.9993	1.8338	0.0124	0.0244
14.4654	2.2979	-0.0050	-0.0195
24.5312	2.7142	-0.0155	-0.0501
40.8632	3.0823	-0.0005	-0.0396
58.7950	3.3415	0.0011	-0.0266
74.5937	3.5026	0.0032	-0.0089
91.4856	3.6347	0.0057	0.0127
106.3110	3.7324	0.0035	0.0283
124.9494	3.8448	-0.0095	0.0378
AVERAGE DN		0.0069	0.0281

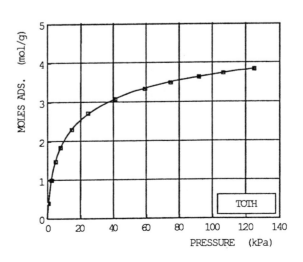

TEMP   (K) =   333.15
PSAT (kPa) = 2111.1799

TOTH		UNILAN		UNIT
B =	1.8364	C =	1082.5300	kPa
M =	0.4178	S =	6.7810	
NI =	5.6453	M =	9.8329	mmol/g
DG =	−57.3584	DG =	−58.9574	J/g
BIS =	3.65049+00	BIS =	1.63433+00	1/g
QIS =	42.4385	QIS =	38.4125	KJ/mol

P(kPa)	N(mmol/g)	DN TOTH	DN UNILAN
0.3466	0.3159	−0.0931	−0.1357
1.3732	0.6385	−0.0696	−0.0946
3.2531	0.9432	−0.0106	−0.0051
5.7862	1.2221	0.0207	0.0409
10.1991	1.5603	0.0340	0.0565
17.0386	1.9226	0.0198	0.0343
29.3842	2.3206	0.0096	0.0106
32.5972	2.3791	0.0259	0.0245
43.9296	2.5946	0.0254	0.0185
44.0896	2.6263	−0.0036	−0.0106
55.4886	2.7927	−0.0055	−0.0144
64.3279	2.8681	0.0238	0.0152
72.6338	2.9930	−0.0159	−0.0233
89.9923	3.1514	−0.0261	−0.0286
93.8587	3.1680	−0.0139	−0.0150
108.8441	3.2701	−0.0161	−0.0111
130.7222	3.3915	−0.0164	−0.0011
AVERAGE DN		0.0253	0.0318

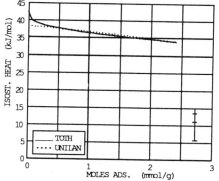

TEMP   (K) =   363.15
PSAT (kPa) = 3768.76

TOTH		UNILAN		UNIT
B =	3.2653	C =	1308.2400	kPa
M =	0.4242	S =	5.6409	
NI =	6.0224	M =	8.8755	mmol/g
DG =	−56.5003	DG =	−57.1469	J/g
BIS =	1.11731+00	BIS =	.511507+00	1/g
QIS =	42.4385	QIS =	38.4125	KJ/mol

P(kPa)	N(mmol/g)	DN TOTH	DN UNILAN
7.2527	0.7242	0.0338	0.0159
7.9193	0.7666	0.0311	0.0164
8.1993	0.8130	0.0008	−0.0127
9.0259	0.8701	−0.0108	−0.0207
10.2258	0.9316	−0.0107	−0.0161
11.6390	0.9995	−0.0115	−0.0125
13.3722	1.0762	−0.0131	−0.0097
15.4787	1.1654	−0.0195	−0.0119
18.9984	1.2801	−0.0119	0.0002
23.5980	1.4104	−0.0056	0.0098
30.3708	1.5728	−0.0001	0.0164
40.7165	1.7754	0.0029	0.0175
55.6753	2.0025	0.0055	0.0146
72.2605	2.1930	0.0123	0.0151
81.5397	2.2956	0.0025	0.0021
93.3654	2.3956	0.0073	0.0033
97.7250	2.4291	0.0093	0.0041
106.6443	2.4978	0.0085	0.0011
132.2954	2.7039	−0.0297	−0.0420
AVERAGE DN		0.0119	0.0127

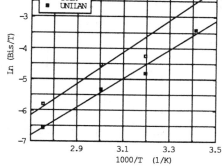

PROPANE
_____

ACTIVATED CARBON
_____

ADSORBENT : NUXIT-AL, (1.5 mm DIAMETER, 3-4 mm LONG)
EXP.   : STATIC VOLUMETRIC
REF.   : SZEPESY, L., ILLES, V., *Acta Chim. Hung.*,**35**, 53(1963)

TEMP   (K) =   293.15
PSAT (kPa) =   832.82

TOTH		UNILAN		UNIT
B =	1.1603	C =	1374.2500	kPa
M =	0.4249	S =	11.8757	
NI =	5.9823	M =	10.8594	mmol/g
DG =	-56.6702	DG =	-74.3686	J/g
BIS =	1.02760+01	BIS =	1.16552+02	1/g
QIS =	31.0564	QIS =	70.4261	KJ/mol

P (kPa)	N (mmol/g)	DN TOTH	DN UNILAN
102.9462	4.2085	-0.0066	0.0364
151.5822	4.3843	0.0342	0.0375
179.4466	4.5458	-0.0392	-0.0469
204.7778	4.5984	-0.0255	-0.0391
236.0872	4.6524	-0.0107	-0.0281
269.5245	4.6997	0.0037	-0.0148
304.8869	4.7434	0.0153	-0.0022
341.5666	4.7867	0.0213	0.0065
402.7668	4.8546	0.0220	0.0140
450.0856	4.8992	0.0217	0.0202
585.3545	5.0312	-0.0115	0.0082
678.6748	5.1102	-0.0384	-0.0031
	AVERAGE DN	0.0208	0.0214

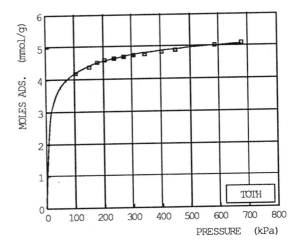

TEMP   (K) =   313.15
PSAT (kPa) =   1363.76

TOTH		UNILAN		UNIT
B =	1.3399	C =	63.7129	kPa
M =	.4069	S =	6.8628	
NI =	5.8550	M =	7.0210	mmol/g
DG =	-58.0569	DG =	-67.3283	J/g
BIS =	.742712+01	BIS =	1.99848+01	1/g
QIS =	31.0564	QIS =	70.4261	KJ/mol

P (kPa)	N (mmol/g)	DN TOTH	DN UNILAN
103.4528	3.7230	-0.0046	0.0349
136.7887	3.8854	0.0040	0.0150
154.9259	3.9698	-0.0070	-0.0058
174.7856	4.0389	-0.0068	-0.0135
191.4029	4.0978	-0.0146	-0.0261
204.8792	4.1219	-0.0011	-0.0155
265.6741	4.2544	0.0049	-0.0156
307.1161	4.3236	0.0094	-0.0110
347.6461	4.3829	0.0111	-0.0073
389.8986	4.4316	0.0172	0.0024
459.5089	4.5056	0.0188	0.0118
549.8907	4.5989	0.0045	0.0096
595.3857	4.6497	-0.0126	-0.0010
655.0661	4.6716	0.0051	0.0255
692.0497	4.7376	-0.0387	-0.0127
	AVERAGE DN	0.0107	0.0138

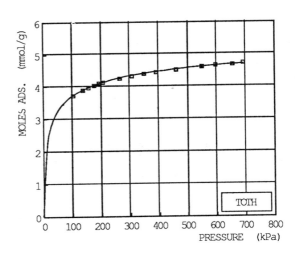

```
TEMP    (K) =    333.15
PSAT  (kPa) =   2111.18
```

TOTH		UNILAN		UNIT
B =	1.9360	C =	443.3470	kPa
M =	0.4138	S =	6.6705	
NI =	5.9042	M =	8.3235	mmol/g
DG =	−57.4117	DG =	−61.2674	J/g
BIS=3.31349+00		BIS=3.07473+00		1/g
QIS =	31.0564	QIS =	70.4261	KJ/mol

P (kPa)	N (mmol/g)	DN TOTH	DN UNILAN
102.7436	3.2273	−0.0048	0.0255
117.3343	3.3179	−0.0012	0.0173
139.2205	3.4450	−0.0091	−0.0037
165.4637	3.5606	−0.0070	−0.0119
185.4247	3.6311	−0.0014	−0.0116
204.4738	3.6904	0.0035	−0.0102
226.6640	3.7556	0.0049	−0.0112
258.9867	3.8417	0.0031	−0.0144
295.4637	3.9238	0.0024	−0.0146
331.4341	3.9836	0.0119	−0.0029
362.7435	4.0403	0.0085	−0.0034
437.6227	4.1478	0.0087	0.0059
506.7263	4.2357	0.0018	0.0092
586.6718	4.3258	−0.0101	0.0103
644.0217	4.3847	−0.0206	0.0094
AVERAGE DN		0.0066	0.0108

```
TEMP    (K) =    363.15
PSAT  (kPa) =   3768.76
```

TOTH		UNILAN		UNIT
B =	3.1730	C =	1201.4500	kPa
M =	0.4209	S =	5.7704	
NI =	5.9614	M =	8.5568	mmol/g
DG =	−56.0721	DG =	−57.1233	J/g
BIS=1.15834+00		BIS=.597497+00		1/g
QIS =	31.0564	QIS =	70.4261	KJ/mol

P (kPa)	N (mmol/g)	DN TOTH	DN UNILAN
106.8979	2.4947	−0.0048	0.0152
138.1060	2.6923	−0.0062	0.0017
161.4107	2.8146	−0.0093	−0.0078
176.2042	2.8824	−0.0104	−0.0120
198.4957	2.9587	0.0036	−0.0018
225.2455	3.0524	0.0050	−0.0034
264.3569	3.1697	0.0069	−0.0039
302.4551	3.2608	0.0147	0.0035
351.5977	3.3768	0.0077	−0.0023
413.5073	3.4968	0.0029	−0.0033
494.4660	3.6222	0.0016	0.0026
577.5525	3.7382	−0.0093	0.0008
660.3350	3.8341	−0.0167	0.0034
AVERAGE DN		0.0076	0.0048

DN = (NCAL−NEXP)

PROPANE
```
-------------------------------------------------------------------------------
CARBON MOLECULAR SIEVE
-------------------------------------------------------------------------------
ADSORBENT : MSC-5A;TAKEDA CHEMICAL Co.;0.56 cc/g;650 Sqm/g
EXP.      : STATIC GRAVIMETRIC (McBAIN QUARTZ SPRING BALANCE)
REF.      : NAKAHARA,T.,HIRATA,M.and OMORI,T.;J.Chem.Eng.Data;19,310(1974)
```

```
                    TEMP (K)=    278.55
                    PSAT (kPa)=  555.51
```

	TOTH		UNILAN	UNITS
B =	.2664	C =	.5169	kPa
M =	.3222	S =	3.6939	
NI =	2.5414	M =	2.1011	mmol/g
DG =	-35.1285	DG =	-32.5779	J/g
BIS=	.35685+003	BIS=	.51196+002	1/g
QIS=	33.1947	QIS=	28.2668	KJ/mol

P(kPa)	N(mmol/g)	DN TOTH	DN UNILAN
.0400	.3424	.1038	.0591
.2267	.8186	.0193	.0102
.4267	1.0340	-.0339	-.0353
.7200	1.2130	-.0769	-.0728
2.0530	1.4290	-.0272	-.0112
4.9330	1.6100	-.0037	.0223
9.2670	1.7300	.0091	.0370
16.2700	1.8300	.0163	.0373
29.0700	1.9390	.0063	.0088
51.8700	2.0160	.0166	-.0106
83.2000	2.0700	.0257	-.0323
AVERAGE DN		.0308	.0306

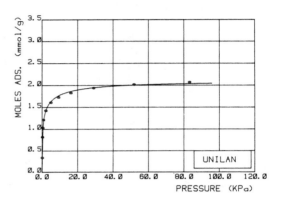

```
                    TEMP (K)=    303.15
                    PSAT (kPa)=  1074.33
```

	TOTH		UNILAN	UNITS
B =	.3505	C =	.8558	kPa
M =	.3389	S =	3.0645	
NI =	2.5077	M =	1.9983	mmol/g
DG =	-37.8121	DG =	-34.7258	J/g
BIS=	.13935+003	BIS=	.20527+002	1/g
QIS=	33.1947	QIS=	28.2668	KJ/mol

P(kPa)	N(mmol/g)	DN TOTH	DN UNILAN
.0800	.4217	.0033	-.0646
.2267	.5646	.0862	.0503
.8933	.9546	.0485	.0573
1.6530	1.2380	-.0701	-.0446
4.0000	1.4630	-.0653	-.0222
8.0000	1.6120	-.0469	-.0005
13.6000	1.7230	-.0400	-.0021
24.0000	1.8030	-.0051	.0108
39.4700	1.8590	.0296	.0152
57.6700	1.9000	.0516	.0085
84.0000	1.9550	.0538	-.0209
AVERAGE DN		.0455	.0270

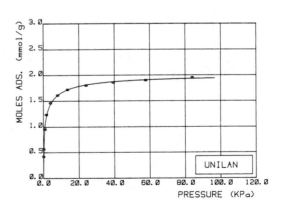

TEMP (K)=   323.15
PSAT (kPa)= 1707.32

	TOTH		UNILAN	UNITS
B =	.4774	C =	1.8759	kPa
M =	.3476	S =	3.2229	
NI =	2.4966	M =	1.9808	mmol/g
DG =	-38.1289	DG =	-35.2030	J/g
BIS=	.56280+002	BIS=	.11030+002	1/g
QIS=	33.1947	QIS=	28.2668	KJ/mol

P(kPa)	N(mmol/g)	DN TOTH	DN UNILAN
.0933	.2721	.0278	-.0237
.6667	.6963	.0116	.0045
1.7600	.9477	.0157	.0246
3.4000	1.1610	-.0181	-.0026
6.6670	1.3290	-.0059	.0138
12.0700	1.5170	-.0427	-.0228
21.3300	1.6400	-.0304	-.0163
34.1300	1.7080	.0046	-.0070
52.4500	1.7870	.0123	-.0027
85.3300	1.8370	.0519	.0090
	AVERAGE DN	.0221	.0127

DN=(NCAL-NEXP)

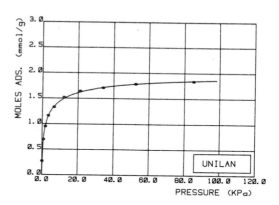

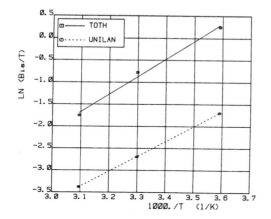

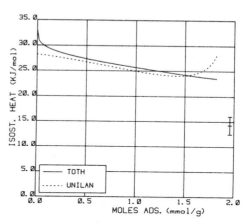

PROPANE

SILICA GEL

ADSORBENT : DAVISON CHEMICAL Co.OF BALTIMORE;751 Sqm/g;Mesh 14x20
EXP.   : STATIC VOLUMETRIC
REF.   : LEWIS,W.K.,GILLILAND,E.R.,CHERTOW,B. and HOFFMAN,W.H.;J.Am.Chem.Soc.;
            72,1153(1950)

TEMP (K)=   298.15
PSAT (kPa)= 947.94

	TOTH		UNILAN		UNITS
B =	8.2120	C =	426.0350		kPa
M =	.3741	S =	2.6191		
NI =	17.0810	M =	5.5586		mmol/g
DG =	-21.6295	DG =	-19.2012		J/g
BIS=	.15221+000	BIS=	.84282-001		1/g

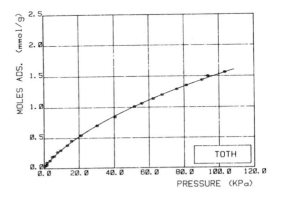

P(kPa)	N(mmol/g)	DN TOTH	DN UNILAN
.6000	.0436	-.0151	-.0234
1.4800	.0564	.0075	-.0072
1.6000	.0931	-.0247	-.0401
3.3330	.1252	.0031	-.0176
4.6800	.1912	-.0212	-.0430
5.8000	.1980	.0046	-.0173
7.6270	.2690	-.0167	-.0373
9.5200	.2986	.0017	-.0165
13.3300	.3850	.0040	-.0083
15.7700	.4560	-.0146	-.0230
21.1900	.5441	.0042	.0042
30.3300	.7020	.0053	.0158
40.5600	.8430	.0200	.0365
51.6000	1.0100	.0023	.0197
55.8700	1.0610	.0049	.0214
62.4000	1.1380	.0062	.0201
67.4500	1.1990	.0030	.0140
75.8700	1.2880	.0056	.0106
81.4800	1.3500	.0017	.0018
90.3700	1.4340	.0056	-.0031
93.7300	1.5010	-.0295	-.0417
103.3000	1.5620	-.0027	-.0260
AVERAGE DN		.0093	.0204

DN=(NCAL-NEXP)

PROPANE

SILICA GEL

ADSORBENT : DAVISON CHEMICAL Co.;REF.GRADE;751 Sqm/g;mesh 14x20
EXP.   : STATIC VOLUMETRIC
REF.   : LEWIS,W.K.,GILLILAND,E.R.,CHERTOW,B. and BAREIS,D.;J.Am.Chem.Soc.;
            72,1160(1950)

TEMP (K)=   273.15
PSAT (kPa)= 472.80

TOTH		UNILAN		UNITS
B =	6.5900	C =	539.4550	kPa
M =	.4323	S =	3.5640	
NI =	11.1695	M =	8.8468	mmol/g
DG =	-21.9210	DG =	-21.0734	J/g
BIS=	.32348+000	BIS=	.18431+000	1/g
QIS=	14.4878	QIS=	19.3398	KJ/mol

P(kPa)	N(mmol/g)	DN TOTH	DN UNILAN
2.2130	.2137	-.0124	-.0460
3.9200	.3721	-.0532	-.0889
5.0270	.3960	-.0092	-.0433
8.5870	.5678	.0086	-.0149
9.7600	.6306	.0012	-.0182
12.4300	.7307	.0176	.0073
14.5900	.9438	-.1091	-.1128
17.2400	.9010	.0318	.0353
29.1200	1.2590	.0402	.0630
35.2800	1.4430	.0135	.0398
39.8400	1.5200	.0427	.0697
57.2500	1.8810	.0289	.0483
66.8100	2.0720	-.0006	.0103
78.2800	2.2410	.0043	.0029
89.1900	2.4200	-.0255	-.0399
101.7000	2.5820	-.0321	-.0624
AVERAGE DN		.0269	.0439

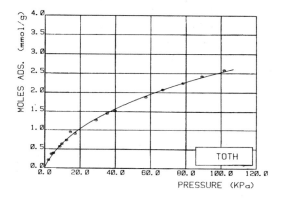

TEMP (K)=   313.15
PSAT (kPa)= 1363.85

TOTH		UNILAN		UNITS
B =	15.8581	C =	1426.9900	kPa
M =	.4827	S =	3.4681	
NI =	8.7026	M =	6.4135	mmol/g
DG =	-19.7183	DG =	-17.9094	J/g
BIS=	.73871-001	BIS=	.54057-001	1/g
QIS=	14.4878	QIS=	19.3398	KJ/mol

P(kPa)	N(mmol/g)	DN TOTH	DN UNILAN
1.3470	.0418	-.0088	-.0142
3.7200	.0900	-.0064	-.0158
5.0670	.0916	.0184	.0080
6.2270	.1407	-.0090	-.0197
12.8300	.2580	-.0153	-.0240
18.2500	.3352	-.0124	-.0177
19.5500	.3276	.0133	.0088
27.2000	.4470	-.0063	-.0064
37.6000	.5680	-.0065	-.0024
39.7300	.5790	.0056	.0103
49.7300	.6871	-.0003	.0057
53.3300	.7068	.0146	.0206
61.6800	.7876	.0102	.0154
72.1100	.8840	.0026	.0054
73.9900	.9040	-.0020	.0002
85.7300	.9908	.0029	.0005
102.5000	1.1280	-.0141	-.0250
AVERAGE DN		.0087	.0118

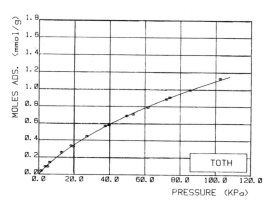

TEMP   (K)=   373.15

	TOTH		UNILAN	UNITS
B =	4.6892	C =	73948.7002	kPa
M =	.2693	S =	8.0593	
NI =	7.8098	M =	3.0949	mmol/g
BIS=	.78020-001	BIS=	.25480-001	1/g
QIS=	14.4878	QIS=	19.3398	KJ/mol

P(kPa)	N(mmol/g)	DN TOTH	DN UNILAN
12.8500	.0531	.0338	.0310
15.8700	.1471	-.0464	-.0476
54.2000	.2087	.0160	.0216
80.2000	.2781	.0064	.0077
100.5000	.3360	-.0115	-.0158
AVERAGE DN		.0228	.0248

DN=(NCAL-NEXP)

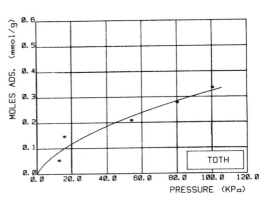

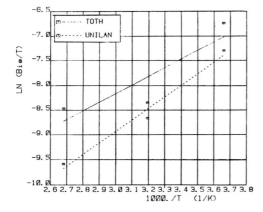

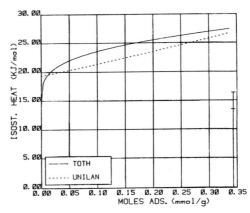

PROPANE

SILICA GEL

ADSORBENT : PA-100;DAVISON CHEMICAL Co.;751 Sqm/g (N2)
EXP.   : STATIC GRAVIMETRIC (McBAIN BALANCE)
REF.   : MASLAN,F.and ABERTH,E.R.;J.Chem.Eng.Data;17,286(1972)

                    TEMP   (K)=   238.15
                    PSAT (kPa)=   137.10

            TOTH   UNILAN   UNITS
        ------------   ------------   ------
B =   1.9411   C =   215.1420   kPa
M =   .2429   S =   3.6250
NI =   38.8389   M =   12.6477   mmol/g
DG =   -24.1230   DG =   -22.8802   J/g
BIS= .50112+001   BIS= .60202+000   1/g
QIS=   39.6669   QIS=   22.2854   KJ/mol

    P(kPa)   N(mmol/g)   DN TOTH   DN UNILAN
    -------- ---------   -------   ---------
     2.0000   .7937   -.0806   -.2721
     7.8670   1.5650   .0156   -.0598
    21.6000   2.4720   .1734   .2479
    40.0000   3.5600   -.0463   .0539
    62.6700   4.3990   -.1395   -.0876
    76.8000   4.7400   -.1107   -.1041
    85.3300   4.8530   -.0243   -.0473
    88.2700   4.8980   -.0041   -.0375
    97.6000   4.8980   .1928   .1256
         AVERAGE DN   .0875   .1151

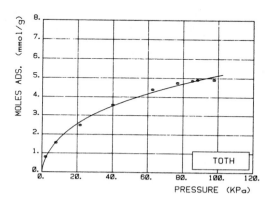

                    TEMP   (K)=   253.15
                    PSAT (kPa)=   244.00

            TOTH   UNILAN   UNITS
        ------------   ------------   ------
B =   2.9901   C =   439.1750   kPa
M =   .2850   S =   3.7737
NI =   31.6963   M =   12.4491   mmol/g
DG =   -24.2952   DG =   -23.1398   J/g
BIS= .14298+001   BIS= .34400+000   1/g
QIS=   39.6669   QIS=   22.2854   KJ/mol

    P(kPa)   N(mmol/g)   DN TOTH   DN UNILAN
    -------- ---------   -------   ---------
     2.2670   .4762   -.0287   -.1420
     8.8000   1.0890   .0075   -.0551
    14.5300   1.5420   -.0609   -.0716
    27.7300   2.1090   .0176   .0684
    41.6000   2.5850   .0437   .1066
    61.2000   3.1290   .0526   .0913
    82.1300   3.6510   .0025   -.0079
   101.6000   4.0820   -.0595   -.1246
         AVERAGE DN   .0341   .0834

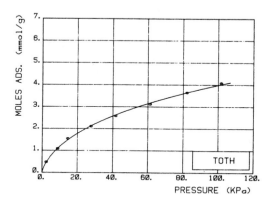

TEMP   (K)=   273.15
PSAT (kPa)=   472.80

	TOTH		UNILAN	UNITS
B =	4.8133	C =	1556.9800	kPa
M =	.3180	S =	4.3831	
NI =	27.0088	M =	12.2800	mmol/g
DG =	-22.7894	DG =	-21.2873	J/g
BIS=	.43844+000	BIS=	.16361+000	1/g
QIS=	39.6669	QIS=	22.2854	KJ/mol

P(kPa)	N(mmol/g)	DN TOTH	DN UNILAN
3.3330	.3175	-.0387	-.0959
9.4670	.6123	-.0117	-.0567
17.2000	.9071	-.0048	-.0195
28.8000	1.2470	.0093	.0258
42.9300	1.5650	.0364	.0675
62.2700	1.9730	.0128	.0376
82.1300	2.3130	.0014	.0021
100.5000	2.6080	-.0301	-.0602
AVERAGE DN		.0181	.0456

DN=(NCAL-NEXP)

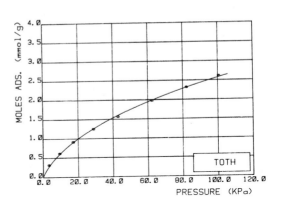

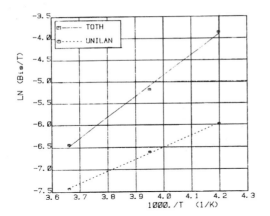

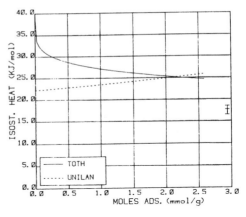

PROPANE

H-MORDENITE

ADSORBENT : Norton Co.; Type Z-900H; 1/16 in. Pellets
EXP.   : Static Volumetric
REF.   : Talu, O., Zwiebel, I.; *AIChE J.* **32**, 1263 (1986)

TEMP (K)   =   283.05
PSAT (kPa) =   638.21

TOTH		UNILAN		UNIT
B =	.3418	C =	268.0926	kPa
M =	.2767	S =	9.7391	
NI =	1.6873	M =	2.7520	mmol/g
DG =	−19.1925	DG =	−19.4013	J/g
BIS =.19224+003		BIS =.21044+002		1/g
QIS =	63.1436	QIS =	39.8266	KJ/mol

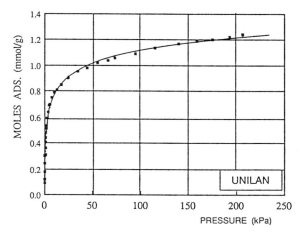

P (kPa)	N (mmol/g)	DN TOTH	DN UNILAN
0.02	0.10	0.0355	0.0156
0.04	0.14	0.0488	0.0383
0.05	0.19	0.0186	0.0116
0.09	0.24	0.0270	0.0287
0.12	0.29	0.0087	0.0139
0.16	0.34	−0.0077	0.0004
0.22	0.39	−0.0182	−0.0081
0.30	0.44	−0.0278	−0.0168
0.43	0.49	−0.0288	−0.0181
0.62	0.55	−0.0370	−0.0280
0.95	0.61	−0.0346	−0.0289
1.44	0.66	−0.0224	−0.0209
2.11	0.71	−0.0147	−0.0174
2.97	0.75	−0.0031	−0.0094
4.04	0.78	0.0132	0.0038
5.35	0.82	0.0149	0.0034
7.03	0.86	0.0150	0.0019
9.26	0.89	0.0246	0.0107
11.00	0.92	0.0190	0.0050
15.32	0.96	0.0249	0.0118
20.23	1.00	0.0222	0.0110
24.92	1.04	0.0095	0.0004
30.49	1.07	0.0053	−0.0011
37.07	1.09	0.0096	0.0065
42.61	1.11	0.0065	0.0062
49.18	1.13	0.0037	0.0064
59.65	1.16	−0.0039	0.0037
71.26	1.19	−0.0137	−0.0012
93.21	1.23	−0.0243	−0.0032
114.96	1.26	−0.0323	−0.0036
	AVERAGE DN	0.0192	0.0112

TEMP (K)   =   303.15
PSAT (kPa) =   1081.14

TOTH		UNILAN		UNIT
B =	.5494	C =	221.0972	kPa
M =	.3412	S =	8.2039	
NI =	1.5351	M =	2.4838	mmol/g
DG =	−18.7299	DG =	−18.9467	J/g
BIS =.22385+002		BIS =.63075+001		1/g
QIS =	63.1436	QIS =	39.8266	KJ/mol

P (kPa)	N (mmol/g)	DN TOTH	DN UNILAN
0.10	0.09	0.0612	0.0577
0.14	0.12	0.0609	0.0614
0.22	0.18	0.0466	0.0522
0.33	0.24	0.0333	0.0423
0.41	0.30	0.0004	0.0105
0.49	0.31	0.0138	0.0243
0.57	0.36	−0.0158	−0.0052
0.77	0.41	−0.0232	−0.0134
0.99	0.44	−0.0160	−0.0079
1.08	0.48	−0.0428	−0.0354
1.47	0.51	−0.0250	−0.0209
1.51	0.53	−0.0408	−0.0370
2.27	0.59	−0.0353	−0.0372
2.40	0.59	−0.0262	−0.0290

3.22	0.64	-0.0280	-0.0355
4.72	0.69	-0.0148	-0.0285
5.06	0.70	-0.0134	-0.0281
7.39	0.75	-0.0013	-0.0213
10.26	0.79	0.0116	-0.0120
12.67	0.81	0.0250	-0.0002
16.70	0.85	0.0277	0.0014
24.81	0.90	0.0366	0.0112
34.28	0.95	0.0326	0.0100
43.85	0.98	0.0361	0.0172
54.62	1.02	0.0249	0.0104
65.79	1.04	0.0284	0.0186
73.19	1.06	0.0215	0.0147
94.66	1.09	0.0221	0.0236
115.89	1.14	-0.0050	0.0042
140.07	1.17	-0.0143	0.0029
158.90	1.19	-0.0210	0.0019
174.76	1.20	-0.0213	0.0063
193.37	1.22	-0.0311	0.0016
206.81	1.24	-0.0445	-0.0082
AVERAGE DN		0.0265	0.0204

$$TEMP(K) = 324.25$$
$$PSAT(kPa) = 1757.20$$

TOTH			UNILAN			UNIT
B	=	.7792	C	=	123.7085	kPa
M	=	.3735	S	=	6.7525	
NI	=	1.4032	M	=	2.0409	mmol/g
DG	=	-18.2205	DG	=	-18.6988	J/g
BIS	=.73778+001		BIS	=.28196+001		1/g
QIS	=	63.1436	QIS	=	39.8266	KJ/mol

P (kPa)	N (mmol/g)	DN TOTH	DN UNILAN
0.51	0.17	0.0488	0.0583
0.67	0.22	0.0299	0.0413
0.91	0.28	0.0078	0.0204
1.21	0.33	-0.0044	0.0082
1.80	0.40	-0.0179	-0.0072
2.63	0.46	-0.0206	-0.0134
4.02	0.53	-0.0237	-0.0221
5.43	0.58	-0.0251	-0.0280
6.92	0.61	-0.0156	-0.0222
8.81	0.64	-0.0063	-0.0164
11.12	0.69	-0.0185	-0.0317
13.69	0.71	-0.0050	-0.0207
16.88	0.74	-0.0018	-0.0193
22.01	0.77	0.0095	-0.0095
28.53	0.81	0.0088	-0.0105
36.22	0.84	0.0139	-0.0046
45.16	0.86	0.0253	0.0086
55.76	0.89	0.0243	0.0103
67.18	0.92	0.0191	0.0084
78.27	0.94	0.0188	0.0114
90.35	0.96	0.0168	0.0131
111.84	0.99	0.0126	0.0152
126.53	1.01	0.0070	0.0138
141.79	1.03	-0.0001	0.0110
159.59	1.05	-0.0069	0.0088
178.21	1.07	-0.0150	0.0055
199.44	1.10	-0.0332	-0.0076
207.51	1.10	-0.0291	-0.0016
AVERAGE DN		0.0166	0.0160

DN= (NCAL–NEXP)

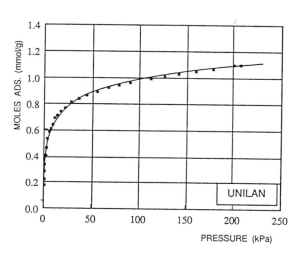

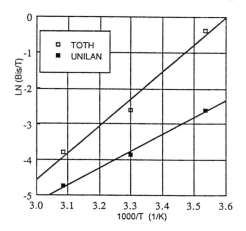

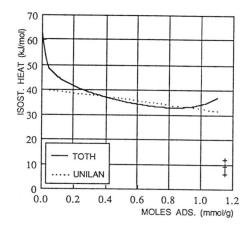

PROPYLENE

ACTIVATED CARBON

ADSORBENT : Fiber Carbon, KF-1500, Toyobo Co., Ltd.
EXP.   : Gravimetric
REF.   : Kuro-Oka, M., Suzuki, T., Nitta, T. and
            Katayama, T., J. Chem. Eng. Japan, **17**, 588
            (1984)

                    TEMP (K)   =   273.15
                    PSAT (kPa) =   585.70

TOTH		UNILAN		UNIT
B =	.5460	C =	5.6035	kPa
M =	.3226	S =	4.3139	
NI =	9.6914	M =	8.3610	mmol/g
DG =	-90.3709	DG =	-89.6384	J/g
BIS =.143639+03		BIS =.293451+02		1/g
QIS =	38.8525	QIS =	36.8174	KJ/mol

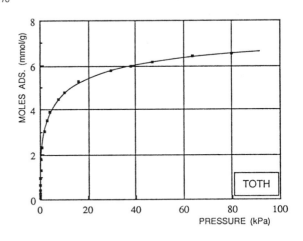

P (kPa)	N (mmol/g)	DN TOTH	DN UNILAN
0.0011	0.0690	-0.0298	-0.0549
0.0016	0.0890	-0.0357	-0.0685
0.0045	0.1680	-0.0478	-0.1115
0.0097	0.2790	-0.0678	-0.1611
0.0198	0.4110	-0.0668	-0.1840
0.0487	0.6520	-0.0475	-0.1671
0.1130	0.9660	0.0010	-0.0756
0.1990	1.2720	0.0125	-0.0167
0.4450	1.7740	0.0663	0.1016
0.8420	2.3100	0.0503	0.1145
1.8620	3.0400	0.0440	0.1068
2.9860	3.5300	0.0116	0.0577
4.0860	3.9100	-0.0587	-0.0274
7.4450	4.4600	-0.0134	-0.0115
10.2800	4.7600	0.0035	-0.0079
16.0100	5.2600	-0.0702	-0.0940
29.6100	5.7700	-0.0144	-0.0400
37.8300	5.9700	0.0006	-0.0208
46.8000	6.1600	-0.0082	-0.0238
63.4900	6.3900	0.0120	0.0072
79.6500	6.5300	0.0505	0.0548
AVERAGE DN		0.0339	0.0718

                    TEMP (K)   =   298.15
                    PSAT (kPa) =   1156.61

TOTH		UNILAN		UNIT
B =	.8999	C =	194.0302	kPa
M =	.3248	S =	6.3770	
NI =	10.5209	M =	13.2738	mmol/g
DG =	-90.1282	DG =	-90.1586	J/g
BIS =.360866+02		BIS =.782022+01		1/g
QIS =	38.8526	QIS =	36.8174	KJ/mol

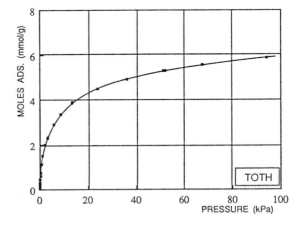

P (kPa)	N (mmol/g)	DN TOTH	DN UNILAN
0.0026	0.0260	-0.0021	-0.0178
0.0116	0.1030	-0.0204	-0.0670
0.0238	0.1780	-0.0340	-0.1055
0.0384	0.2420	-0.0372	-0.1274
0.0552	0.3000	-0.0350	-0.1390
0.1010	0.4500	-0.0513	-0.1720
0.1870	0.6130	-0.0247	-0.1456
0.2830	0.7670	-0.0152	-0.1223
0.6230	1.1360	0.0188	-0.0320
1.1570	1.5250	0.0361	0.0420
2.1810	2.0200	0.0387	0.0923
3.1150	2.3200	0.0520	0.1214
5.6090	2.9100	0.0236	0.0983
8.5270	3.3600	0.0007	0.0642
13.3100	3.8900	-0.0584	-0.0165
23.7200	4.5000	-0.0439	-0.0363
35.9800	4.9300	-0.0231	-0.0377
51.1600	5.2900	-0.0070	-0.0343
51.8900	5.3200	-0.0220	-0.0496
67.5600	5.5600	0.0148	-0.0167

```
94.1200  5.8600  0.0545  0.0267
    AVERAGE DN   0.0292  0.0745
```

```
                    TEMP (K)   =      323.15
                    PSAT (kPa) =     2055.31
```

TOTH		UNILAN		UNIT
B =	1.3320	C =	658.7496	kPa
M =	.3253	S =	6.5130	
NI =	10.8409	M =	13.4324	mmol/g
DG =	−87.1951	DG =	−85.6348	J/g
BIS =	.120660+02	BIS =	.283397+01	1/g
QIS =	38.8525	QIS =	36.8174	KJ/mol

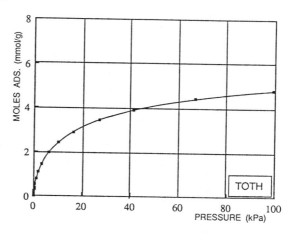

P (kPa)	N (mmol/g)	DN TOTH	DN UNILAN
0.0025	0.0070	0.0012	−0.0044
0.0052	0.0150	0.0008	−0.0095
0.0110	0.0300	0.0002	−0.0185
0.0173	0.0480	−0.0037	−0.0299
0.0294	0.0740	−0.0056	−0.0434
0.0456	0.1020	−0.0050	−0.0550
0.0807	0.1620	−0.0116	−0.0802
0.1200	0.2160	−0.0143	−0.0966
0.1980	0.3060	−0.0183	−0.1158
0.2970	0.3940	−0.0151	−0.1204
0.4680	0.5220	−0.0135	−0.1186
0.9860	0.7990	−0.0028	−0.0798
1.8740	1.1080	0.0229	−0.0041
3.2140	1.4590	0.0210	0.0421
6.2980	1.9840	0.0187	0.0858
10.0900	2.4100	0.0129	0.0923
16.2400	2.9000	−0.0138	0.0580
27.0700	3.4400	−0.0216	0.0212
41.0200	3.8900	−0.0197	−0.0125
66.8100	4.4200	−0.0073	−0.0489
99.2400	4.8300	0.0254	−0.0558
AVERAGE DN		0.0122	0.0568

DN= (NCAL−NEXP)

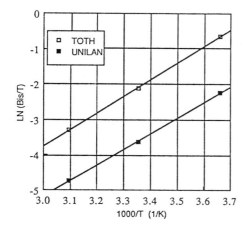

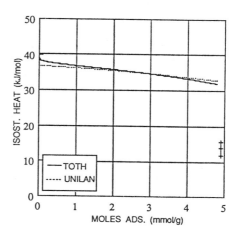

PROPYLENE

ACTIVATED CARBON

ADSORBENT : BPL;PITTSBURGH ACTIVATED CARBON Co;1050-1150 Sqm/g;MESH 20X40
EXP.   : STATIC GRAVIMETRIC (McBAIN QUARTZ SPRING)
REF.   : LAUKHUF,W.L.S. and PLANK,C.A.;J.CHEM.ENG.DATA; 14,48(1969)

TEMP (K)=   303.15
PSAT (kPa)= 1307.15

	TOTH		UNILAN	UNITS
B =	.7731	C =	815.0540	kPa
M =	.2404	S =	7.3705	
NI =	12.1341	M =	13.0710	mmol/g
DG =	-75.3415	DG =	-70.6342	J/g
BIS=	.89188+002	BIS=	.43554+001	1/g
QIS=	123.9940	QIS=	23.4517	KJ/mol

P(kPa)	N(mmol/g)	DN TOTH	DN UNILAN
1.1330	1.2120	-.0291	-.1784
4.0670	1.9530	-.0015	-.0120
7.6670	2.3570	.0615	.0983
8.4000	2.4900	.0000	.0413
16.4700	3.0300	.0164	.0730
25.0700	3.4340	-.0179	.0323
33.4700	3.7020	-.0235	.0161
47.6000	4.1060	-.1008	-.0797
49.8000	4.0400	.0076	.0260
63.6000	4.2420	.0365	.0389
76.6700	4.4440	.0123	.0014
77.3000	4.4440	.0254	.0135
91.0700	4.5790	.0418	.0181
97.9300	4.7120	-.0216	-.0509
100.6000	4.7120	.0041	-.0271
113.5000	4.8480	-.0161	-.0567
AVERAGE DN		.0260	.0477

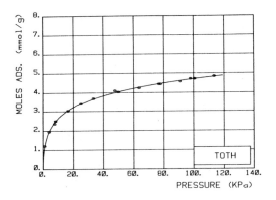

TEMP (K)=   313.15
PSAT (kPa)= 1650.42

	TOTH		UNILAN	UNITS
B =	1.1751	C =	312.2340	kPa
M =	.3551	S =	6.1832	
NI =	7.9355	M =	10.8853	mmol/g
DG =	-73.6442	DG =	-72.4966	J/g
BIS=	.13118+002	BIS=	.35565+001	1/g
QIS=	123.9940	QIS=	23.4517	KJ/mol

P(kPa)	N(mmol/g)	DN TOTH	DN UNILAN
1.4670	1.1430	-.0555	-.0983
1.7330	1.0790	.1027	.0701
4.1330	1.8180	-.0659	-.0546
4.2670	1.6820	.0934	.1057
4.5330	1.8870	-.0670	-.0528
9.7330	2.5240	-.0999	-.0778
13.6000	2.6260	.0815	.0990
15.4700	2.7610	.0577	.0725
19.4700	3.1370	-.1176	-.1083
21.9300	3.0300	.0939	.1002
26.7300	3.2980	.0006	.0019
27.8000	3.4010	-.0678	-.0674
37.5300	3.5000	.0982	.0926
40.0700	3.7070	-.0511	-.0578
42.3300	3.7070	-.0029	-.0102
51.9300	3.7760	.1069	.0982
54.5300	3.9690	-.0436	-.0523
56.6700	3.9760	-.0173	-.0258
67.7300	4.1470	-.0347	-.0416
77.1300	4.2420	-.0189	-.0233
80.8700	4.3130	-.0499	-.0529
82.2000	4.1710	.1059	.1033
90.6000	4.3840	-.0255	-.0247
94.8000	4.4790	-.0827	-.0802
97.8000	4.3840	.0382	.0421
103.6000	4.4440	.0258	.0324
AVERAGE DN		.0617	.0633

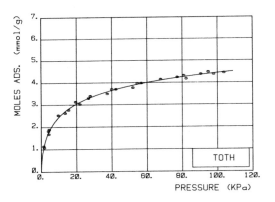

```
         TEMP   (K)=     323.15
         PSAT (kPa)=    2055.31
```

	TOTH		UNILAN	UNITS
B =	1.8534	C =	329.0940	kPa
M =	.4560	S =	5.8715	
NI =	6.5709	M =	10.5687	mmol/g
DG =	-72.1653	DG =	-73.9874	J/g
BIS=	.45628+001	BIS=	.26067+001	1/g
QIS=	123.9940	QIS=	23.4517	KJ/mol

P(kPa)	N(mmol/g)	DN TOTH	DN UNILAN
1.5330	1.0100	-.1482	-.1320
7.1330	1.8870	.0233	.0589
13.2700	2.3570	.0865	.0983
20.6700	2.8280	.0131	.0048
34.0000	3.2750	.0135	-.0097
47.7300	3.5690	.0176	-.0054
61.8000	3.8350	-.0282	-.0429
75.5300	3.9690	.0037	.0011
88.4000	4.1110	-.0117	-.0010
101.4000	4.2420	-.0352	-.0098
AVERAGE DN		.0381	.0364

```
         DN=(NCAL-NEXP)
```

PROPYLENE

ACTIVATED CARBON

ADSORBENT : BLACK PEARLS I;GODFREY L. CABOT Co. OF BOSTON;705 Sqm/g
EXP.   : STATIC VOLUMETRIC
REF.   : LEWIS,W.K.,GILLILAND,E.R.,CHERTOW,B. and HOFFMAN,W.H.;J.Am.Chem.Soc.;
             72,1153(1950)

                      TEMP   (K)=   298.15
                      PSAT (kPa)=   1155.23

                 TOTH   UNILAN   UNITS
         ----------------   ----------------   -----
         B =   2.2428   C =   90.4586   kPa
         M =   .5067   S =   4.3903
         NI =   3.9766   M =   5.3676   mmol/g
         DG = -36.9904   DG = -38.6711   J/g
         BIS= .20019+001   BIS= .13510+001   1/g

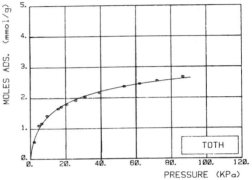

              P(kPa)   N(mmol/g) DN TOTH   DN UNILAN

              2.2000   .5670   .0827   .0966
              4.8000   1.1070   -.1073   -.0902
              6.3600   1.1710   -.0250   -.0114
              9.4670   1.4190   -.0523   -.0473
             15.8000   1.6480   .0182   .0097
             17.4700   1.7030   .0229   .0121
             20.5300   1.7860   .0360   .0220
             25.6000   1.9130   .0400   .0232
             30.8000   2.0410   .0206   .0035
             39.2000   2.1680   .0329   .0186
             53.1300   2.3660   .0048   .0008
             61.9300   2.4560   -.0025   .0020
             71.6700   2.5520   -.0218   -.0070
             86.6000   2.6770   -.0506   -.0192
              AVERAGE DN   .0370   .0260

                 DN=(NCAL-NEXP)

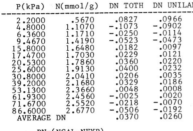

PROPYLENE

ACTIVATED CARBON

ADSORBENT : COLUMBIA GRADE L;MESH 20x40;1152 Sq m/g (N2)
EXP.   : STATIC VOLUMETRIC
REF.   : RAY,G.C.and BOX,E.O.;IND.ENG.CHEM.;42,1315(1950)

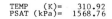

```
          TEMP    (K)=    310.92
          PSAT (kPa)=   1568.76


          TOTH              UNILAN         UNITS
     ---------------   ---------------    -----
B   =      1.2382    C   =    14.2138      kPa
M   =       .4238    S   =     3.3308
NI  =      6.6748    M   =     6.0105    mmol/g
DG  =    -74.0636    DG  =   -73.1613      J/g
BIS=  .10422+002    BIS=  .45823+001      1/g
QIS=    32.1642    QIS=     32.2597    KJ/mol


  P(kPa)   N(mmol/g)  DN TOTH  DN UNILAN
------------------------------------------
   1.8670    1.3790     .0010     .0078
  10.2700    2.7290    -.0023     .0036
  39.7300    3.8670     .0015    -.0088
  76.5300    4.3640     .0030     .0072
  98.7900    4.5490    -.0031     .0099
    AVERAGE DN          .0022     .0075
```

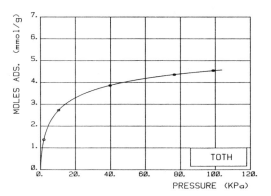

```
          TEMP    (K)=    338.70
          PSAT (kPa)=   2823.90


          TOTH              UNILAN         UNITS
     ---------------   ---------------    -----
B   =      1.3000    C   =738316.0000      kPa
M   =       .3245    S   =    13.4976
NI  =      8.3663    M   =    22.1154    mmol/g
DG  =    -77.0124    DG  =   -75.9257      J/g
BIS=  .10496+002    BIS=  .22736+001      1/g
QIS=    32.1642    QIS=     32.2597    KJ/mol


  P(kPa)   N(mmol/g)  DN TOTH  DN UNILAN
------------------------------------------
   2.1330     .9731    -.0099    -.0456
  15.6000    2.2260     .0162     .0644
  48.1300    3.1890    -.0169    -.0102
  89.1900    3.6960     .0070    -.0196
    AVERAGE DN          .0125     .0350
```

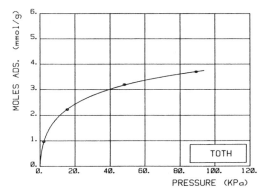

```
          TEMP    (K)=    366.48


          TOTH              UNILAN         UNITS
     ---------------   ---------------    -----
B   =      4.2349    C   =    90.5109      kPa
M   =       .5062    S   =     3.0059
NI  =      6.0381    M   =     5.9234    mmol/g

BIS=  .10630+001    BIS=  .66851+000      1/g
QIS=    32.1642    QIS=     32.2597    KJ/mol


  P(kPa)   N(mmol/g)  DN TOTH  DN UNILAN
------------------------------------------
   4.8000     .7060     .0240     .0090
  20.5300    1.6940    -.0247    -.0106
  52.8000    2.4800     .0015     .0029
  93.1900    2.9860     .0075     .0017
    AVERAGE DN          .0144     .0060
```

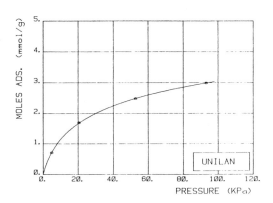

TEMP　(K)=　394.20

	TOTH		UNILAN	UNITS
B　=	2.0183	C　=	83397.5000	kPa
M　=	.2167	S　=	9.3404	
NI　=	30.7681	M　=	16.0166	mmol/g
BIS=	.39475+001	BIS=	.38375+000	1/g
QIS=	32.1642	QIS=	32.2597	KJ/mol

P(kPa)	N(mmol/g)	DN TOTH	DN UNILAN
5.8660	.6602	-.0924	-.1557
18.0000	.9583	.0926	.1055
38.6600	1.5340	.0041	.0413
56.5300	1.8480	-.0111	.0088
88.3900	2.2560	-.0147	-.0522
AVERAGE DN		.0430	.0727

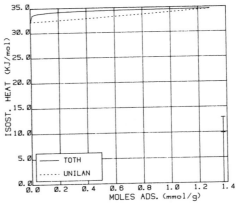

TEMP　(K)=　449.80

	TOTH		UNILAN	UNITS
B　=	6.5799	C　=	96679.0996	kPa
M　=	.4376	S　=	8.7641	
NI　=	5.6176	M　=	11.3963	mmol/g
BIS=	.28362+000	BIS=	.16097+000	1/g
QIS=	32.1642	QIS=	32.2597	KJ/mol

P(kPa)	N(mmol/g)	DN TOTH	DN UNILAN
9.3330	.3229	.0031	-.0101
42.8000	.8613	.0006	.0124
65.9900	1.0950	-.0078	-.0023
94.9300	1.2950	.0055	-.0039
AVERAGE DN		.0043	.0072

DN=(NCAL-NEXP)

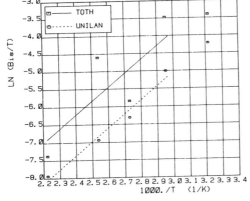

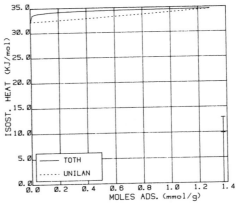

PROPYLENE

ACTIVATED CARBON

ADSORBENT : NUXIT-AL, (1.5 mm DIAMETER, 3-4 mm LONG)
EXP.   : STATIC VOLUMETRIC
REF.   : SZEPESY, L., ILLES, V., *Acta Chim. Hung.*,**35**, 37(1963)

TEMP   (K) =   293.15
PSAT (kPa) = 1019.07

TOTH		UNILAN		UNIT
B = 0.8475		C = 841.0930		kPa
M = 0.3541		S = 8.2164		
NI = 6.9183		M = 12.1799		mmol/g
DG = -65.8509		DG = -66.6979		J/g
BIS=2.69066+01		BIS=.794962+01		1/g
QIS= 42.8765		QIS= 37.6335		KJ/mol

P (kPa)	N (mmol/g)	DN TOTH	DN UNILAN
0.2800	0.6202	0.0143	-0.0251
1.0799	1.2899	-0.0242	0.0068
3.2131	1.9851	-0.0166	0.0289
7.1994	2.5250	0.0384	0.0594
14.2921	3.0765	0.0125	0.0048
24.0513	3.4946	-0.0097	-0.0322
25.3845	3.5155	0.0098	-0.0136
29.9575	3.6775	-0.0289	-0.0538
38.2901	3.8225	0.0057	-0.0181
41.1565	3.8926	-0.0123	-0.0350
55.9552	4.1094	-0.0121	-0.0252
62.5280	4.1701	0.0037	-0.0039
98.6049	4.4731	0.0024	0.0298
129.0157	4.6377	0.0059	0.0639
AVERAGE DN		0.0140	0.0286

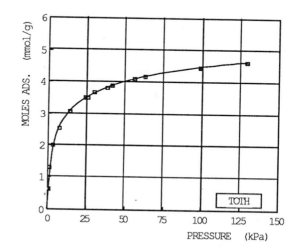

TEMP   (K) =   313.15
PSAT (kPa) = 1650.32

TOTH		UNILAN		UNIT
B = 1.1754		C = 972.2880		kPa
M = 0.3544		S = 7.3562		
NI = 6.6889		M = 10.7861		mmol/g
DG = -61.8805		DG = -62.3955		J/g
BIS=1.10381+01		BIS=3.07416+00		1/g
QIS= 42.8765		QIS= 37.6335		KJ/mol

P (kPa)	N (mmol/g)	DN TOTH	DN UNILAN
0.6133	0.6059	-0.0388	-0.1023
2.5598	1.2377	-0.0450	-0.0401
6.2128	1.7040	0.0249	0.0544
12.0523	2.1578	0.0279	0.0534
19.3050	2.5411	-0.0126	0.0019
30.0374	2.8137	0.0427	0.0451
30.3974	2.8864	-0.0212	-0.0191
38.6767	3.0176	0.0265	0.0232
43.2763	3.0863	0.0409	0.0356
53.1555	3.2639	0.0146	0.0069
53.4621	3.2541	0.0286	0.0209
65.1278	3.4107	0.0157	0.0074
75.9935	3.6597	-0.1224	-0.1294
77.8867	3.5539	0.0010	-0.0058
100.6981	3.7484	-0.0124	-0.0133
127.9491	3.9046	-0.0038	0.0051
AVERAGE DN		0.0299	0.0352

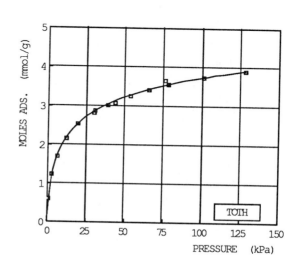

TEMP (K) = 333.15
PSAT (kPa) = 2528.62

TOTH		UNILAN		UNIT
B =	1.8307	C =	1016.1800	kPa
M =	0.3704	S =	6.2904	
NI =	7.1743	M =	10.3971	mmol/g
DG =	-62.8870	DG =	-63.0700	J/g
BIS=	3.88364+00	BIS=	1.21503+00	1/g
QIS =	42.8765	QIS =	37.6335	KJ/mol

P(kPa)	N(mmol/g)	DN TOTH	DN UNILAN
0.8266	0.4913	-0.1097	-0.1906
5.7595	1.1373	0.0334	0.0200
13.2122	1.6647	0.0376	0.0551
24.9712	2.1475	0.0275	0.0483
43.3830	2.6615	-0.0455	-0.0342
44.1296	2.6017	0.0283	0.0391
61.1681	2.9373	-0.0388	-0.0361
62.4347	2.8993	0.0161	0.0182
85.8460	3.1653	0.0137	0.0088
88.8991	3.2376	-0.0296	-0.0352
111.6305	3.3835	0.0122	0.0035
131.3488	3.5365	-0.0076	-0.0172
AVERAGE DN		0.0333	0.0422

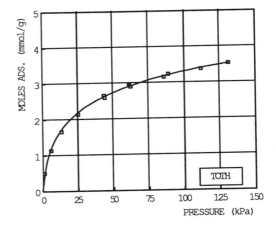

TEMP (K) = 363.15
PSAT (kPa) = 4447.38

TOTH		UNILAN		UNIT
B =	3.1716	C =	1236.9100	kPa
M =	0.4236	S =	5.6075	
NI =	5.8935	M =	8.7317	mmol/g
DG =	-58.5072	DG =	-59.5485	J/g
BIS=	1.16657+00	BIS=	.517818+00	1/g
QIS =	42.8765	QIS =	37.6335	KJ/mol

P(kPa)	N(mmol/g)	DN TOTH	DN UNILAN
4.9462	0.5466	0.0648	0.0272
15.6653	1.1146	0.0482	0.0478
18.3984	1.3185	-0.0607	-0.0573
22.9047	1.4488	-0.0550	-0.0480
29.9041	1.6157	-0.0470	-0.0380
31.3973	1.5376	0.0640	0.0731
41.5831	1.8213	-0.0244	-0.0164
51.5556	1.9289	0.0229	0.0285
57.7018	2.0766	-0.0421	-0.0383
83.4462	2.3715	-0.0609	-0.0638
85.2728	2.2234	0.1036	0.1004
108.8174	2.4799	0.0323	0.0250
130.7889	2.6776	-0.0253	-0.0348
AVERAGE DN		0.0501	0.0460

DN= (NCAL-NEXP)

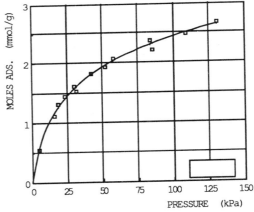

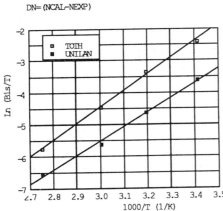

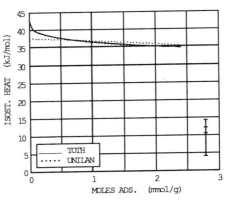

PROPYLENE

ACTIVATED CARBON

ADSORBENT : NUXIT-AL, (1.5 mm DIAMETER, 3-4 mm LONG)
EXP.   : STATIC VOLUMETRIC
REF.   : SZEPESY, L., ILLES, V., *Acta Chim. Hung.*,**35**, 53(1963)

TEMP (K)   =   293.15
PSAT (kPa) = 1019.07

TOTH		UNILAN		UNIT
B =	0.2662	C =	1273.0200	kPa
M =	0.1801	S =	12.9266	
NI =	8.3335	M =	11.3039	mmol/g
DG =	-84.5539	DG =	-88.5345	J/g
BIS =	3.15691+04	BIS =	3.44154+02	1/g
QIS =	128.7515	QIS =	88.3544	KJ/mol

P(kPa)	N(mmol/g)	DN TOTH	DN UNILAN
118.8542	4.5735	0.0350	0.0417
134.2556	4.6627	0.0023	0.0058
168.5035	4.7912	-0.0218	-0.0234
208.2229	4.8925	-0.0272	-0.0321
265.9781	4.9924	-0.0181	-0.0250
319.6804	5.0656	-0.0107	-0.0178
438.0280	5.1740	0.0163	0.0115
529.7271	5.2427	0.0275	0.0259
622.4395	5.3007	0.0364	0.0384
764.4971	5.4605	-0.0395	-0.0315
AVERAGE DN		0.0235	0.0253

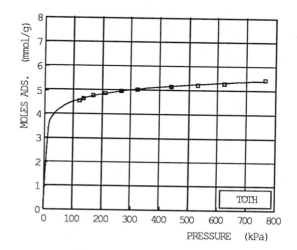

TEMP (K)   =   313.15
PSAT (kPa) = 1650.32

TOTH		UNILAN		UNIT
B =	0.4358	C =	1309.4000	kPa
M =	0.2174	S =	10.4495	
NI =	7.5282	M =	10.1312	mmol/g
DG =	-72.8309	DG =	-73.2039	J/g
BIS =	8.94314+02	BIS =	.332818+02	1/g
QIS =	128.7515	QIS =	88.3544	KJ/mol

P(kPa)	N(mmol/g)	DN TOTH	DN UNILAN
118.8542	3.8524	0.0394	0.0502
143.3749	3.9814	0.0075	0.0121
154.5206	4.0197	0.0076	0.0100
191.3016	4.1482	-0.0124	-0.0150
239.6336	4.2754	-0.0267	-0.0330
304.2790	4.3994	-0.0331	-0.0412
372.1667	4.4936	-0.0299	-0.0378
469.9453	4.5882	-0.0138	-0.0193
561.1378	4.6547	0.0023	0.0002
679.1815	4.7368	0.0076	0.0106
803.5072	4.7711	0.0487	0.0578
AVERAGE DN		0.0208	0.0261

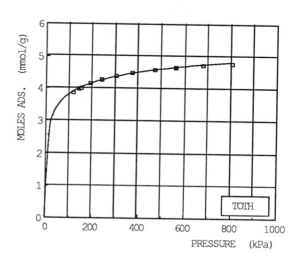

TEMP   (K) =   363.15
PSAT (kPa) =   4447.38

TOTH		UNILAN		UNIT
B =	2.7771	C =	165.3100	kPa
M =	0.3767	S =	3.1630	
NI =	7.0049	M =	5.6776	mmol/g
DG =	-60.1351	DG =	-58.4023	J/g
BIS =1.40522+00		BIS =.386854+00		1/g
QIS =	128.7515	QIS =	88.3544	KJ/mol

P(kPa)	N(mmol/g)	DN TOTH	DN UNILAN
120.8807	2.5879	-0.0051	-0.0069
159.5869	2.8065	0.0025	0.0032
204.4738	3.0033	0.0088	0.0107
254.3257	3.1965	-0.0056	-0.0033
306.8121	3.3487	-0.0047	-0.0025
372.1667	3.5008	-0.0006	0.0014
433.1644	3.6199	0.0019	0.0034
500.5455	3.7538	-0.0174	-0.0164
602.3771	3.8787	0.0023	0.0023
700.4597	3.9872	0.0097	0.0085
784.2554	4.0840	-0.0016	-0.0040
AVERAGE DN		0.0055	0.0057

DN = (NCAL-NEXP)

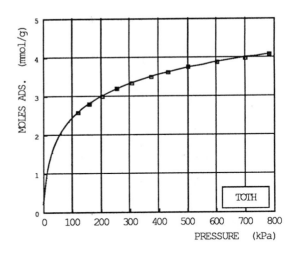

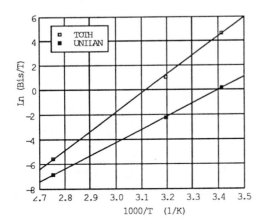

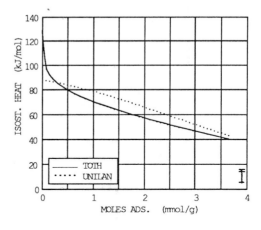

PROPYLENE

CARBON MOLECULAR SIEVE

ADSORBENT : MSC-5A;TAKEDA CHEMICAL Co.;0.56 cc/g;650 Sqm/g
EXP.   : STATIC GRAVIMETRIC (McBAIN QUARTZ SPRING BALANCE)
REF.   : NAKAHARA,T.,HIRATA,M.and OMORI,T.;J.Chem.Eng.Data;19,310(1974)

TEMP (K)= 278.65
PSAT (kPa)= 687.74

	TOTH		UNILAN	UNITS
B =	.2021	C =	.2516	kPa
M =	.3237	S =	3.5017	
NI =	2.9491	M =	2.4974	mmol/g
DG =	-47.5482	DG =	-42.8867	J/g
BIS=	.95442+003	BIS=	.10882+003	1/g
QIS=	57.5690	QIS=	36.9818	KJ/mol

P(kPa)	N(mmol/g)	DN TOTH	DN UNILAN
.0267	.4801	.1439	.0564
.1867	1.1100	.0625	.0386
.2533	1.2260	.0390	.0250
.3600	1.4020	-.0307	-.0332
.4133	1.4780	-.0652	-.0631
.5333	1.5810	-.0924	-.0815
.8933	1.7060	-.0677	-.0387
1.5470	1.8250	-.0352	.0125
4.0000	2.0390	-.0118	.0572
8.5330	2.1890	.0021	.0655
13.8000	2.2770	.0060	.0519
24.3300	2.3700	.0097	.0223
39.2700	2.4310	.0209	-.0023
57.4700	2.4670	.0366	-.0179
74.6700	2.4900	.0463	-.0303
90.8000	2.5170	.0424	-.0509
AVERAGE DN		.0445	.0405

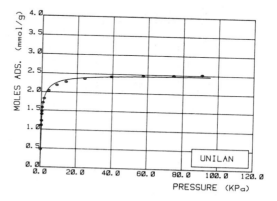

TEMP (K)= 303.15
PSAT (kPa)= 1307.15

	TOTH		UNILAN	UNITS
B =	.3095	C =	.7929	kPa
M =	.3552	S =	3.9604	
NI =	2.8154	M =	2.4778	mmol/g
DG =	-47.2227	DG =	-43.5894	J/g
BIS=	.19277+003	BIS=	.52168+002	1/g
QIS=	57.5690	QIS=	36.9818	KJ/mol

P(kPa)	N(mmol/g)	DN TOTH	DN UNILAN
.2000	.8318	-.0093	-.0025
.6000	1.1500	.0079	.0050
2.9330	1.6300	.0116	-.0016
8.2670	1.9200	-.0025	-.0038
14.2700	2.0560	-.0117	-.0049
24.2700	2.1580	-.0036	.0076
38.4000	2.2460	-.0067	.0023
54.2700	2.2960	.0013	.0039
68.1300	2.3290	.0036	-.0002
90.0000	2.3650	.0080	-.0060
AVERAGE DN		.0066	.0038

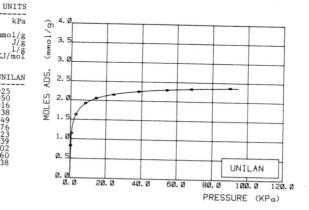

TEMP (K)=   323.15
PSAT (kPa)= 2055.31

	TOTH		UNILAN	UNITS
B =	.5362	C =	2.1787	kPa
M =	.4021	S =	3.4141	
NI =	2.7429	M =	2.4115	mmol/g
DG =	-45.3976	DG =	-42.9179	J/g
BIS=	.34723+002	BIS=	.13221+002	1/g
QIS=	57.5690	QIS=	36.9818	KJ/mol

P(kPa)	N(mmol/g)	DN TOTH	DN UNILAN
.2667	.5418	.0051	.0048
3.8670	1.3930	.0048	.0018
8.5330	1.6640	-.0132	-.0160
13.3300	1.8010	-.0188	-.0185
22.2700	1.9110	.0099	.0144
36.9300	2.0170	.0264	.0324
58.0000	2.1390	.0017	.0040
74.1300	2.1940	-.0051	-.0076
87.4700	2.2270	-.0074	-.0143
AVERAGE DN		.0103	.0126

DN=(NCAL-NEXP)

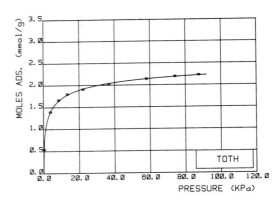

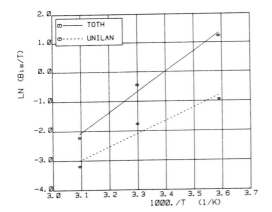

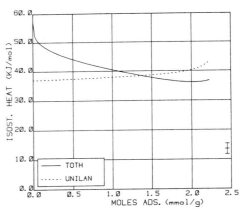

PROPYLENE

CARBON MOLECULAR SIEVE

ADSORBENT : MSC-5A;TAKEDA CHEMICAL Co.;0.56 cc/g;650 Sqm/g
EXP.   : STATIC GRAVIMETRIC (McBAIN QUARTZ SPRING BALANCE)
REF.   : NAKAHARA,T.,HIRATA,M.and MORI,H.;J.Chem.Eng.Data;27,317(1982)

TEMP (K)=   274.85
PSAT (kPa)=   615.96

TOTH		UNILAN		UNITS
B =	.1451	C =	5.2762	kPa
M =	.2528	S =	11.4378	
NI =	3.3026	M =	4.5576	mmol/g
DG =	-55.0011	DG =	-49.0634	J/g
BIS=	.15617+005	BIS=	.80043+004	1/g
QIS=	76.2118	QIS=	87.8520	KJ/mol

P(kPa)	N(mmol/g)	DN TOTH	DN UNILAN
.7600	1.8630	.0010	.0298
.8000	1.8840	-.0072	.0190
1.8400	2.0750	.0018	-.0061
2.7470	2.1850	-.0182	-.0362
4.3470	2.2520	.0123	-.0118
6.1600	2.3490	-.0146	-.0393
8.3330	2.3730	.0193	-.0031
9.5200	2.4160	.0010	-.0196
14.8700	2.4810	.0149	.0042
15.1200	2.5040	-.0052	-.0154
21.8300	2.5230	.0363	.0387
23.3200	2.5350	.0348	.0399
28.1700	2.5900	.0092	.0225
30.3500	2.5580	.0525	.0694
34.5700	2.6310	-.0012	.0223
38.2900	2.7840	-.1393	-.1103
AVERAGE DN		.0230	.0305

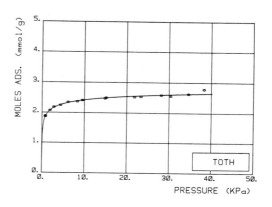

TEMP (K)=   303.15
PSAT (kPa)=   1307.15

TOTH		UNILAN		UNITS
B =	.2710	C =	428.1030	kPa
M =	.3079	S =	11.4093	
NI =	3.0000	M =	5.9324	mmol/g
DG =	-49.4957	DG =	-48.1991	J/g
BIS=	.52503+003	BIS=	.13799+003	1/g
QIS=	76.2118	QIS=	87.8520	KJ/mol

P(kPa)	N(mmol/g)	DN TOTH	DN UNILAN
.8000	1.2850	.0264	.0494
.8800	1.3460	-.0065	.0130
1.3730	1.4480	.0216	.0262
1.5070	1.5160	-.0194	-.0177
2.2530	1.6280	-.0169	-.0254
2.7730	1.6640	.0049	-.0075
4.0800	1.7700	.0034	-.0133
4.6800	1.8360	-.0265	-.0436
6.7070	1.8920	.0092	-.0062
7.7470	1.9640	-.0272	-.0407
9.1070	1.9710	.0048	-.0057
11.3700	2.0450	-.0171	-.0220
13.9700	2.0580	.0165	.0185
15.0900	2.1090	-.0174	-.0124
19.8400	2.1320	.0184	.0357
22.2300	2.1850	-.0111	.0123
23.0400	2.1640	.0173	.0426
AVERAGE DN		.0156	.0231

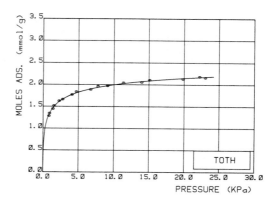

TEMP (K)=   323.15
PSAT (kPa)= 2055.31

	TOTH		UNILAN	UNITS
B =	.3944	C =	590.8490	kPa
M =	.3283	S =	10.1492	
NI =	2.9431	M =	5.8187	mmol/g
DG =	-48.4550	DG =	-48.5922	J/g
BIS=	.13448+003	BIS=	.33331+002	1/g
QIS=	76.2118	QIS=	87.8520	KJ/mol

P(kPa)	N(mmol/g)	DN TOTH	DN UNILAN
.7330	1.0080	-.0320	-.0081
.9600	1.0120	.0447	.0631
1.8130	1.2620	-.0114	-.0078
2.8930	1.3760	.0173	.0108
3.4270	1.4600	-.0156	-.0250
4.8400	1.5750	-.0277	-.0416
5.1730	1.5480	.0188	.0044
7.3870	1.6810	-.0114	-.0268
7.7470	1.6780	.0050	-.0102
11.0300	1.7630	.0178	.0058
12.4000	1.8140	-.0018	-.0117
15.3900	1.8440	.0250	.0201
16.9500	1.9110	-.0172	-.0193
21.0700	1.9640	-.0156	-.0100
21.9300	1.9570	.0013	.0085
28.1300	2.0230	-.0049	.0138
34.0400	2.0540	.0082	.0374
	AVERAGE DN	.0162	.0191

DN=(NCAL-NEXP)

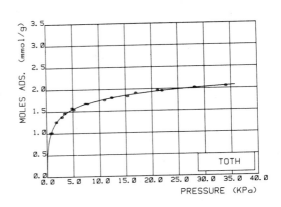

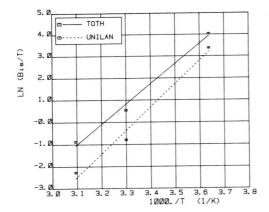

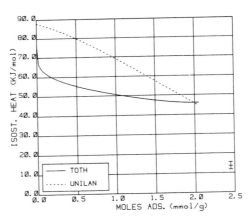

PROPYLENE

SILICA GEL

ADSORBENT : DAVISON CHEMICAL Co.OF BALTIMORE;751 Sqm/g;Mesh 14x20
EXP.   : STATIC VOLUMETRIC
REF.   : LEWIS,W.K.,GILLILAND,E.R.,CHERTOW,B. and HOFFMAN,W.H.;J.Am.Chem.Soc.;
            72,1153(1950)

TEMP (K)=   298.15
PSAT (kPa)= 1155.23

	TOTH		UNILAN	UNITS
B =	6.0181	C =	110.8690	kPa
M =	.5356	S =	2.7026	
NI =	4.6869	M =	4.4776	mmol/g
DG =	-28.3392	DG =	-28.2584	J/g
BIS=	.40731+000	BIS=	.27506+000	1/g

P(kPa)	N(mmol/g)	DN TOTH	DN UNILAN
.5867	.1090	-.0316	-.0463
4.5600	.3738	.0399	.0203
9.5200	.7227	-.0371	-.0444
9.8930	.7596	-.0569	-.0634
12.2100	.7472	.0543	.0518
20.7200	1.1250	-.0373	-.0320
25.9100	1.1290	.0949	.1016
26.4400	1.1680	.0686	.0754
33.7600	1.5620	-.1668	-.1601
36.2000	1.4010	.0410	.0473
37.5700	1.4730	-.0058	.0003
47.4100	1.6330	-.0047	-.0007
61.0500	1.8080	.0008	.0017
61.2000	1.8140	-.0034	-.0026
73.4300	1.9180	.0251	.0234
74.0300	1.9280	.0210	.0192
86.6700	2.0730	-.0085	-.0123
101.4000	2.1840	-.0043	-.0098
102.4000	2.1970	-.0101	-.0157
AVERAGE DN		.0375	.0383

DN=(NCAL-NEXP)

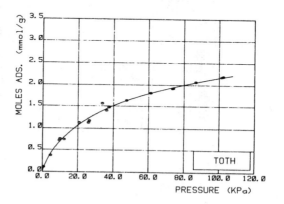

PROPYLENE

SILICA GEL

ADSORBENT : DAVISON CHEMICAL Co.;REF.GRADE;751 Sqm/g;mesh 14x20
EXP.   : STATIC VOLUMETRIC
REF.   : LEWIS,W.K.,GILLILAND,E.R.,CHERTOW,B. and BAREIS,D.;J.Am.Chem.Soc.;
            72,1160(1950)

TEMP (K)=   273.15
PSAT (kPa)=   585.05

	TOTH		UNILAN	UNITS
B =	1.7557	C =	102.4580	kPa
M =	.3072	S =	3.5176	
NI =	10.1528	M =	6.3618	mmol/g
DG =	-33.0446	DG =	-31.4262	J/g
BIS=	.36896+001	BIS=	.67493+000	1/g
QIS=	29.9484	QIS=	22.2620	KJ/mol

P(kPa)	N(mmol/g)	DN TOTH	DN UNILAN
6.8130	1.0520	.0576	.0096
12.6900	1.4680	.0180	.0151
20.2300	1.9030	-.0885	-.0678
28.5100	2.0650	.0150	.0433
47.3200	2.5480	-.0425	-.0217
57.1200	2.6400	.0325	.0446
75.3200	2.9350	-.0096	-.0161
101.6000	3.1790	.0287	-.0053
AVERAGE DN		.0366	.0279

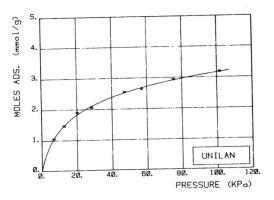

TEMP (K)=   313.15
PSAT (kPa)=   1648.35

	TOTH		UNILAN	UNITS
B =	2.9593	C =	738.3470	kPa
M =	.3173	S =	4.5420	
NI =	9.2099	M =	6.0709	mmol/g
DG =	-31.3791	DG =	-27.6729	J/g
BIS=	.78481+000	BIS=	.22121+000	1/g
QIS=	29.9484	QIS=	22.2620	KJ/mol

P(kPa)	N(mmol/g)	DN TOTH	DN UNILAN
7.0530	.4586	-.0010	-.0308
9.9600	.5911	-.0280	-.0444
12.5300	.6300	.0132	.0067
24.5500	.9260	.0012	.0202
33.2500	1.0590	.0218	.0460
54.1200	1.3660	-.0045	.0132
67.0900	1.4990	-.0004	.0070
80.7200	1.6260	-.0032	-.0084
102.0000	1.7910	-.0029	-.0298
AVERAGE DN		.0085	.0230

DN=(NCAL-NEXP)

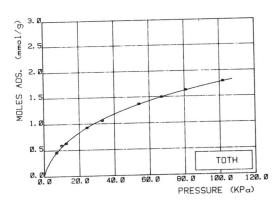

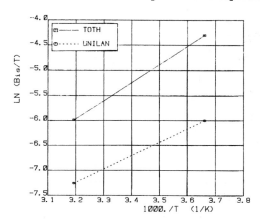

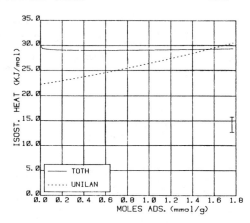

PROPYLENE

SILICA GEL

ADSORBENT : PA-100;DAVISON CHEMICAL Co.;751 Sqm/g (N2)
EXP.   : STATIC GRAVIMETRIC (McBAIN BALANCE)
REF.   : MASLAN,F.and ABERTH,E.R.;J.Chem.Eng.Data;17,286(1972)

TEMP (K)=   238.15
PSAT (kPa)=   174.10

	TOTH		UNILAN	UNITS
B =	.7970	C =	1771.1700	kPa
M =	.1733	S =	7.5646	
NI =	32.5635	M =	17.4294	mmol/g
DG =	-37.6079	DG =	-35.0977	J/g
BIS=	.23889+003	BIS=	.24839+001	1/g
QIS=	50.9889	QIS=	24.8896	KJ/mol

P(kPa)	N(mmol/g)	DN TOTH	DN UNILAN
.8000	.9981	.0021	-.2763
4.9330	2.1620	-.0357	-.0286
12.8000	2.9470	.0432	.1680
23.8700	3.7070	-.0412	.0896
40.8000	4.2540	.0633	.1423
60.6700	4.8480	-.0076	-.0029
81.2000	5.2760	-.0298	-.0995
99.4700	5.5370	.0021	-.1291
AVERAGE DN		.0281	.1170

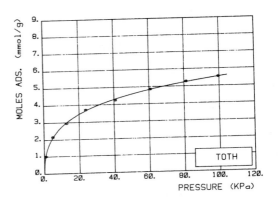

TEMP (K)=   253.15
PSAT (kPa)=   306.20

	TOTH		UNILAN	UNITS
B =	1.0142	C =	1943.3100	kPa
M =	.1873	S =	6.9873	
NI =	28.9262	M =	14.6297	mmol/g
DG =	-35.1586	DG =	-32.6965	J/g
BIS=	.56452+002	BIS=	.12277+001	1/g
QIS=	50.9889	QIS=	24.8896	KJ/mol

P(kPa)	N(mmol/g)	DN TOTH	DN UNILAN
.6667	.6416	-.0822	-.3109
6.0000	1.6160	-.0407	-.0786
14.5300	2.2340	.0105	.0773
27.8700	2.7800	.0693	.1566
42.5300	3.2560	.0382	.1010
62.4000	3.6600	.0734	.0847
82.6700	4.0640	.0129	-.0320
99.3300	4.4440	-.1334	-.2236
AVERAGE DN		.0576	.1331

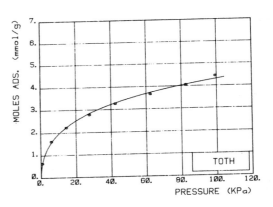

TEMP   (K)=   273.15
PSAT (kPa)=   585.05

	TOTH		UNILAN	UNITS
B =	1.4414	C =	1937.8800	kPa
M =	.2098	S =	6.2526	
NI =	25.4561	M =	11.6801	mmol/g
DG =	-33.1459	DG =	-30.5622	J/g
BIS=	.10119+002	BIS=	.56846+000	1/g
QIS=	50.9889	QIS=	24.8896	KJ/mol

P(kPa)	N(mmol/g)	DN TOTH	DN UNILAN
2.1330	.5941	-.0369	-.1718
8.9330	1.1410	.0224	.0004
14.8000	1.4730	-.0029	.0239
28.1300	1.9250	.0172	.0781
41.6000	2.2810	-.0033	.0514
62.5300	2.6620	.0050	.0250
82.8000	2.9700	-.0109	-.0338
101.3000	3.1840	-.0031	-.0668
AVERAGE DN		.0127	.0564

DN=(NCAL-NEXP)

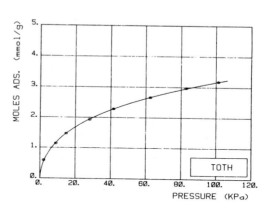

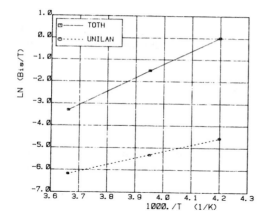

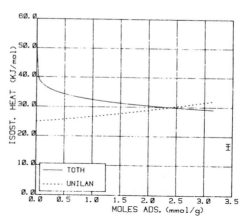

CHAPTER 3

ADSORPTION OF GAS MIXTURES

Design of columns for thermal swing adsorption (TSA) and pressure swing adsorption (PSA) require the simultaneous solution of the partial differential equations for the material, energy, and momentum balances describing the dynamics of adsorption in columns, in conjunction with the kinetic and equilibrium properties of the adsorbates[Si69(85)]. This chapter is devoted to adsorption equilibria of gas mixtures.

A compilation of references to experimental data is given in Appendix E. Only those systems accompanied by tabulated data for the single–gas adsorption isotherms are included in this chapter.

GUIDE TO TABLES

Ordering of Systems

Adsorbates are presented in alphabetical order for component no. 1, which is defined as the most strongly adsorbed species (at low coverage, if an azeotrope exists). Mixtures tabulated in this chapter are indexed in Appendix B.

Heading

The heading is the same as that described in Chapter 2, except that there are two adsorbates instead of one.

Adsorption Isotherm

For adsorption from binary gas mixtures, there are three degrees of freedom. Usually the independent variables are $\{T, P, y_1\}$ and the measurements are $\{n, x_1\}$ or $\{n_1, n_2\}$. Experimental data at constant T are tabulated as a function of both P and y_1. In some cases, either P or y_1 is held constant for a set of isothermal measurements.

A sample isotherm for the adsorption of ethylene(1) and methane(2) on activated carbon is:

TEMP	(K)=	293.15		
P (KPa)	Y1	X1	N(mmol/g)	Sexp
98.8316	.1100	.6340	1.3299	14.0154
98.2983	.1120	.6510	1.3522	14.7894
99.2849	.2920	.8510	1.8930	13.8482
98.7649	.3030	.8330	1.8327	11.4741
99.4182	.5070	.9300	2.2954	12.9188
99.2716	.5340	.9390	2.3440	13.4332
99.4315	.7310	.9710	2.6197	12.3213

In this case P is nearly but not exactly constant. Each point gives the pressure P in units of kPa, the mole fraction y_1 of ethylene (component number 1) in the gas phase, the mole fraction x_1 of ethylene in the adsorbed phase, the total amount adsorbed in mol/kg [= mmol/g], and the selectivity of the adsorbent $s_{1,2}$ for ethylene relative to methane:

$$s_{1,2} = \frac{x_1/y_1}{x_2/y_2} \qquad (3\text{-}1)$$

The correspondence between the symbols in the computer printout and the symbols in the above equations is given in Table 3-1.

Table 3-1. Explanation of Symbols Used in Computer Printout

Name of Variable or Constant	Symbol for Computer Printout	Symbol Used Elsewhere	Units
Pressure	P	P	kPa
Mole fraction comp. no. 1 in vapor phase	Y1	y_1	–
Mole fraction comp. no. 1 in adsorbed phase	X1	x_1	–
Total amount adsorbed	N	n_t	mol/kg
Selectivity	Sexp	$s_{1,2}$	–

CALCULATION OF MIXED–GAS ADSORPTION
FROM SINGLE–GAS ISOTHERMS

Any adsorption isotherm equation for single gases such as the Toth equation (2-1) or the UNILAN equation (2-2) can be used to predict adsorption from mixtures using ideal–adsorbed–solution (IAS) theory [My(121)65)], which is thermodynamically consistent and exact at the limit of zero pressure. The success of calculations of this type depends to a large extent upon the ability of the equation to fit the pure–gas adsorption data accurately. Small errors in the fitting of the experimental data, particularly at low surface coverage, may generate large errors in the calculated value of the selectivity. Other sources of error are the neglect of energetic heterogeneity of adsorption sites and adsorbate–adsorbate interactions, both of which can cause the mixture equilibria to exhibit nonideal behavior [My691(83)]. As a rule, however, predictions by means of IAS theory are accurate when the amount adsorbed is less than half of the saturation capacity of the adsorbent. At higher coverage, negative deviations from Raoult's law have been observed for some systems; these are caused primarily by energetic heterogeneity and by adsorbed-phase nonidealities.

Let the adsorption isotherm of pure ith adsorbate be represented by the function $n_i^\circ(P_i^\circ)$. For a multicomponent system containing N adsorbates, the IAS equations for a perfect gas phase are:

$$Py_i = P_i^\circ x_i \qquad \{i = 1, 2, \ldots, N\} \tag{3-2}$$

$$\psi_1^\circ(P_1^\circ) = \psi_2^\circ(P_2^\circ) = \cdots = \psi_N^\circ(P_N^\circ) \tag{3-3}$$

$$\sum_{i=1}^{N} x_i = 1 \tag{3-4}$$

The function $\psi_i^\circ(P^\circ)$ is the integral for the spreading pressure $(\Pi A/RT)$ from Eq. (1-17) or (1-18) for the adsorption of pure ith adsorbate at the same T as the mixture. Given the usual set of independent variables $\{T, P, y_1 \ldots y_{N-1}\}$, there are $2N$ unknowns $\{P_i^\circ, x_i\}$ and $2N$ equations (3-2) to (3-4). After the above equations are solved, the total amount adsorbed n_t is found from:

$$\frac{1}{n_t} = \sum_{i=1}^{N} \frac{x_i}{n_i^\circ} \tag{3-5}$$

The amount of ith component adsorbed is given by:

$$n_i = n_t x_i \tag{3-6}$$

General Algorithm for IAS Theory

The solution algorithm [My392(86)] for Eqs. (3-2) to (3-6) is based upon three functions. The first of these is the single–gas adsorption equation:

$$n_i^\circ = n_i^\circ(P_i^\circ) \tag{3-7}$$

The second function is the integral for $\psi = \Pi A / RT$ by Eq. (1-17) or (1-18):

$$\psi_i^\circ = \psi_i^\circ(P_i^\circ) \tag{3-8}$$

and the third function is the inverse of Eq. (3-8):

$$P_i^\circ = P_i^\circ(\psi_i^\circ) \tag{3-9}$$

For isothermal adsorption of a mixture of N components on a particular adsorbent, an algorithm for solving Eqs. (3-2) to (3-6) is given in Fig. 3-1. The single equation

$$\mathcal{F}(\psi) = \sum_{i=1}^{N} x_i - 1 = \sum_{i=1}^{N} \left[\frac{P y_i}{P_i^\circ} \right] - 1 = 0$$

is solved for ψ using Newton's method.

In general the function $\psi_i^\circ(P_i^\circ)$ in Eq. (3-8) is a numerical integration, but in some cases the integral can be simplified or represented in terms of a series expansion. Its inverse function $P_i^\circ(\psi_i^\circ)$ is also evaluated numerically by the algorithm in Fig. 3-2, which implements Newton's method to solve the equation $G(P_i^\circ) = 0$ for P_i°.

The IAS algorithm in Figs. 3-1 and 3-2 is for arbitrary single-gas adsorption isotherms. Specific Eqs. (3-7), (3-8), and (3-9) are given for the Toth and UNILAN adsorption isotherms so that mixed-gas adsorption can be predicted from the constants reported in Chapter 2.

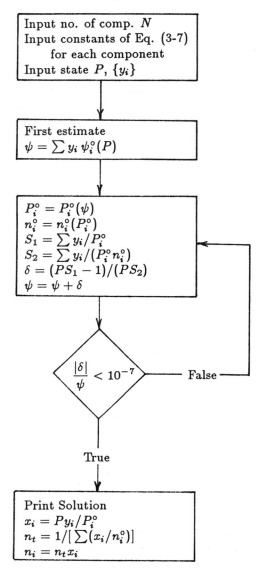

Fig. 3-1. General algorithm for solving IAS equations. For notation see Eqs. (3-7) to (3-9). Summations are over the N adsorbates present in the gas mixture. The equality sign ($=$) stands for an assignment as defined in *FORTRAN* or *C* programming languages.

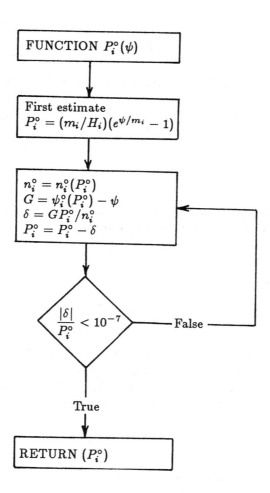

Fig. 3-2. Subprogram for computing the inverse function $P_i^\circ = P_i^\circ(\psi_i^\circ)$ of Eq. (3-9). m_i is the saturation capacity of the ith component and $H_i = B_{iS}/RT$ is the Henry constant given by Eq. (1-1). The equality sign $(=)$ stands for assignment to the variable on the left-hand side of the equation.

Toth Equation

For the Toth equation (2-1),

$$n = n(P) = \frac{mP}{(b + P^t)^{1/t}} \qquad (3\text{-}10)$$

$\Pi A/RT$ may be expressed as an infinite series according to Eq. (2-6):

$$\psi = \psi(P) = m \left[\theta - \frac{\theta}{t} \ln(1 - \theta^t) - \sum_{j=1}^{\infty} \frac{\theta^{jt+1}}{jt(jt + 1)} \right] \qquad (3\text{-}11)$$

where

$$\theta = \frac{P}{(b + P^t)^{1/t}} \qquad (3\text{-}12)$$

To simplify notation, the symbols n_i^o, P_i^o, and ψ_i^o in Eqs. (3-7) through (3-9) are replaced by n, P, and ψ in Eqs. (3-10) to (3-14).

UNILAN Equation

For the UNILAN equation (2-2),

$$n = n(P) = \frac{m}{2s} \ln \left[\frac{c + Pe^{+s}}{c + Pe^{-s}} \right] \qquad (3\text{-}13)$$

the straightforward method of calculating $\psi = \Pi A/RT$ is Eq. (1-17). Since the UNILAN equation is itself an integral over a uniform distribution of adsorption energies, it is possible to integrate first for spreading pressure and then integrate with respect to the energy distribution according to Eq. (2-10):

$$\psi = \psi(P) = \frac{m}{2s} \int_{-s}^{+s} \ln[1 + (P/c)e^z]\, dz \qquad (3\text{-}14)$$

This integral is evaluated numerically, e.g., by Romberg's method [Pr114(86)] and converges more rapidly than Eq. (1-17).

The inverse function $P(\psi)$ is calculated by the subprogram in Fig. 3-2. The Henry constant (H) for the Toth and UNILAN equations is given by Eqs. (2-7) and (2-11).

Up to this point, equations in this chapter are written for the case when the mixture, as well as the single gases, obeys the perfect gas law. For a real gas, the use of the rigorous Eq. (1-16) generates values of $\Pi A/RT$ that differ significantly from the approximate values of Eq. (1-17) when the pressure is 500 kPa or greater. However, precise calculations of mixed–gas adsorption equilibria based upon Eqs. (1-16) and (1-35) are very nearly the same as the results obtained by assuming a perfect gas, because of a partial cancellation of errors. Therefore the effect of gas phase imperfections is usually neglected. An example for a real gas is given on page 218.

SAMPLE CALCULATION

Mixed–gas adsorption calculations are illustrated for the adsorption of a binary mixture of propylene(1) and ethylene(2) on a carbon molecular sieve at 323.15 K. Experimental data are on page 248 for the mixture, and on pages 106 and 200 for the single gases.

Constants for the Toth and UNILAN equations are reproduced in Tables 3-2 and 3-3.

Table 3-2. Constants for Toth Equation.

Gas	m, [mol/kg]	b, $[\mathrm{kPa}]^t$	t
propylene	2.9431	0.3944	0.3283
ethylene	4.7087	2.1941	0.3984

Table 3-3. Constants for UNILAN Equation.

Gas	m, [mol/kg]	c, kPa	s
propylene	5.8187	590.849	10.1492
ethylene	5.6264	359.111	5.2416

Using these constants for the experimental point at $T = 323.15$ K, $P = 13.4133$ kPa and $y_1 = 0.046$, results of the mixed–gas adsorption calculation are given in Table 3-4. Note that the predictions of the Toth and UNILAN equation are in good agreement. In fact, IAS predictions are independent of the particular equation adopted, to the extent that the equation fits the experimental data for the single-gas isotherms. In this case, both equations fit the data within experimental error.

Table 3-4. Comparison of Predictions of Toth and UNILAN Equations with Experimental Data for Adsorption of Propylene(1) and Ethylene(2) on Carbon Molecular Sieve at $T = 323.15$ K, $P = 13.4133$ kPa and $y_1 = 0.046$.

Variable	Toth	UNILAN	Expt.
x_1	0.5014	0.5147	0.4960
n_t, mol/kg	1.268	1.270	1.295
$s_{1,2}$	20.86	22.00	20.41
P_1°, kPa	1.2304	1.1987	–
P_2°, kPa	25.667	26.370	–
n_1°, mol/kg	1.1320	1.1374	–
n_2°, mol/kg	1.4422	1.4488	–
$\Pi A/RT$, mol/kg	2.746	2.701	–

RAPID CALCULATIONS
OF MULTICOMPONENT ADSORPTION EQUILIBRIA

Prediction of adsorption equilibria for a system of N components from single-gas isotherms calls for the simultaneous solution of N algebraic equations. These are Eqs. (3-3) to (3-4), which upon elimination of the x_i by Eq. (3-2) contain N unknowns (P_i°). The algorithm in Fig. 3-1 reduces the problem to solving one equation in one unknown (ψ). Only 5 to 10 iterations by Newton's method are needed, but at each step a numerical integration for ψ must be performed several times for each component. This procedure may not be fast enough to be incorporated as a subroutine in numerical calculations of column dynamics.

A series expansion [OB1467(84)] of the usual adsorption integral equation in terms of the central moments of the adsorption energy distribution yields, following truncation after the second term of the series:

$$n = m\left[\frac{\eta}{1+\eta} + \frac{\sigma^2\eta(1-\eta)}{2(1+\eta)^3}\right] \tag{3-15}$$

where η is dimensionless pressure (KP). This isotherm contains three constants: m, σ, and K. The integral for $\psi = \Pi A/RT$ is:

$$\psi = m\left[\ln(1+\eta) + \frac{\sigma^2\eta}{2(1+\eta)^2}\right] \tag{3-16}$$

Since Eq. (3-16) is analytical, the N Eqs. (3-3) and (3-4) may be solved simultaneously by the Newton-Raphson method for their N unknowns (P_i°). This algorithm, called FastIAS[OB1188(85)], is 25 times faster than the algorithm in Fig. 3-1. Its use is restricted to equations like (3-15) or (3-17) that can be integrated analytically for spreading pressure. An additional factor of 4 in speed can be realized[OB(88)] by exploiting the fact that the Jacobian matrix for the Newton-Raphson solution is sparse and structured, allowing a solution for which computer time is linear in the number of components. Computation time for a system of 10 components is 0.18 s on a personal computer with an 8088/8087 CPU operated at 4.77 MHz. Specifications for the FastIAS algorithm may be $\{P, \mathbf{y}\}$, $\{P, \mathbf{x}\}$, or $\{n_t, \mathbf{x}\}$.

ADSORPTION IN ZEOLITES

The isotherm equation for adsorption of a single gas in a zeolite containing m moles of identical cavities per kilogram is[Ru696(72)]:

$$\frac{n}{m} = \frac{\eta + \sum_i \left[\frac{\eta^i (1 - ib)^i}{(i-1)!} \right]}{1 + \eta + \sum_i \left[\frac{\eta^i (1 - ib)^i}{i!} \right]} \tag{3-17}$$

$\eta = KP$ and b is the ratio of the effective co-volume of the adsorbate molecule to the volume of the cavity. Summations are over $2 \le i \le t$, and t is the number (integer) of molecules that can fit into a cavity: $t \le (1/b)$. Since m is a physical property of the zeolite, Eq. (3-17) contains only two constants: b and the Henry constant K, and these are usually found by fitting Eq. (3-17) to experimental data.

Another form[Ru696(72)] of Eq. (3-17) contains an exponential factor in each term of the summation for adsorbate-adsorbate attractive forces. This factor is omitted in Eq. (3-17) because it is less important[Ru78(84)] than the free-volume reduction incorporated in the constant b.

The generalization of Eq. (3-17) for adsorption of gas mixtures[Ru701(73)] is explicit in pressure and gas-phase composition; for a binary mixture of components nos. 1 and 2:

$$\frac{n_1}{m} = \frac{\eta_1 + \sum_i \sum_j \left[\frac{\eta_1^i \eta_2^j (1 - ib_1 - jb_2)^{i+j}}{(i-1)! j!} \right]}{1 + \eta_1 + \eta_2 + \sum_i \sum_j \left[\frac{\eta_1^i \eta_2^j (1 - ib_1 - jb_2)^{i+j}}{i! j!} \right]} \tag{3-18}$$

with a similar expression for component 2. The symbols $\eta_1 = K_1 P y_1$ and $\eta_2 = K_2 P y_2$ are dimensionless partial pressures. The summation in the denominator is over all integer values $0, 1, 2, \ldots$ of i and j satisfying the restrictions $i + j \ge 2$ and $ib_1 + jb_2 \le 1$. The summation in the numerator has the same restrictions plus one more: $i > 0$.

Since Eq. (3-18) is based upon a statistical model, it is thermodynamically consistent and satisfies Eq. (1-29). It also satisfies Eq. (1-27): the selectivity at the limit of zero pressure is the ratio of Henry constants: $\lim_{P\to 0}(s_{1,2}) = (K_1/K_2)$. If the molecules are the same size ($b_1 = b_2$) then the selectivity is constant: $s_{1,2} = (K_1/K_2)$. Let molecule no. 1 be more strongly adsorbed than molecule 2 at low pressure, so that $K_1 > K_2$. If molecule 1 is the larger one, which is the usual case when dispersion forces dominate, then $s_{1,2}$ decreases with pressure according to Eq. (3-18). Otherwise, if molecule 1 is smaller than molecule 2, then the selectivity increases with pressure; this is an unusual but not impossible case.

Predictions of mixed-gas adsorption equilibria from Eq. (3-18) and from IAS theory based on Eq. (3-17) differ by only a few percent. However Eq. (3-18) is analytical and therefore calculations are fast. A problem is that Eq. (3-17), having only two undetermined constants, lacks flexibility and does not fit single-gas isotherms as well as Eq. (2-1) or (2-2). Predictions of selectivity are very sensitive to the single-gas isotherms. Small errors in the single-gas isotherms, especially at low pressure, generate large errors in selectivity.

THERMODYNAMIC CONSISTENCY

The integral of Eq. (1-29) provides a quantitative test of thermodynamic consistency of isobaric binary data. Some data sets do not contain enough points for an accurate numerical integration. In this case the data can be checked by the intersection rule [My1(87)]: At constant T and P any pair of thermodynamically consistent x–y curves cross at least once in the region $0 < x_1 < 1$. Fig. 3-3 is an example for which the experimental data intersect the theoretical IAS curve at about $y_1 = 0.13$, $x_1 = 0.64$. Therefore these experimental data are nonideal, but thermodynamically consistent with the intersection rule.

NONIDEAL ADSORBED SOLUTIONS

Nonideal adsorbed solutions such as the experimental data in Fig. 3-3 can be described in terms of adsorbed-phase activity coefficients. Let $\mathcal{G}(x_1)$ be an equation for excess Gibbs free energy of a liquid phase such as the Wilson, UNIQUAC or NRTL equation [Pr233(86)]. For example, for the particular case of the Wilson equation: $\mathcal{G} = -RT \sum_i [x_i \ln(\sum_j (x_j \Lambda_{ij}))]$. The corresponding expression for excess free energy of an adsorbed phase, Eq. (1-36), is [Ko(88)]:

$$g^e = (1 - e^{-C\psi})\mathcal{G} \tag{3-19}$$

where $\psi = (\Pi A/RT)$ and C is a constant for each adsorbent with units of [kg/mol]. At the limit of zero surface coverage, $\Pi = 0$ and the adsorbed phase is ideal ($g^e = 0$).

The Wilson equation by itself can be used only at constant Π. Experimental data are often taken at constant P but never at constant Π. The excess reciprocal amount adsorbed σ^e in Eq. (1-37) is obtained from Eqs. (1-38) and (3-19):

$$\sigma^e = C\left(\frac{\mathcal{G}}{RT}\right)e^{-C\psi} \tag{3-20}$$

Unlike g^e, the limit of σ^e as $P \rightarrow 0$ is non-zero as predicted by theory [OB451(87)].

Explicit equations for activity coefficients in the adsorbed phase for use in Eq. (1-35) follow from Eqs. (1-39) and (3-19). For a binary mixture, there are three constants $(C, \Lambda_{12}, \Lambda_{21})$ determined by fitting the experimental data with Eq. (3-19). Since C is a property of the adsorbent, Eq. (3-19) can be applied to as many components as desired without introducing any constants other than those obtained from binary data.

Experimental data can be explained by nonidealities in the adsorbed phase provided the data intersect the prediction for an ideal adsorbed solution, as shown on Fig. 3-3.

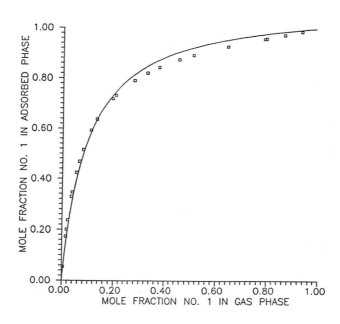

Fig. 3-3. Mole fraction in adsorbed phase (x_1) versus mole fraction in gas phase (y_1). Solid line is IAS theory. Experimental data intersect IAS curve at $y_1 = 0.13$, $x_1 = 0.64$. These data are nonideal but thermodynamically consistent.

FUGACITY CALCULATIONS FOR REAL GAS

Eq. (1-35) is the fundamental equation of equilibrium for adsorption:

$$P y_i \phi_i = P_i^\circ \phi_i^\circ \gamma_i x_i \tag{1-35}$$

ϕ_i is the fugacity coefficient of ith component at the pressure, temperature, and composition of the gas mixture; ϕ_i° is the fugacity coefficient of pure i at its reference pressure P_i°. In this chapter, the prime symbol on x and γ is omitted to simplify notation. Here the focus is on gas-phase imperfections and it is assumed that the adsorbed phase is ideal ($\gamma_i = 1$). Therefore Eqs. (3-3) through (3-6) are unchanged. However in the case of a real gas the spreading pressure in the single-gas reference state is obtained from Eq. (1-16):

$$\psi_i^\circ = \int_0^{f_i^\circ} n_i^\circ \, d\ln f_i^\circ \tag{3-21}$$

Assuming that the fugacity coefficient can be represented by the virial equation of state terminated after the second virial coefficient:

$$\ln \phi_i = \frac{P}{RT} \Big[B_{ii} + (1 - y_i)^2 \delta_{12} \Big] \tag{3-22}$$

for a binary mixture. For pure adsorbate at its reference state:

$$\ln \phi_i^\circ = \frac{B_{ii} P_i^\circ}{RT} \tag{3-23}$$

where $\delta_{12} \equiv 2B_{12} - B_{11} - B_{22}$. It follows that:

$$\psi_i^\circ = \int_0^{P_i^\circ} \frac{n_i^\circ}{P} dP \; + \; \frac{B_{ii}}{RT} \int_0^{P_i^\circ} n_i^\circ dP \tag{3-24}$$

The second integral in Eq. (3-24) is the correction for gas-phase imperfections. For most adsorption equations, both integrals must be calculated numerically.

It is instructive to compare results obtained from the above equations with approximate results for a perfect gas ($B_{ii} = 0$). Consider the adsorption of a binary gas mixture of ethylene (1) and methane (2) on BPL activated carbon at $y_1 = 0.235$, $T = 301.4$ K, and $P = 1430.6$ kPa. Second virial coefficients at 301.4 K are $B_{11} = -137.0$ cm^3/mol, $B_{12} = -76.0$ cm^3/mol, and $B_{22} = -42.2$ cm^3/mol. Table 3-5 compares IAS calculations for an imperfect gas to approximate results obtained by assuming a perfect gas. Single-gas adsorption isotherms for this system are tabulated on pages 98 and 136.

Table 3-5. Comparison of perfect-gas approximation with calculation for real gas. Adsorption of binary mixture of ethylene (1) and methane (2) on BPL activated carbon at 301.4 K, 1430.6 kPa and $y_1 = 0.235$.

Quantity	Imperfect gas	Perfect gas	Error, %
P_1°, kPa	449.3	462.8	+3.0
P_2°, kPa	4024.2	3999.6	−0.6
P_2°/P_1°	8.956	8.641	−3.5
n_1°, mol/kg	4.0087	4.0394	+0.8
n_2°, mol/kg	4.3385	4.3325	−0.1
ϕ_1	0.9332	1.0	+7.2
ϕ_2	0.9770	1.0	+2.4
ϕ_1°	0.9757	1.0	+2.5
ϕ_2°	0.9345	1.0	+7.0
x_1	0.7156	0.7264	+1.5
x_2	0.2844	0.2736	−3.8
n_1, mol/kg	2.9322	2.9894	+2.0
n_2, mol/kg	1.1650	1.1262	−3.3
n_t, mol/kg	4.0972	4.1156	+0.4
ψ, mol/kg	8.8392	9.0324	+2.2
$s_{1,2}$	8.193	8.641	+5.5

Note that the equilibrium value of ψ for an imperfect gas is obtained by solving Eqs. (1-35), (3-3) through (3-6), and (3-24) simultaneously.

Errors in calculated amounts adsorbed introduced by assuming a perfect gas are only a few percent, even when fugacity coefficients deviate from unity by 7 percent. The selectivity from Eqs. (1-35) and (3-1) is:

$$s_{1,2} = \frac{P_2^\circ \phi_2^\circ \phi_1}{P_1^\circ \phi_1^\circ \phi_2} \qquad (3\text{-}25)$$

There is a partial compensation of errors: the perfect-gas approximation for the weakly adsorbed species ($B_{22} = 0$) lowers the vapor-pressure ratio P_2°/P_1°, while the perfect-gas approximation for the strongly adsorbed species ($B_{11} = 0$) raises the value of P_2°/P_1°.

The experimental results[Re336(80)] are $n_t = 4.527$ mol/kg and $x_1 = 0.612$, which is in poor agreement with both calculations in Table 3-5. The discrepancy must be due to nonidealities in the adsorbed phase. When the pressure is high enough to require corrections for gas-phase imperfections, the surface coverage is also high and adsorbed phase nonidealities are dominant. In most cases, vapor-phase imperfections may be safely ignored unless the pressure is greater than 500 kPa and experimental error is less than a few percent.

(1) ACETYLENE

----

(2) ETHYLENE

SILICA GEL

ADSORBENT : DAVISON CHEMICAL Co.;REF.GRADE;751 Sqm/g;mesh 14x20
EXP   : STATIC VOLUMETRIC
REF   : LEWIS,W.K.,GILLILAND,E.R.,CHERTOW,B.,and MILLIKEN,W.;J.Am.Chem.soc.;
            72,1157(1950)

                   TEMP   (K)=   298.15

P (KPa)	Y1	X1	N(mmol/g)	Sexp
101.7910	.0680	.1360	1.0780	2.1574
101.8050	.1620	.3700	1.1700	3.0380
101.5650	.1860	.4080	1.1930	3.0161
101.7780	.2860	.5420	1.2980	2.9544
101.4580	.4380	.7080	1.3970	3.1111
101.5780	.7578	.9314	1.6220	4.3394

->

(1) ISOBUTANE
----
(2) ETHANE

ZEOLITE MOLECULAR SIEVE

ADSORBENT : 13X;LINDE(UNION CARBIDE);.3 cc/g;525 Sqm/g (N2)
EXP   : STATIC VOLUMETRIC
REF   : HYUN,S.H.and DANNER,R.P.;J.Chem.Eng.Data;27,196(1982)

                    TEMP   =   298.15 (K)
                    PRESSURE =   137.80 (KPa)

        Y1   X1   N(mmol/g)   Sexp

        .0030   .0829   2.2600   30.041
        .0269   .2194   2.1800   10.167
        .0996   .4514   2.0500   7.438
        .1270   .6540   1.9600   12.993
        .2298   .7903   1.9500   12.631
        .3184   .8487   1.9600   12.008
        .4868   .8654   1.9300   6.778
        .5223   .8947   1.9300   7.771
        .9200   .9790   1.9100   4.054
        .9632   .9976   1.9100   15.881

                    TEMP   =   323.15 (K)
                    PRESSURE =   137.80 (KPa)

        Y1   X1   N(mmol/g)   Sexp

        .0347   .3625   1.8400   15.818
        .0369   .2771   1.8300   10.005
        .1223   .6353   1.7500   12.502
        .2114   .8017   1.6800   15.081
        .6529   .9218   1.6700   6.267
        .8190   .9742   1.6800   8.345
        .8905   .9946   1.6700   22.648
END OF FILE
->

(1) ISOBUTANE
---
(2) ETHYLENE

ZEOLITE MOLECULAR SIEVE

ADSORBENT : 13X;LINDE(UNION CARBIDE);.3 cc/g;525 Sqm/g (N2)
EXP   : STATIC VOLUMETRIC
REF   : HYUN,S.H.and DANNER,R.P.;J.Chem.Eng.Data;27,196(1982)

TEMP   =   298.15 (K)
PRESSURE =   137.80 (KPa)

Y1	X1	N(mmol/g)	Sexp
.0450	.1223	2.6800	2.957
.1168	.2085	2.6300	1.992
.1515	.3086	2.4600	2.500
.2163	.4237	2.3900	2.664
.3249	.5161	2.3000	2.216
.3485	.5378	2.2700	2.175
.5310	.6897	2.1900	1.963
.7302	.7274	2.1400	.986
.8047	.7660	2.0700	.794
.8745	.8218	2.0500	.662

TEMP   =   323.15 (K)
PRESSURE =   137.80 (KPa)

Y1	X1	N(mmol/g)	Sexp
.0776	.1863	2.3700	2.721
.2843	.4575	2.1300	2.123
.3237	.4855	2.0000	1.972
.5675	.6406	1.9000	1.358
.6533	.7116	1.9200	1.309
.7223	.7316	1.9300	1.048
.8566	.7959	1.8700	.653
.9493	.9178	1.6700	.596

TEMP   =   373.15 (K)
PRESSURE =   137.80 (KPa)

Y1	X1	N(mmol/g)	Sexp
.0339	.1021	1.7400	3.241
.1876	.4227	1.6800	3.171
.4425	.6616	1.5700	2.463
.7279	.7644	1.4700	1.213
.8839	.8962	1.4000	1.134
.9617	.9832	1.3800	2.331

END OF FILE
->

(1) BUTANE
----
(2) METHANE

ACTIVATED CARBON

ADSORBENT : BPL;PITTSBURGH ACTIVATED CARBON Co.;1040 Sqm/g
EXP   : FLOW NON-CHROMATOGRAPHIC
REF   : GRANT,R.J. and MANES,M.;Ind.Eng.Chem.Fundam.;5,490(1966)

                    TEMP   =   298.15 (K)
                    Y1   =   .002760

         P (KPa)   X1   N(mmol/g)   Sexp

         101.325   .8470   2.0321   2000.242
        1135.530   .9520   2.5161   7166.156
        1962.890   .6160   3.9046   579.616
        3548.670   .6450   3.5176   656.481
        5272.350   .4480   4.6374   293.244

                    TEMP   =   298.15 (K)
                    Y1   =   .005610

         P (KPa)   X1   N(mmol/g)   Sexp

         101.325   .8920   2.3139   1463.979
        1135.530   .8030   3.4556   722.509
        1962.890   .7140   3.9263   442.513
        3548.670   .5120   4.8633   185.970
        5272.350   .3680   5.9249   103.211

->

(1) BUTANE
----
(2) PROPANE

ACTIVATED CARBON

ADSORBENT : NUXIT-AL (1.5 mm DIAMETER ,3-4 mm LONG)
EXP   : STATIC VOLUMETRIC
REF   : SZEPESY,L.,ILLES,V.;ACTA CHIM. HUNG.;35,245(1963)

TEMP   (K)=   293.15

P (KPa)	Y1	X1	N(mmol/g)	Sexp
100.1110	.0500	.1320	4.1629	2.8894
100.7380	.0830	.2650	4.1196	3.9834
100.1110	.1490	.3780	4.0688	3.4709
100.1380	.1840	.4730	4.0291	3.9804
99.5115	.2750	.5630	3.9956	3.3965
98.7516	.3720	.6610	3.9737	3.2917
98.7516	.5050	.7370	3.9505	2.7468

->

(1) CARBON DIOXIDE
----
(2) ETHYLENE

ZEOLITE MOLECULAR SIEVE

ADSORBENT : 13X;LINDE(UNION CARBIDE);.3 cc/g;525 Sqm/g (N2)
EXP   : STATIC VOLUMETRIC
REF   : HYUN,S.H.and DANNER,R.P.;J.Chem.Eng.Data;27,196(1982)

TEMP   =   298.15 (K)
PRESSURE =   137.80 (KPa)

Y1	X1	N(mmol/g)	Sexp
.1106	.2723	3.0700	3.009
.3032	.5181	3.3500	2.471
.5070	.7037	3.6500	2.309
.7068	.8103	3.7800	1.772
.9099	.9099	3.9300	1.000
.9344	.9248	3.9800	.863

TEMP   =   323.15 (K)
PRESSURE =   137.80 (KPa)

Y1	X1	N(mmol/g)	Sexp
.0996	.1659	2.6600	1.798
.3177	.4200	2.8800	1.555
.3981	.5265	3.0000	1.681
.6957	.7339	3.2900	1.206
.9057	.9134	3.4100	1.098

```
CARBON DIOXIDE _____
- - - - - - - - - - - - - - - - - - - - - - - - - - - - - - - - - -
PROPANE _____
- - - - - - - - - - - - - - - - - - - - - - - - - - - - - - - - - -
ZEOLITE _____
- - - - - - - - - - - - - - - - - - - - - - - - - - - - - - - - - -
ADSORBENT: H-mordenite, Norton Co.;type Z-900H;1/16 in. Pellets
EXP.     : Static Volumetric
REF.     : Talu, O. and Zwiebel, I.; AIChE J. 32, 1263 (1986)
```

T (K)= 303.15

P(kPa)	Y1	X1	N(mmol/g)	Sexp
1.29	0.167	0.288	0.662	2.0176
10.11	0.165	0.349	1.093	2.7129
21.41	0.163	0.374	1.287	3.0678
41.03	0.167	0.406	1.475	3.4093
61.19	0.178	0.434	1.595	3.5410
40.86	0.035	0.273	1.271	10.3535
40.95	0.338	0.490	1.611	1.8818
41.55	0.633	0.577	1.705	0.7908
40.88	0.958	0.801	1.670	0.1764

(1) ETHANE
----
(2) ETHYLENE

ACTIVATED CARBON

ADSORBENT : BPL;PITTSBURGH CHEMICAL Co.;988 Sqm/g
EXP   : VOLUMETRIC
REF   : REICH,R.,ZIEGLER,W.T.and ROGERS,K.A.;Ind.Eng.Chem.Process Des.Dev.
            19,336(1980)

                    TEMP   =   301.40 (K)
                    Y1   =   .2400

        P (KPa)   X1   N(mmol/g)   Sexp

        137.894   .2810   2.9880   1.2376
        308.193   .2900   3.8490   1.2934
        737.043   .2860   4.7190   1.2684
        1341.020   .2750   5.3780   1.2011
        1981.540   .2720   5.9380   1.1832

                    TEMP   =   212.70 (K)
                    Y1   =   .2400

        P (KPa)   X1   N(mmol/g)   Sexp

        139.962   .3460   6.7320   1.6753
        224.078   .3330   7.0940   1.5810
        405.408   .3250   7.7840   1.5247

                    TEMP   =   301.40 (K)
                    Y1   =   .4720

        P (KPa)   X1   N(mmol/g)   Sexp

        217.873   .5590   3.5930   1.4180
        549.508   .5330   4.5820   1.2767
        1132.800   .5330   5.3800   1.2767

                    TEMP   =   301.40 (K)
                    Y1   =   .6820

        P (KPa)   X1   N(mmol/g)   Sexp

        144.099   .7620   3.2670   1.4929
        346.114   .7480   4.0760   1.3840
        692.917   .7460   4.8700   1.3695
        1367.910   .7270   5.6060   1.2417

->

```
           TEMP      =    212.70 (K)
           Y1        =      .6820


      P (KPa)        X1      N(mmol/g)   Sexp
   -------------------------------------------------
      137.205       .7900     6.6750     1.7541
      240.625       .7790     7.1880     1.6436
->    343.356       .7740     7.5140     1.5969
```

(1) ETHANE

(2) ETHYLENE

ACTIVATED CARBON

ADSORBENT : NUXIT-AL (1.5 mm DIAMETER ,3-4 mm LONG)
EXP : STATIC VOLUMETRIC
REF : SZEPESY,L.,ILLES,V.;ACTA CHIM. HUNG.;35,245(1963)

TEMP (K)= 293.15

P (KPa)	Y1	X1	N(mmol/g)	Sexp
101.2580	.1150	.1620	2.9284	1.4877
100.9250	.1970	.2680	2.9454	1.4924
101.2580	.2880	.3810	2.9896	1.5217
101.2580	.3880	.4880	3.0244	1.5034
101.2580	.4780	.5780	3.0556	1.4957
101.2580	.5630	.6720	3.0895	1.5903
101.2580	.6460	.7360	3.1145	1.5277
101.2580	.7080	.7910	3.1364	1.5609
101.2580	.7660	.8340	3.1533	1.5348
101.2580	.8050	.8630	3.1926	1.5259

TEMP (K)= 333.15

P (KPa)	Y1	X1	N(mmol/g)	Sexp
100.1780	.1330	.1710	1.7439	1.3446
99.9648	.1670	.2180	1.7529	1.3905
100.1510	.2160	.2780	1.7676	1.3976
99.9782	.2800	.3590	1.7779	1.4402
100.0710	.3690	.4590	1.8028	1.4508
100.4980	.4890	.5790	1.8457	1.4372
100.3510	.6870	.7500	1.8956	1.3668

->

(1) ETHANE ——

(2) METHANE ———

ACTIVATED CARBON ——

ADSORBENT : BPL;PITTSBURGH CHEMICAL Co.;988 Sqm/g
EXP : VOLUMETRIC
REF : REICH,R.,ZIEGLER,W.T.and ROGERS,K.A.;Ind.Eng.Chem.Process Des.Dev.
        19,336(1980)

$$\begin{array}{lll} \text{TEMP} & = & 301.40 \ (K) \\ Y1 & = & .7330 \end{array}$$

P (KPa)	X1	N(mmol/g)	Sexp
129.620	.9740	2.8930	13.6456
350.940	.9420	3.9080	5.9160
684.644	.9250	4.5200	4.4925
1234.150	.8940	5.2630	3.0721
2005.670	.8720	5.8110	2.4815

$$\begin{array}{lll} \text{TEMP} & = & 212.70 \ (K) \\ Y1 & = & .7330 \end{array}$$

P (KPa)	X1	N(mmol/g)	Sexp
130.310	.9880	6.4330	29.9904
244.072	.9740	6.7370	13.6456
353.009	.9680	7.2400	11.0188

$$\begin{array}{lll} \text{TEMP} & = & 301.40 \ (K) \\ Y1 & = & .5010 \end{array}$$

P (KPa)	X1	N(mmol/g)	Sexp
131.689	.9290	2.5960	13.0323
344.046	.8930	3.5910	8.3125
692.917	.8670	4.3120	6.4928
1385.150	.7930	5.4250	3.8156

$$\begin{array}{lll} \text{TEMP} & = & 212.70 \ (K) \\ Y1 & = & .5010 \end{array}$$

P (KPa)	X1	N(mmol/g)	Sexp
131.689	.9680	6.1280	30.1292
347.493	.9440	6.9100	16.7898
694.986	.9280	7.8390	12.8374

->

```
TEMP      =    301.40 (K)
Y1        =     .2550
```

P (KPa)	X1	N(mmol/g)	Sexp
129.620	.8330	1.9940	14.5729
349.561	.7850	3.0560	10.6671
687.402	.7520	3.8160	8.8589
1373.420	.7040	4.6800	6.9486
1378.250	.7090	4.6860	7.1182

```
TEMP      =    260.20 (K)
Y1        =     .2550
```

P (KPa)	X1	N(mmol/g)	Sexp
124.794	.8750	3.4880	20.4510
411.614	.8320	4.6760	14.4687
941.127	.7980	5.5940	11.5416

```
TEMP      =    212.70 (K)
Y1        =     .2550
```

P (KPa)	X1	N(mmol/g)	Sexp
131.689	.9240	5.6860	35.5201
436.435	.8910	6.6660	23.8818
874.937	.8570	7.2300	17.5090

```
END OF FILE
->
```

(1) ETHANE
_____

(2) METHANE
_____

ACTIVATED CARBON
_____

ADSORBENT : NUXIT-AL (1.5 mm DIAMETER ,3-4 mm LONG)
EXP   : STATIC VOLUMETRIC
REF   : SZEPESY,L.,ILLES,V.;ACTA CHIM. HUNG.;35,245(1963)

TEMP   (K)=   293.15

P (KPa)	Y1	X1	N(mmol/g)	Sexp
102.1910	.0660	.6050	1.3456	21.6751
99.7249	.0830	.6580	1.4116	21.2565
102.4850	.2450	.8520	2.0652	17.7402
99.7249	.2510	.8590	2.0540	18.1795
99.7249	.4890	.9410	2.6255	16.6667
102.5380	.5190	.9530	2.7692	18.7920
99.7515	.7310	.9750	2.9887	14.3516

->

(1) ETHYLENE

----

(2) ACETYLENE

ACTIVATED CARBON

ADSORBENT : EY-51-C;PITTSBURGH COKE and CHEMICAL Co.;805 Sqm/g
EXP   : STATIC VOLUMETRIC
REF   : LEWIS,W.K.,GILLILAND,E.R.,CHERTOW,B.,and MILLIKEN,W.;J.Am.Chem.soc.;
            72,1157(1950)

TEMP   (K)=   298.15

P (KPa)	Y1	X1	N(mmol/g)	Sexp
101.5780	.0903	.0852	2.0570	.9383
101.4050	.1120	.1400	2.0260	1.2907
101.2310	.1366	.1242	2.0330	.8964
101.5510	.1860	.2320	2.0310	1.3220
101.1650	.2890	.3440	2.0680	1.2901
101.4710	.3450	.4280	2.0300	1.4206
102.3910	.4530	.5670	2.0700	1.5812
101.2180	.5510	.6530	2.0850	1.5335
100.9650	.5530	.6610	1.9920	1.5761
101.4310	.8420	.9030	2.0780	1.7469

->

(1) ETHYLENE
----
(2) CARBON DIOXIDE

ACTIVATED CARBON

ADSORBENT : NUXIT-AL (1.5 mm DIAMETER ,3-4 mm LONG)
EXP   : STATIC VOLUMETRIC
REF   : SZEPESY,L.,ILLES,V.;ACTA CHIM. HUNG.;35,245(1963)

                    TEMP   (K)=   293.15

P (KPa)	Y1	X1	N(mmol/g)	Sexp
101.0580	.0840	.2240	2.1504	3.1478
101.3250	.2190	.4540	2.3552	2.9653
101.3650	.3910	.6210	2.5193	2.5521
101.4580	.5140	.7390	2.6362	2.6772
101.1110	.6430	.8250	2.7179	2.6174
101.4450	.7410	.8810	2.7973	2.5877

->

ETHYLENE

ETHANE

ZEOLITE

ADSORBENT : 13X, UNION CARBIDE; 1/16 in. Pellets
EXP.   : Static Volumetric
REF.   : KAUL, B. K., *IND. ENG. CHEM. RES.* 26, 928 (1987)

TEMP   = 323.15 (K)
PRESSURE = 137.89 (kPa)

Y1	X1	N(mmol/g)	Sexp
0.953	0.990	2.476	4.882
0.832	0.959	2.445	4.723
0.771	0.942	2.418	4.824
0.197	0.615	2.222	6.511
0.132	0.505	2.151	6.709
0.123	0.508	2.155	7.362
0.094	0.411	2.097	6.726

TEMP   = 423.15 (K)
PRESSURE = 137.89 (kPa)

Y1	X1	N(mmol/g)	Sexp
0.860	0.931	0.9289	2.196
0.773	0.900	0.8942	2.643
0.732	0.870	0.8634	2.450
0.644	0.833	0.8339	2.757
0.539	0.782	0.7607	3.068
0.538	0.783	0.7585	3.098
0.470	0.744	0.7402	3.277
0.360	0.672	0.6644	3.642
0.310	0.590	0.6215	3.203
0.250	0.530	0.6032	3.383
0.160	0.465	0.5653	4.563
0.150	0.450	0.5488	4.636
0.120	0.384	0.5220	4.571

(1) ETHYLENE
----
(2) METHANE

ACTIVATED CARBON

ADSORBENT : BPL;PITTSBURGH CHEMICAL Co.;988 Sqm/g
EXP   : VOLUMETRIC
REF   : REICH,R.,ZIEGLER,W.T.and ROGERS,K.A.;Ind.Eng.Chem.Process Des.Dev.
            19,336(1980)

              TEMP   =   301.40 (K)
              Yl   =   .7400

     P (KPa)   Xl   N(mmol/g)   Sexp

     130.999   .9640   2.4530   9.4084
     344.735   .9320   3.6040   4.8156
     682.575   .9210   4.3480   4.0961
    1225.880   .9120   5.1360   3.6413
    2000.840   .8910   5.7310   2.8721

              TEMP   =   212.70 (K)
              Yl   =   .7400

     P (KPa)   Xl   N(mmol/g)   Sexp

     135.826   .9830   6.3380   20.3164
     240.625   .9720   6.7820   12.1969
     414.371   .9690   7.3070   10.9826
     680.507   .9620   7.9160   8.8947

              TEMP   =   301.40 (K)
              Yl   =   .4640

     P (KPa)   Xl   N(mmol/g)   Sexp

     122.036   .8940   2.0530   9.7427
     344.046   .8460   3.1770   6.3459
     676.370   .8260   3.9340   5.4837
    1240.360   .7900   4.7780   4.3456
    2031.180   .7530   5.3910   3.5216

              TEMP   =   212.70 (K)
              Yl   =   .4640

     P (KPa)   Xl   N(mmol/g)   Sexp

     615.697   .9100   7.3540   11.6801
    1041.100   .8890   8.0770   9.2518
    1043.860   .8940   8.0360   9.7427

```
                    TEMP      =    301.40 (K)
                    Y1        =      .4660
```

P (KPa)	X1	N(mmol/g)	Sexp
519.171	.8320	3.7870	5.6750

```
                    TEMP      =    212.70 (K)
                    Y1        =      .4660
```

P (KPa)	X1	N(mmol/g)	Sexp
319.914	.9270	6.9180	14.5516
725.322	.9020	7.6220	10.5472

```
                    TEMP      =    301.40 (K)
                    Y1        =      .2350
```

P (KPa)	X1	N(mmol/g)	Sexp
128.931	.7540	1.7160	9.9777
443.329	.6910	2.9940	7.2797
693.607	.6640	3.5370	6.4331
697.054	.6620	3.5730	6.3758
1430.650	.6120	4.5270	5.1347

```
                    TEMP      =    260.20 (K)
                    Y1        =      .2350
```

P (KPa)	X1	N(mmol/g)	Sexp
145.478	.7920	3.2170	12.3953
350.251	.7500	4.2350	9.7660
683.954	.7300	4.9850	8.8014
683.954	.7330	4.9880	8.9369
1411.350	.6760	5.9350	6.7920

```
                    TEMP      =    212.70 (K)
                    Y1        =      .2350
```

P (KPa)	X1	N(mmol/g)	Sexp
135.826	.8630	5.4680	20.5061
441.261	.8260	6.6540	15.4534
442.640	.8240	6.6690	15.2408
896.311	.8040	7.3640	13.3535

```
END OF FILE
->
```

(1) ETHYLENE
----
(2) METHANE

ACTIVATED CARBON

ADSORBENT : NUXIT-AL (1.5 mm DIAMETER ,3-4 mm LONG)
EXP   : STATIC VOLUMETRIC
REF   : SZEPESY,L.,ILLES,V.;ACTA CHIM. HUNG.;35,245(1963)

                    TEMP   (K)=   293.15

        P (KPa)   Y1   X1   N(mmol/g)   Sexp

        98.8316   .1100   .6340   1.3299   14.0154
        98.2983   .1120   .6510   1.3522   14.7894
        99.2849   .2920   .8510   1.8930   13.8482
        98.7649   .3030   .8330   1.8327   11.4741
        99.4182   .5070   .9300   2.2954   12.9188
        99.2716   .5340   .9390   2.3440   13.4332
        99.4315   .7310   .9710   2.6197   12.3213
->

(1) HEXANE
----
(2) METHANE

ACTIVATED CARBON

ADSORBENT : BPL;PITTSBURGH ACTIVATED CARBON Co.;1040 Sqm/g
EXP   : FLOW NON-CHROMATOGRAPHIC
REF   : GRANT,R.J. and MANES,M.;Ind.Eng.Chem.Fundam.;5,490(1966)

$$\text{TEMP} = 298.15 \ (K)$$
$$Y1 = .000795$$

P (KPa)	X1	N(mmol/g)	Sexp
101.325	.9890	2.6885	113003.2
1135.530	.9580	2.9096	28668.4
1962.890	.8620	3.1748	7850.8
3548.670	.7830	3.4517	4535.1
5272.350	.6110	4.1735	1974.1

->

HYDROGEN SULFIDE _____

CARBON DIOXIDE _____

ZEOLITE _____

ADSORBENT : H-mordenite, Norton Co.;type Z-900H;1/16 in. Pellets
EXP.   : Static Volumetric
REF.   : Talu, O. and Zwiebel, I.; *AIChE J.* **32,** 1263 (1986)

T   (K)= 303.15

P(kPa)	Y1	X1	N(mmol/g)	Sexp
0.60	0.229	0.813	0.738	14.6375
0.62	0.216	0.812	0.645	15.6769
7.00	0.212	0.789	1.378	13.8990
12.17	0.205	0.783	1.548	13.9931
15.77	0.066	0.538	1.436	16.4795
15.55	0.181	0.735	1.614	12.5501
15.69	0.464	0.872	1.826	7.8696
14.93	0.638	0.922	1.886	6.7069
14.38	0.840	0.968	1.945	5.7619

HYDROGEN SULFIDE

PROPANE

ZEOLITE

ADSORBENT : H-mordenite, Norton Co.;type Z-900H;1/16 in. Pellets
EXP.   : Static Volumetric
REF.   : Talu, O. and Zwiebel, I.; *AIChE J.* **32,** 1263 (1986)

T   (K) = 303.15

P (kPa)	Y1	X1	N (mmol/g)	Sexp
1.85	0.041	0.475	0.960	21.1626
4.87	0.031	0.485	1.181	29.4372
8.11	0.040	0.524	1.357	26.4202
18.32	0.035	0.537	1.580	31.9781
7.97	0.103	0.600	1.558	13.0631
8.30	0.279	0.701	1.762	6.0587
8.14	0.584	0.782	1.863	2.5552
8.13	0.936	0.916	1.851	0.7456

(1) PROPANE
----
(2) ETHANE

ACTIVATED CARBON

ADSORBENT : NUXIT-AL (1.5 mm DIAMETER ,3-4 mm LONG)
EXP   : STATIC VOLUMETRIC
REF   : SZEPESY,L.,ILLES,V.;ACTA CHIM. HUNG.;35,245(1963)

TEMP   (K)=   293.15

P (KPa)	Y1	X1	N(mmol/g)	Sexp
101.4980	.0304	.1848	3.4027	7.2303
101.3250	.0617	.3451	3.5231	8.0136
101.0450	.1061	.4898	3.6276	8.0882
101.3510	.1703	.6156	3.7672	7.8023
101.1510	.2584	.7145	3.8769	7.1825
101.2980	.3274	.7859	3.8885	7.5410
101.3110	.4907	.8658	4.1326	6.6961
101.3650	.6534	.9158	4.2383	5.7695
101.2980	.7838	.9481	4.2352	5.0389
101.4310	.8668	.9684	4.2615	4.7093

TEMP   (K)=   333.15

P (KPa)	Y1	X1	N(mmol/g)	Sexp
101.9910	.1420	.5340	2.4221	6.9240
102.2310	.2820	.7220	2.6728	6.6125
102.3510	.4800	.8440	2.9030	5.8611
101.5910	.6590	.9130	3.0485	5.4303
101.6050	.8020	.9500	3.1457	4.6908

->

(1) PROPANE

---

(2) ETHANE

CARBON MOLECULAR SIEVE

ADSORBENT : MSC-5A;TAKEDA CHEMICAL Co.;0.56 cc/g;650 Sqm/g
EXP   : STATIC VOLUMETRIC
REF   : NAKAHARA,T.,HIRATA,M.,KOMATSU,S.;J.Chem.Eng.Data;26,161(1981)

TEMP   (K)=   278.15

P (KPa)	Y1	X1	N(mmol/g)	Sexp
13.1863	.0000	.0000	2.6220	
13.2397	.0740	.5740	1.9670	16.8609
13.3063	.1850	.7330	2.0170	12.0942
13.2530	.3340	.8260	2.0250	9.4658
13.2663	.5050	.8810	2.0580	7.2568
13.1863	.6370	.9090	2.0390	5.6923
13.3863	.7960	.9360	2.0180	3.7481
13.5997	1.0000	1.0000	1.9900	

TEMP   (K)=   303.15

P (KPa)	Y1	X1	N(mmol/g)	Sexp
13.3863	.0000	.0000	1.4570	
13.4797	.0480	.3110	1.5030	8.9523
13.5197	.0530	.4810	1.4760	16.5597
12.8797	.0900	.5040	1.4800	10.2742
13.3330	.1770	.7180	1.6560	11.8386
13.2397	.2850	.8130	1.6810	10.9071
13.2663	.6750	.9200	1.7650	5.5370
13.3863	.8320	.9450	1.7520	3.4694
13.1863	.8990	.9690	1.7040	3.5118
13.4530	1.0000	1.0000	1.7410	

TEMP   (K)=   303.15

P (KPa)	Y1	X1	N(mmol/g)	Sexp
40.1323	.0000	.0000	1.9290	
40.0523	.1160	.5520	1.8500	9.3898
40.4790	.2240	.7310	1.8350	9.4141
40.2523	.4740	.8610	1.8590	6.8738
39.9990	1.0000	1.0000	1.8590	

->

TEMP (K)= 323.15

P (KPa)	Y1	X1	N(mmol/g)	Sexp
13.3330	.0000	.0000	1.1410	
13.6663	.0510	.4300	1.2380	14.0375
13.4263	.0970	.5770	1.3000	12.6985
13.5330	.1930	.7420	1.3660	12.0254
13.4663	.3200	.8470	1.4020	11.7639
13.3063	.4800	.8690	1.4810	7.1864
13.6797	.6290	.8560	1.5760	3.5062
13.5463	.7680	.9420	1.5540	4.9063
13.3330	1.0000	1.0000	1.5580	

->

(1) PROPANE
----

(2) ETHYLENE

SILICA GEL

ADSORBENT : DAVISON CHEMICAL Co.;REF.GRADE;751 Sqm/g;mesh 14x20
EXP   : STATIC VOLUMETRIC
REF   : LEWIS,W.K.,GILLILAND,E.R.,CHERTOW,B.,and BAREIS,D.;J.Am.Chem.soc.;
            72,1160(1950)

TEMP   (K)=   273.15

P (KPa)	Y1	X1	N(mmol/g)	Sexp
102.2050	.0220	.0450	1.6200	2.0947
101.5110	.0830	.1790	1.7470	2.4088
101.2050	.1750	.3010	1.8040	2.0300
102.2050	.2100	.3750	1.8880	2.2571
102.1910	.3160	.4950	1.9900	2.1217
102.2180	.4550	.6390	2.1160	2.1202
102.3250	.4580	.6140	2.1080	1.8824
101.5110	.5990	.7490	2.2730	1.9977
101.8180	.6780	.7860	2.3750	1.7444
101.6850	.7960	.8770	2.4520	1.8273
101.8180	.9010	.9240	2.5750	1.3359
101.6450	.9571	.9730	2.5720	1.6153

TEMP   (K)=   298.15

P (KPa)	Y1	X1	N(mmol/g)	Sexp
101.9110	.0470	.0810	.9900	1.7872
103.1380	.1320	.2310	1.0780	1.9753
101.6310	.2380	.3740	1.1120	1.9128
103.0850	.3390	.5090	1.1710	2.0213
101.6310	.5120	.6520	1.2620	1.7857
103.1380	.5990	.7350	1.3290	1.8568
102.3650	.6190	.7670	1.3230	2.0262
101.8310	.7630	.8420	1.3920	1.6553
102.4050	.9190	.9530	1.4660	1.7872

TEMP   (K)=   313.15

P (KPa)	Y1	X1	N(mmol/g)	Sexp
103.4580	.0690	.1160	.7160	1.7705
103.4180	.2250	.3650	.7850	1.9799
103.3780	.3520	.5060	.8380	1.8856
103.0050	.5060	.6600	.9000	1.8951
102.8180	.5080	.6540	.9000	1.8306
102.5910	.8410	.8970	1.0130	1.6465

->

(1) PROPANE
----
(2) METHANE

ACTIVATED CARBON

ADSORBENT : BPL;PITTSBURGH ACTIVATED CARBON Co.;1040 Sqm/g
EXP   : FLOW NON-CHROMATOGRAPHIC
REF   : GRANT,R.J. and MANES,M.;Ind.Eng.Chem.Fundam.;5,490(1966)

$$TEMP = 298.15 \text{ (K)}$$
$$Y1 = .010530$$

P (KPa)	X1	N(mmol/g)	Sexp
101.325	.7670	1.5986	309.3240
1135.530	.5630	3.7069	121.0602
1962.890	.4690	4.5853	82.9951
3548.670	.4380	4.7666	73.2339
5272.350	.3540	5.3105	51.4926
6996.020	.3650	5.0158	54.0124

$$TEMP = 298.15 \text{ (K)}$$
$$Y1 = .020000$$

P (KPa)	X1	N(mmol/g)	Sexp
101.325	.8370	1.8991	251.6135
1962.890	.6310	4.3873	83.7913
3548.670	.5530	4.8802	60.6197
5272.350	.4690	5.2768	43.2787
6996.020	.4190	5.6847	35.3373

->

(1) PROPYLENE

----

(2) ETHYLENE

ACTIVATED CARBON

ADSORBENT : NUXIT-AL (1.5 mm DIAMETER ,3-4 mm LONG)
EXP   : STATIC VOLUMETRIC
REF   : SZEPESY,L.,ILLES,V.;ACTA CHIM. HUNG.;35,245(1963)

TEMP   (K)=   293.15

P (KPa)	Y1	X1	N(mmol/g)	Sexp
101.2050	.0420	.3650	3.1538	13.1110
101.3380	.0950	.5330	3.3429	10.8726
101.3250	.1400	.6660	3.5423	12.2489
101.3250	.2410	.8000	3.8163	12.5975
101.3650	.4000	.8910	4.0643	12.2615

->

(1) PROPYLENE ———

(2) ETHYLENE ——

CARBON MOLECULAR SIEVE ————————————————————————————————————

ADSORBENT : MSC-5A;TAKEDA CHEMICAL Co.;0.56 cc/g;650 Sqm/g.
EXP   : STATIC VOLUMETRIC
REF   : NAKAHARA,T.,HIRATA,M.and MORI,H.;J Chem.Eng.Data;27,317(1982)

TEMP   (K)=   274.85

P (KPa)	Y1	X1	N(mmol/g)	Sexp
13.8133	.0240	.2660	2.1640	14.7375
13.2933	.0850	.7550	2.1280	33.1729
13.3200	.2550	.8890	2.2580	23.3989
13.3600	.3300	.8840	2.2220	15.4723
13.4133	.4510	.9300	2.3550	16.1726
13.3200	.5060	.9400	2.3190	15.2951
13.4400	.6310	.9530	2.3550	11.8575
13.0267	.7410	.9520	2.3780	6.9323
13.2400	.8190	.9430	2.4170	3.6562
13.5067	.8720	.9600	2.4630	3.5229
13.0133	.8860	.9750	2.4740	5.0181

TEMP   (K)=   303.15

P (KPa)	Y1	X1	N(mmol/g)	Sexp
13.4133	.0990	.6230	1.6700	15.0396
13.5600	.2900	.8640	1.9000	15.5538
13.4933	.5240	.9350	2.0030	13.0669
13.5333	.6320	.9440	2.0160	9.8155
13.3200	.6330	.9620	2.0010	14.6776
13.0400	.7580	.9550	2.0810	6.7754
13.6267	.8880	.9600	2.1350	3.0270
13.4667	.9120	.9670	2.1460	2.8275

TEMP   (K)=   323.15

P (KPa)	Y1	X1	N(mmol/g)	Sexp
13.4133	.0460	.4960	1.2950	20.4099
13.7600	.1170	.6810	1.3520	16.1113
13.8533	.2920	.8820	1.6230	18.1233
14.0133	.4050	.9020	1.7070	13.5220
13.4400	.4060	.8980	1.6450	12.8806
13.4533	.5330	.9310	1.8330	11.8220
13.3867	.6400	.9740	1.6580	21.0721
13.4933	.6550	.9490	1.8060	9.8011
13.5067	.7550	.9710	1.7750	10.8653
13.4267	.8760	.9900	1.7860	14.0137

(1) PROPYLENE

----

(2) ETHYLENE

SILICA GEL

```
ADSORBENT : DAVISON CHEMICAL Co.;REF.GRADE;751 Sqm/g;mesh 14x20
EXP       : STATIC VOLUMETRIC
REF       : LEWIS,W.K.,GILLILAND,E.R.,CHERTOW,B.,and BAREIS,D.;J.Am.Chem.soc.;
            72,1160(1950)
```

TEMP   (K)=   273.15

P (KPa)	Y1	X1	N(mmol/g)	Sexp
102.5380	.0110	.1150	1.6620	11.6831
102.5510	.0520	.2550	1.8070	6.2401
103.2580	.0970	.4590	1.9810	7.8983
103.3780	.1860	.6380	2.2280	7.7130
102.7110	.2890	.7930	2.5040	9.4249
101.2580	.3180	.7620	2.3720	6.8665
102.7910	.4210	.8550	2.6110	8.1095
102.8980	.6360	.9338	2.8900	8.0731

TEMP   (K)=   298.15

P (KPa)	Y1	X1	N(mmol/g)	Sexp
102.8180	.0320	.1930	1.0490	7.2345
102.9650	.0920	.4030	1.2080	6.6624
103.0980	.1830	.6030	1.4050	6.7811
102.3110	.3720	.8050	1.6800	6.9691
102.2710	.4740	.8770	1.7630	7.9123
102.4050	.7520	.9631	2.0250	8.6075

TEMP   (K)=   313.15

P (KPa)	Y1	X1	N(mmol/g)	Sexp
102.0850	.0650	.2950	.8240	6.0191
102.0050	.1480	.5290	.9420	6.4657
102.2450	.2970	.7300	1.1430	6.3997
102.4180	.4370	.8910	1.3050	10.5312
102.6050	.6790	.9398	1.4810	7.3803

->

(1) PROPYLENE ——

(2) PROPANE ——

ACTIVATED CARBON
———
ADSORBENT : BLACK PEARL I;GODFREY L. CABOT Co OF BOSTON;705 Sqm/g
EXP   : STATIC VOLUMETRIC
REF   : LEWIS,W.K.,GILLILAND,E.R.,CHERTOW,B.,and HOFFMAN,W.H.;J.Am.Chem.soc.;
            72,1153(1950)

TEMP   (K)=   298.15

P (KPa)	Y1	X1	N(mmol/g)	Sexp
102.9780	.0472	.0483	2.6700	1.0245
101.4980	.0842	.1008	2.7450	1.2193
101.7250	.1590	.1685	2.7680	1.0719
100.4050	.2305	.2609	2.7250	1.1784
102.7910	.2423	.2523	2.7700	1.0552
102.2580	.3264	.3312	2.8200	1.0220
100.9110	.3893	.5888	2.6050	2.2463
100.8580	.3990	.4075	2.7770	1.0360
101.1910	.4017	.4135	2.7930	1.0501
100.8310	.4540	.3325	2.5600	.5991
100.6580	.4780	.4964	2.7630	1.0764
102.7910	.4812	.5181	2.8310	1.1591
102.5250	.5090	.5425	2.9270	1.1439
102.1250	.5649	.5923	2.8380	1.1190
100.7510	.5672	.5853	2.7720	1.0769
101.5910	.6041	.6272	2.7940	1.1026
100.9250	.6201	.6325	2.9070	1.0544
101.3250	.7184	.7404	2.8230	1.1180
101.1910	.8119	.8387	2.8310	1.2046
100.7380	.8896	.9007	2.8190	1.1257

->

(1) PROPYLENE
----
(2) PROPANE

ACTIVATED CARBON

ADSORBENT : NUXIT-AL (1.5 mm DIAMETER ,3-4 mm LONG)
EXP   : STATIC VOLUMETRIC
REF   : SZEPESY,L.,ILLES,V.;ACTA CHIM. HUNG.;35,245(1963)

                TEMP   (K)=   293.15

P (KPa)	Y1	X1	N(mmol/g)	Sexp
101.1250	.1390	.1300	4.2450	.9256
101.0580	.2350	.2640	4.3084	1.1677
101.3510	.3620	.3720	4.4038	1.0440
101.2850	.4690	.4800	4.4551	1.0451
101.3780	.5360	.5670	4.4864	1.1336
101.5510	.6210	.6330	4.4872	1.0527
101.4580	.6810	.6890	4.5042	1.0378

->

(1) PROPYLENE _____

(2) PROPANE _____

SILICA GEL _____

ADSORBENT : DAVISON CHEMICAL Co.;REF.GRADE;751 Sqm/g;mesh 14x20
EXP   : STATIC VOLUMETRIC
REF   : LEWIS,W.K.,GILLILAND,E.R.,CHERTOW,B.,and HOFFMAN,W.H.;J.Am.Chem.soc.;
            72,1153(1950)

TEMP   (K)=   298.15

P (KPa)	Y1	X1	N(mmol/g)	Sexp
101.3250	.0790	.5990	1.4260	17.4147
101.3250	.1036	.7470	1.9930	25.5471
101.3250	.2520	.2770	1.6860	1.1372
100.5250	.3748	.2993	1.7010	.7125
100.4710	.3780	.4450	1.8510	1.3194
100.5250	.3860	.6409	1.9740	2.8389
102.1650	.4644	.6880	1.9670	2.5432
100.4710	.4667	.6345	1.9630	1.9837
101.4580	.4700	.7184	2.0410	2.8768
102.3650	.5960	.7044	2.0520	1.6153
101.4450	.7010	.7424	2.0130	1.2293
102.5510	.7555	.8922	2.1970	2.6785

->

CHAPTER 4

ADSORPTION OF LIQUID MIXTURES

A compilation of references to experimental data for adsorption of liquid mixtures is given in Appendix F. Only those binary systems accompanied by tabulated data for the surface excess (n_1^e) are included in this chapter.

GUIDE TO TABLES

Ordering of Systems

Adsorbates are listed in alphabetical order for component number 1, which is defined as the more strongly adsorbed species. Mixtures tabulated in this chapter are indexed in Appendix C.

Heading

The heading includes the names of the two adsorbates, the type of adsorbent, and the reference. In the case of U-shaped isotherms, adsorbate no. 1 is defined as the more strongly adsorbed species so that n_1^e is positive over the entire range of bulk concentration. S-shaped isotherms indicate reversal of selectivity at the point where n_1^e changes sign.

Adsorption Isotherm

There are three degrees of freedom for adsorption of a binary liquid mixture. However, the liquid and solid phases are both condensed and the effect of pressure may be neglected for the systems listed in this chapter. The two remaining independent variables are $\{T, x_1\}$, where x_1 is the mole fraction of adsorbate no. 1 in the bulk liquid phase at equilibrium. The measurement of surface excess (n_1^e) as a function of x_1 along a isotherm is based on the relation[Si535(86)]

$$n_1^e = n^\circ(x_1^\circ - x_1) \tag{4-1}$$

where x_1^o is the mole fraction of component no. 1 in the bulk liquid solution before it is contacted with the adsorbent, and n^o is the total number of moles of bulk liquid solution per unit mass of adsorbent. The values of n^o and x_1^o are usually not reported because in principle they have no effect on the function $n_1^e(x_1)$. However, the accuracy of the measurement of n_1^e improves with the magnitude of the difference $(x_1^o - x_1)$.

The definition of surface excess for a binary solution is:

$$n_1^e = n_1 x_2 - n_2 x_1 \tag{4-2}$$

n_i is the specific amount (mol/kg) of ith component adsorbed, but for a non–porous adsorbent the absolute amount adsorbed (n_1 or n_2) cannot be measured experimentally because of uncertainty in the location of the dividing surface between the bulk liquid and the adsorbed phase. The experimental variable is the change in concentration of the liquid solution upon contact with a solid adsorbent, Eq. (4-1).

A sample pair of isotherms from the data bank is the adsorption of methyl acetate(1) and benzene(2) on silica gel:

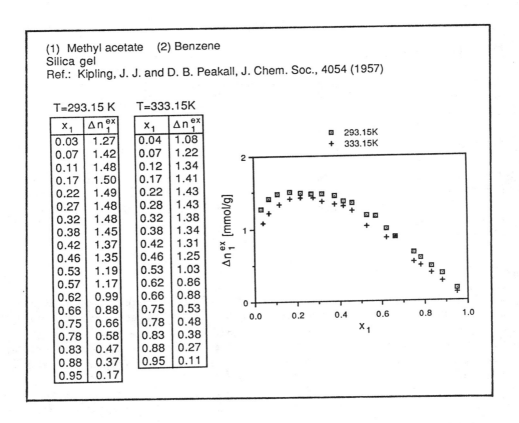

(1) Methyl acetate   (2) Benzene
Silica gel
Ref.: Kipling, J. J. and D. B. Peakall, J. Chem. Soc., 4054 (1957)

| T=293.15 K | | T=333.15K | |
x_1	Δn_1^{ex}	x_1	Δn_1^{ex}
0.03	1.27	0.04	1.08
0.07	1.42	0.07	1.22
0.11	1.48	0.12	1.34
0.17	1.50	0.17	1.41
0.22	1.49	0.22	1.43
0.27	1.48	0.28	1.43
0.32	1.48	0.32	1.38
0.38	1.45	0.38	1.34
0.42	1.37	0.42	1.31
0.46	1.35	0.46	1.25
0.53	1.19	0.53	1.03
0.57	1.17	0.62	0.86
0.62	0.99	0.66	0.88
0.66	0.88	0.75	0.53
0.75	0.66	0.78	0.48
0.78	0.58	0.83	0.38
0.83	0.47	0.88	0.27
0.88	0.37	0.95	0.11
0.95	0.17		

Surface excess $\left(n_1^e\right)$ of methyl acetate in units of mol/kg [= mmol/g] is a function of the equilibrium mole fraction of methyl acetate (x_1) in the bulk liquid phase. The computer printout uses the symbol Δn_1^{ex} for n_1^e. The isotherms pass through the points $(x_1, n_1^e) = (0,0)$ and $(1,0)$. The slope of the isotherm in the limit as $x_1 \to 0$ is finite but often very steep. Extrapolation from the experimental data to the point $(0,0)$ may be difficult.

These isotherms show the usual decrease of n_1^e with increasing temperature. The temperature coefficient $\left(\partial n_1^e / \partial T\right)_{x_1}$ is related to the heat of immersion of the adsorbent by Eq. (1-34).

ADSORPTION EQUATIONS

U-shaped adsorption isotherms (but not S-shaped isotherms) can be fit approximately by the following equation [La1025(71)]:

$$n_1^e = \frac{m_1\left(sa_1x_2 - a_2x_1\right)}{sa_1 + \beta a_2} \tag{4-3}$$

$a_1 = \gamma_1 x_1$ and $a_2 = \gamma_2 x_2$ are activities in the bulk liquid solution. Activity coefficients γ_i are obtained independently, for example, from vapor–liquid equilibrium data. For an ideal liquid mixture, $a_1 = x_1$ and $a_2 = x_2$.

m_i is the saturation capacity of ith adsorbate in units of mol/kg. For non-porous adsorbents:

$$m_i = A/\tilde{a}_i \tag{4-4}$$

and for microporous adsorbents such as high–area activated carbons or zeolites:

$$m_i = V/v_i \tag{4-5}$$

where A is the specific surface area of the adsorbent (m^2/kg) and V is the specific pore volume (m^3/kg). $\tilde{a}_i$ is molar area (m^2/mol) of the adsorbate [Mc577(67)] and v_i is the molar volume of the adsorbate in the liquid state (m^3/mol).

β is the ratio of saturation capacities:

$$\beta = \frac{m_1}{m_2} = \frac{\tilde{a}_2}{\tilde{a}_1} = \frac{v_2}{v_1} \tag{4-6}$$

and may be estimated by the ratio of molar areas or molar volumes, depending on whether the adsorbent is nonporous or microporous.

The selectivity s in Eq. (4-3) is defined as:

$$s = \frac{x_1' a_2}{x_2' a_1} \tag{4-7}$$

where x_i' is the mole fraction of ith component in the adsorbed phase. Except for the case of adsorbate molecules of equal size ($\beta = 1$), s is a function of composition:

$$s = s_0(sa_1 + a_2)^{(\beta-1)/\beta} \tag{4-8}$$

s_0 is a constant; $s = s_0$ at $x_1 = 0$ and $s = s_0^\beta$ at $x_1 = 1$.

To summarize, n_1^e in Eq. (4-3) is not explicit in x_1 because s varies with composition. Given the mole fraction (x_i) and activity (a_i) in the bulk solution, as well as the constants β, s_0 and m_1, Eq. (4-8) is solved numerically for s, which is substituted into Eq. (4-3) to obtain n_1^e.

For an ideal bulk solution, Eq. (4-3) reduces to:

$$n_1^e = \frac{m_1(s-1)x_1 x_2}{sx_1 + \beta x_2} \tag{4-9}$$

but s is still a function of composition through Eq. (4-8).

For $\beta = 1$, $m_1 = m_2 = m$, Eq. (4-8) gives $s = s_0 = $ constant, and Eq. (4-3) reduces to the explicit form [Si2828(70)]:

$$n_1^e = \frac{m(s_0 a_1 x_2 - a_2 x_1)}{s_0 a_1 + a_2} \tag{4-10}$$

If $\beta = 1$ and the bulk solution is ideal [Ev1803(64)]:

$$n_1^e = \frac{m(s_0 - 1)x_1 x_2}{s_0 x_1 + x_2} \tag{4-11}$$

It is recommended that the ratio β be estimated by Eq. (4-6). Given bulk–phase activity coefficients (γ_i), Eqs. (4-3), (4-9), (4-10) and (4-11) contain only two parameters: m_1 and s_0. In the case of U-shaped isotherms, these constants are usually extracted from the experimental data.

Take as an illustration the isotherm for adsorption of 1,2-dichloroethane + benzene on silica gel reported on page 276. The silica gel is microporous, and from Eq. (4-6), $\beta = v_2/v_1 = (88.26)/(79.17) = 1.11$. The liquids form an ideal solution so Eq. (4-9) applies. Values of constants optimized for a best fit of the experimental data are $m_1 = 0.947$ mol/kg and $s_0 = 2.970$. The solid line in Fig. 4-1 is Eq. (4-9) and the experimental points are from page 276.

A sample calculation of n_1^e is illustrated as follows for the composition $x_1 = 0.2$. From Eq. (4-8) for an ideal solution:

$$s = s_0(sx_1 + x_2)^{(\beta-1)/\beta}$$

The solution at $x_1 = 0.2$ and $x_2 = 0.8$ is $s = 3.074$. From Eq. (4-9):

$$n_1^e = \frac{(0.947)(3.074 - 1)(0.2)(0.8)}{(3.074)(0.2) + (1.11)(0.8)} = 0.209 \text{ mol/kg}$$

This family of two–parameter equations fits experimental data for U-shaped isotherms fairly well, but the goodness of fit should not be taken as evidence for adequacy of the theory. For example, Eq. (4-11) with the constants $m_1 = m_2 = m = 1.099$ mol/kg and $s_o = 2.554$ fit the experimental data for the system 1,2-dichloroethane + benzene + silica gel as well as Eq. (4-9) used with the parameters $m_1 = 0.947$, $m_2 = 0.853$, $\beta = 1.11$ and $s_o = 2.970$.

Eq. (4-3) and its simplified versions are derived for the special case when the adsorbed phase (but not necessarily the liquid phase) forms an ideal solution. Moreover, Eq. (4-3) does not take into account the energetic heterogeneity of the surface. As a result, constants of Eq. (4-3) derived from experimental data cannot be interpreted literally. In some cases, particularly for S-shaped isotherms, Eq. (4-3) does not fit experimental data very well. The theory of an ideal, energetically homogeneous adsorbed phase should be extended to account for energetic heterogeneity and adsorbed-phase nonidealities. So far, attempts to do so have led to equations that are either too complicated or contain too many undetermined constants. Research is needed to develop an isotherm equation that provides quantitative agreement with experiment using constants that can be determined by independent methods.

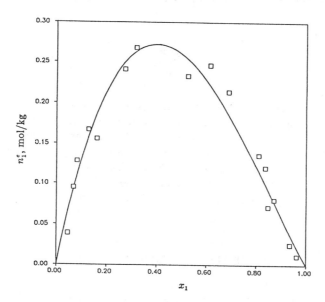

Fig. 4-1. Surface excess (n_1^e) as a function of mole fraction in bulk liquid solution (x_1) for 1,2-dichloroethane(1) + benzene(2) on silica gel at 303 K. Solid line is Eq. (4-9) or Eq. (4-11).

FREE ENERGY OF IMMERSION

The free energy of immersion ($\phi = \Delta G_{im}$) is defined as the change in Gibbs free energy accompanying the contact of a clean adsorbent with a liquid. Since adsorption is spontaneous, ϕ is a negative quantity with units of J/kg. There is an exact relationship [Si186(71)] between the *difference* in free energies of immersion and the surface excess:

$$\frac{\phi_1^\circ - \phi_2^\circ}{RT} = -\int_{x_1=0}^{1} \frac{n_1^e}{x_1 x_2 \gamma_1} d(\gamma_1 x_1) \tag{4-12}$$

This integration requires data on the bulk–phase activity coefficient γ_1.

Free Energy of Immersion from Adsorption Equations

The constant s_o in Eqs. (4-8), (4-10), and (4-11) is connected to the free energies of immersion by:

$$s_o = \exp\left(\frac{\phi_2^\circ - \phi_1^\circ}{m_1 RT}\right) \tag{4-13}$$

Obtaining free energies of immersion from Eq. (4-13) by fitting the experimental data with any of the equations in the previous section is more convenient than numerical integration by Eq. (4-12). The integral is a physical quantity whose value is nearly independent of the constants in the equations, as long as they fit the experimental data. To illustrate this point, it was shown in the last section that for the system 1,2-dichloroethane + benzene + silica gel at 303.15 K, the constants of Eq. (4-9) are: $m_1 = 0.947$ mol/kg and $s_o = 2.970$. From Eq. (4-13):

$$\phi_2^\circ - \phi_1^\circ = m_1 RT \ln(s_o)$$
$$= (0.947)(8.3145)(303.15) \ln(2.97) = 2598. \text{ J/kg}$$

For the same data, the constants of Eq. (4-11) are $m = 1.099$ mol/kg and $s_o = 2.554$. Again, from Eq. (4-13):

$$\phi_2^\circ - \phi_1^\circ = m_1 RT \ln(s_o)$$
$$= (1.099)(8.3145)(303.15) \ln(2.554) = 2597. \text{ J/kg}$$

THERMODYNAMIC CONSISTENCY

The difference in free energies of immersion provides the basis for a powerful test of thermodynamic consistency of binary adsorption isotherms [Si186(71)]. Consider a set of three liquids (nos. 1, 2, and 3) and the isothermal adsorption of their binary pairs (1–2, 1–3, and 2–3) on the same adsorbent. The test of consistency is:

$$(\phi_1^o - \phi_2^o) + (\phi_2^o - \phi_3^o) + (\phi_3^o - \phi_1^o) = 0 \qquad (4\text{-}14)$$

The three integrals of surface excess in Eq. (4-12) for the three binary pairs should sum to zero. A non–zero sum indicates either non–equilibrium conditions or experimental error. For example, data is tabulated in this chapter for binary combinations of benzene(1) + ethyl acetate(2) + cyclohexane(3) adsorbed on activated carbon at 303.15 K. The binary integrals of Eq. (4-12) were evaluated [Mi453(73)] using known values of bulk–phase activity coefficients, yielding:

$$(\phi_1^o - \phi_2^o) + (\phi_2^o - \phi_3^o) + (\phi_3^o - \phi_1^o) = -(14.77) - (9.25) + (23.76)$$
$$= -0.26 \text{ J/g}$$

The error is only one percent of the largest difference in free energies, so these data exhibit a high degree of thermodynamic consistency.

(1) Acetic acid   (2) Water
Activated carbon
Ref: Foti, G., L. G. Nagy and G. Schay, Acta Chim. Acad. Sci. Hung., 80, 25 (1974)

T = 298.15 K

x_1	Δn_1^{ex}
0.031	2.96
0.116	3.16
0.176	2.47
0.246	2.07
0.320	1.34
0.389	0.83
0.545	0.35
0.724	0.28
0.880	0.11

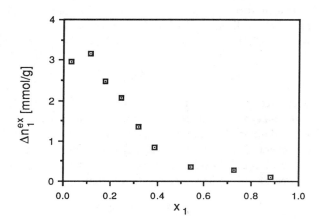

(1) Benzene   (2) Chlorobenzene
Silica gel, 420 m2./g (BET)
Ref: Peisen, L. and T. Gu, Scientia Sinica, 22, 1384 (1979)

T = 293.15 K

x_1	Δn_1^{ex}
0.0547	0.1570
0.1102	0.2130
0.2185	0.3518
0.3246	0.4474
0.4260	0.4302
0.5285	0.3827
0.7236	0.2771
0.8177	0.1828
0.9097	0.1014
0.9555	0.0478

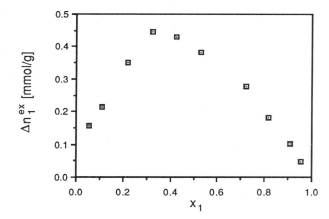

(1) Benzene   (2) Cyclohexane
Activated carbon, BPL, Pittsburgh Activated Carbon Co., 4x10, 1100 m2/g
Ref: Minka, C. and A. L. Myers, AIChE J., 19, 453 (1973)

T= 303.15 K

x_1	Δn_1^{ex}
0.019	0.642
0.021	0.723
0.077	1.051
0.079	1.014
0.143	1.233
0.152	1.239
0.246	1.264
0.346	1.207
0.470	1.061
0.565	0.872
0.667	0.669
0.692	0.609
0.783	0.441
0.888	0.233
0.945	0.107

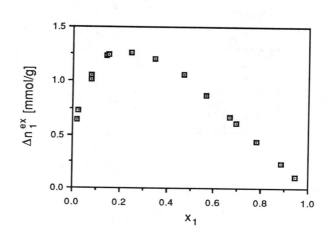

(1) Benzene   (2) Cyclohexane
Graphitized Carbon Black, Graphon, Cabot Co., 89 m2/g (BET)
Ref: Matayo, D. R. and J. P. Wightman, J. Colloid Interface Sci.,
     44, 162 (1973)

T = 303.15 K

x_1	Δn_1^{ex}
0.1	0.0415
0.2	0.0692
0.3	0.0724
0.4	0.0692
0.5	0.0798
0.6	0.0742
0.7	0.0395
0.8	0.0386
0.9	0.0177

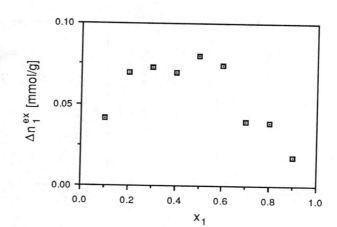

(1) Benzene   (2) Cyclohexane
Silica gel, Cab-O-Sil, Cabot Corp., 223 m2/g
Ref.: Matayo, D. R. and J. P. Wightman, J. Colloid Interface Sci., 44, 162 (1973)

T = 303.15 K

x_1	Δn_1^{ex}
0.1	0.368
0.2	0.439
0.3	0.366
0.4	0.392
0.5	0.308
0.6	0.162
0.7	0.116
0.8	0.086

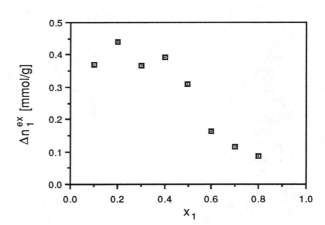

(1) Benzene   (2) Cyclohexane
Silica Gel, Hi-Sil 215, PPG Industries, Inc., 143 m2/g
Ref.: Matayo, D. R. and J. P. Wightman, J. Colloid Interface Sci., 44, 162 (1973)

T = 303.15 K

x_1	Δn_1^{ex}
0.1	0.136
0.2	0.183
0.3	0.275
0.4	0.243
0.5	0.256
0.6	0.179
0.7	0.114
0.8	0.080
0.9	0.031

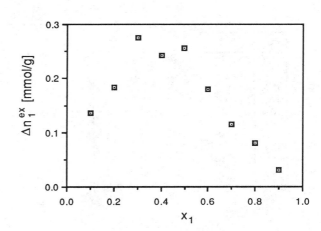

(1) Benzene (2) Cyclohexane
Silica gel, PA 400, Davison Chemical Company, 8-12 mesh, 740 m2/g
Ref.: Sircar, S. and A. L. Myers, J. Phys. Chem., 74, 2828 (1970)

T = 303.15 K

x_1	Δn_1^{ex}	x_1	Δn_1^{ex}
0.011	0.674	0.549	1.096
0.030	1.008	0.630	0.908
0.033	0.986	0.652	0.900
0.074	1.274	0.703	0.785
0.078	1.361	0.708	0.802
0.107	1.425	0.740	0.683
0.145	1.454	0.755	0.680
0.202	1.579	0.810	0.488
0.240	1.552	0.811	0.460
0.344	1.555	0.844	0.451
0.398	1.410	0.906	0.261
0.434	1.415	0.927	0.191
0.443	1.362	0.930	0.130
0.508	1.240	0.945	0.173
0.536	1.192		

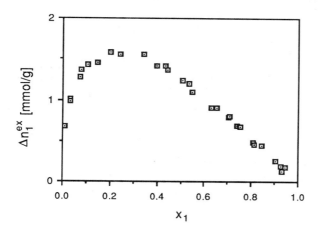

(1) Benzene   (2) Cyclohexane
Silica gel, Davison Chemical Co., grade 37, mesh 12-42, 350 m2/g
Ref.: Sircar, S., J. Novosad and A. L. Myers, Ind. Eng. Chem. Fundam.,
      11, 249 (1972)

T = 273.15 K

x_1	Δn_1^{ex}
0.045	1.30
0.111	1.75
0.180	1.92
0.263	1.86
0.353	1.71
0.390	1.62
0.542	1.34
0.793	0.67
0.900	0.33

T = 303.15 K

x_1	Δn_1^{ex}
0.051	1.05
0.117	1.47
0.187	1.61
0.268	1.59
0.357	1.50
0.395	1.41
0.544	1.20
0.794	0.59
0.900	0.30

T = 333.15 K

x_1	Δn_1^{ex}
0.056	0.83
0.123	1.19
0.193	1.33
0.272	1.35
0.360	1.28
0.390	1.23
0.547	1.06
0.795	0.52
0.900	0.27

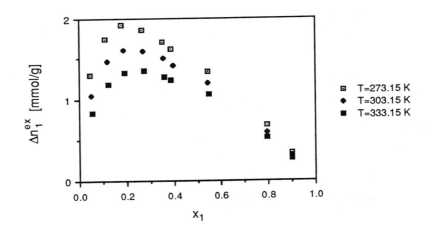

(1) Benzene   (2) 1,2-Dichloroethane
Activated carbon, BPL, Pittsburgh Activated Carbon Co., 4x10, number 23
Ref.: Sircar, S. and A. L. Myers, J. Phys. Chem. 74, 2828 (1970)

T = 303.15 K

x_1	Δn_1^{ex}	x_1	Δn_1^{ex}
0.074	0.529	0.551	0.298
0.102	0.575	0.556	0.342
0.209	0.568	0.687	0.234
0.221	0.572	0.713	0.186
0.359	0.538	0.803	0.151
0.364	0.519	0.939	0.047
0.471	0.429		

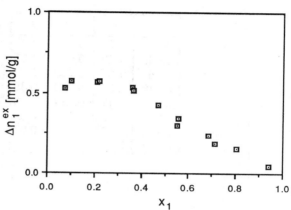

(1) Benzene   (2) Ethanol
Activated carbon
Ref.: Foti, G., L. G. Nagy and G. Schay, Acta Chim. Acad. Sci. Hung. 80, 25 (1974)

T = 298.15 K

x_1	Δn_1^{ex}
0.054	1.12
0.108	1.66
0.342	1.82
0.491	1.30
0.593	0.78
0.717	0.31
0.802	-0.24
0.918	-0.58
0.976	-0.41

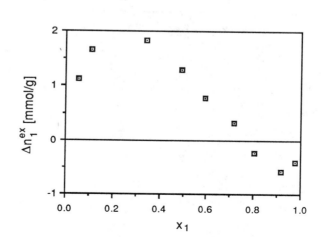

(1) Benzene   (2) Ethanol
Activated carbon, BPL, Pittsburgh Activated Carbon Co. BPL, 4x10, number 23
Ref.: Sircar, S. and A. L. Myers, J. Phys. Chem. 74, 2828 (1970)

T = 303.15 K

x_1	Δn_1^{ex}	x_1	Δn_1^{ex}
0.014	0.74	0.581	1.43
0.036	1.29	0.671	1.05
0.061	1.66	0.758	0.78
0.118	2.00	0.850	0.35
0.177	2.11	0.914	0.13
0.205	2.13	0.940	0.02
0.273	2.18	0.958	0.03
0.429	1.99		

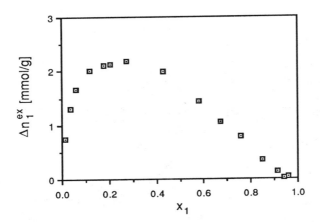

(1) Benzene   (2) Ethanol
Graphitized Carbon Black, Graphon, Cabot Corp., 86 m2/g (BET)
Ref.: Brown, C. E., D. H. Everett and C. J. Morgan, Trans. Faraday Soc.,
      71, 883 (1975)

x_1	Δn_1^{ex} mmol/g							
	T [K]							
	283	288	293	298	303	313	323	333
0.0670	0.080			0.078		0.077		0.074
0.1220	0.136	0.133	0.131	0.129	0.126	0.122	0.119	0.118
0.1630	0.167	0.166	0.165	0.164	0.162	0.158	0.154	0.149
0.2020	0.192			0.187	0.185	0.181	0.177	0.173
0.2585	0.226	0.224	0.223	0.220	0.218	0.212	0.206	0.199
0.3310	0.256	0.254	0.251	0.248	0.244	0.236	0.227	0.218
0.3840	0.258	0.255	0.251	0.249	0.244	0.236	0.227	0.217
0.4380	0.257	0.254	0.250	0.245	0.244	0.236	0.220	0.217
0.4910	0.254	0.249	0.243	0.239	0.232	0.220	0.208	0.195
0.5915	0.225	0.218	0.212	0.206	0.200	0.186	0.174	0.161
0.6420	0.202	0.196	0.190	0.183	0.177	0.164	0.152	0.141
0.7030	0.173	0.168	0.162	0.156	0.145	0.133	0.123	0.112
0.7570	0.142	0.136	0.130	0.125	0.120	0.110	0.101	0.093
0.8190	0.101	0.095	0.091	0.087	0.084	0.078	0.071	0.065
0.9000	0.046			0.039				0.029

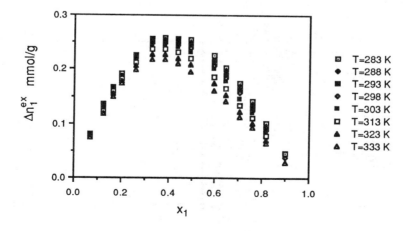

(1) Benzene   (2) Ethanol
Graphitized Carbon Black, Graphon, Cabot Co., 89 m2/g
Ref.: Matayo, D. R. and J. P. Wightman, J. Colloid Interface Sci.,
     44, 162 (1973)

T = 303.15 K

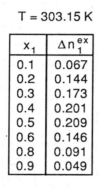

x_1	Δn_1^{ex}
0.1	0.067
0.2	0.144
0.3	0.173
0.4	0.201
0.5	0.209
0.6	0.146
0.8	0.091
0.9	0.049

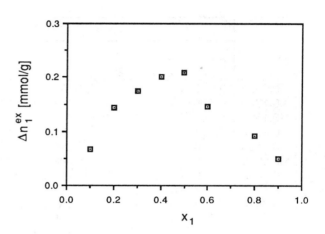

(1) Benzene   (2) Ethanol
Silica Gel, Cab-O-Sil, Cabot Corp., 223 m2/g
Ref.: Matayo, D. R. and J. P. Wightman, J. Colloid Interface Sci.,
     44, 162 (1973)

T = 303.15 K

x_1	Δn_1^{ex}
0.05	1.269
0.10	1.407
0.20	1.195
0.30	0.977
0.40	0.874
0.50	0.718
0.70	0.578
0.80	0.348
0.90	0.239

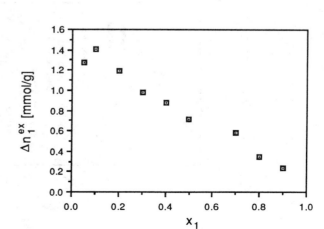

(1) Benzene   (2) Ethanol
Silica gel, Hi-Sil 215, PPG Industries, Inc., 143 m2/g
Ref.: Matayo, D. R. and J. P. Wightman, J. Colloid Interface Sci.,
    44, 162 (1973)

T = 303.15 K

x_1	Δn_1^{ex}
0.1	0.914
0.2	0.807
0.3	0.755
0.4	0.642
0.5	0.518
0.6	0.410
0.7	0.295
0.8	0.243

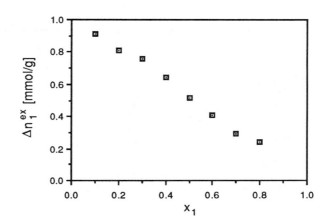

(1) Benzene   (2) Ethyl acetate
Activated carbon, BPL, Pittsburgh Activated Carbon Co., 4x10, 1100 m2/g
Ref.: Minka, C. and A. L. Myers, AIChE J., 19, 453 (1973)

T = 303.15 K

x_1	Δn_1^{ex}
0.039	0.394
0.080	0.561
0.183	0.809
0.278	0.832
0.376	0.782
0.479	0.686
0.592	0.603
0.688	0.416
0.790	0.281
0.898	0.143
0.947	0.079

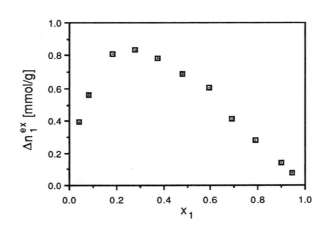

(1) Benzene   (2) n-Heptane
Silica gel, 360 m2/g
Ref.: Foti, G., L. G. Nagy and G. Schay, Acta Chim. Acad. Sci. Hung.,
      76, 269 (1973)

T = 298.15 K

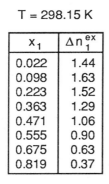

x_1	Δn_1^{ex}
0.022	1.44
0.098	1.63
0.223	1.52
0.363	1.29
0.471	1.06
0.555	0.90
0.675	0.63
0.819	0.37

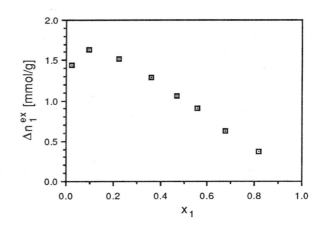

(1) Benzene   (2) n-Heptane
Silica gel, PA 400, Davison Chemical Company, 8-12 mesh, 740 m2/g
Ref.: Sircar, S. and A. L. Myers, J. Phys. Chem., 74, 2828 (1970)

T = 303.15 K

x_1	Δn_1^{ex}	x_1	Δn_1^{ex}
0.006	0.466	0.396	1.464
0.022	0.695	0.405	1.429
0.044	1.005	0.480	1.352
0.071	1.273	0.542	1.230
0.076	1.180	0.596	1.122
0.126	1.440	0.675	0.915
0.153	1.608	0.693	0.963
0.169	1.572	0.709	0.836
0.174	1.546	0.769	0.703
0.224	1.641	0.806	0.560
0.278	1.615	0.819	0.563
0.295	1.672	0.866	0.333
0.340	1.624	0.955	0.124

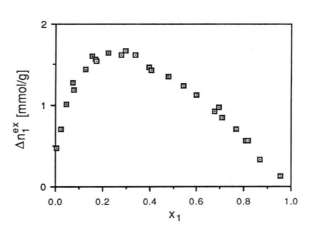

(1) Benzene   (2) Nitrobenzene
Alumina, 65.2 m2/g
Ref.: Suri, S. K. and V. Ramakrishna, Acta Chim. Acad. Sci. Hung.,
      63, 301 (1970)

T = 303.15 K

x_1	Δn_1^{ex}
0.1252	0.1963
0.1880	0.2210
0.2376	0.1892
0.3388	0.1128
0.3664	0.1190
0.4365	0.1108
0.5379	0.1230
0.6313	-0.0290
0.7014	-0.0800
0.8607	-0.1085
0.9088	-0.0890

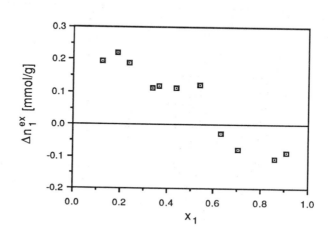

(1) Benzene   (2) Nitrobenzene
Silica gel, 412   m2/g
Ref.: Suri, S. K. and V. Ramakrishna, Acta Chim. Acad. Sci. Hung.,
      63, 301(1970)

T = 293.15 K

x_1	Δn_1^{ex}
0.1213	0.2864
0.2288	0.1160
0.3389	-0.1187
0.4441	-0.3079
0.5478	-0.4798
0.6452	-0.5850
0.7431	-0.6856
0.8237	-0.7395
0.9165	-0.5321

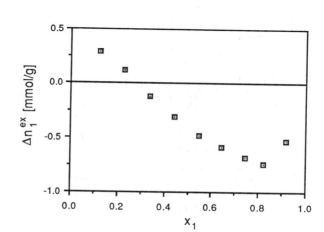

(1) Benzene   (2) Toluene
Silica gel, 420 m2/g
Ref.: Peisen, L. and T. Gu, Scientia Sinica, 22, 1384 (1979)

T = 293.15 K

x_1	Δn_1^{ex}
0.1150	0.2014
0.2269	0.2285
0.3349	0.2326
0.4386	0.2097
0.5411	0.1899
0.6395	0.1361
0.7350	0.1085
0.8258	0.0680
0.9157	0.0382
0.9578	0.0187

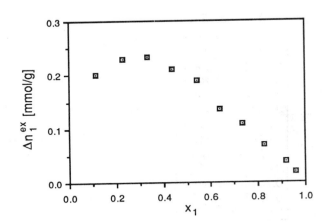

(1) Chlorobenzene   (2) Bromobenzene
Silica gel, 420 m2/g
Ref.: Peisen, L. and T. Gu, Scientia Sinica, 22, 1384 (1979)

T = 293.15 K

x_1	Δn_1^{ex}
0.0680	0.0366
0.1312	0.0460
0.2598	0.0499
0.4238	0.0487
0.7081	0.0366
0.7611	0.0295
0.8501	0.0205
0.8707	0.0107
0.9028	0.0194
0.9284	0.0117
0.9342	0.0105
0.9515	0.0097
0.9672	0.0094

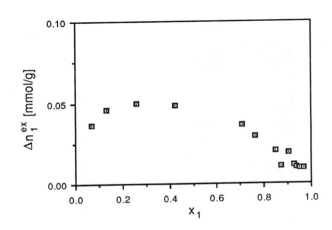

(1) Cyclohexane   (2) Ethanol
Activated carbon, BPL, Pittsburgh Activated Carbon Co., 4x10, number 23
Ref.: Sircar, S. and A. L. Myers, J. Phys. Chem., 74, 2828 (1970)

T = 303.15 K

x_1	Δn_1^{ex}	x_1	Δn_1^{ex}
0.042	0.295	0.440	0.065
0.051	0.485	0.470	0.000
0.072	0.517	0.521	-0.129
0.148	0.586	0.537	-0.362
0.160	0.669	0.610	-0.643
0.213	0.661	0.756	-1.230
0.216	0.583	0.848	-1.310
0.249	0.595	0.893	-1.180
0.286	0.532	0.920	-1.230
0.341	0.383	0.953	-0.996
0.391	0.192	0.974	-0.470

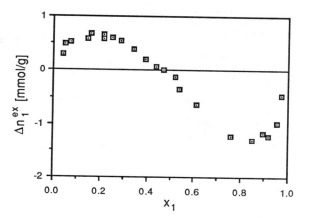

(1) Cyclohexane   (2) Ethanol
Graphitized Carbon Black, Graphon, Cabot Co., 89 m2/g
Ref.: Matayo, D. R. and J. P. Wightman, J. Colloid Interface Sci.,
      44, 162 (1973)

T = 303.15 K

x_1	Δn_1^{ex}
0.1	-0.0064
0.2	0.0630
0.3	0.0359
0.4	0.1326
0.5	0.1290
0.6	0.1006
0.7	0.0809
0.8	0.0367
0.9	0.0527

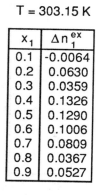

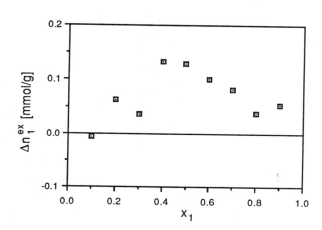

(1) Cyclohexane   (2) Ethanol
Silica gel, Cab-O-Sil, Cabot Corp., 223 m2/g
Ref.: Matayo, D. R. and J. P. Wightman, J. Colloid Interface Sci.,
      44, 162 (1973)

T = 303.15 K

x_1	Δn_1^{ex}
0.1	1.751
0.2	1.780
0.4	1.320
0.5	1.140
0.6	0.803
0.7	0.736
0.9	0.464

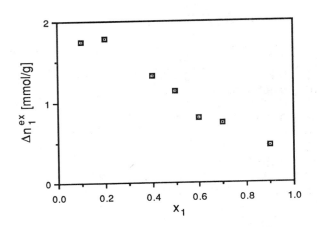

(1) Cyclohexane   (2) Ethanol
Silica Gel, Hi-Sil 215, PPG Industries, Inc., 143 m2/g
Ref.: Matayo, D. R. and J. P. Wightman, J. Colloid Interface Sci.,
      44, 162 (1973)

T = 303.15 K

x_1	Δn_1^{ex}
0.1	1.004
0.2	1.144
0.3	1.005
0.4	0.997
0.5	0.789
0.6	0.599
0.7	0.259
0.8	0.362
0.9	0.162

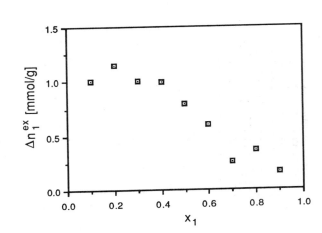

(1) Cyclohexane   (2) Heptane-n
Silica gel, PA 400, Davison Chemical Co., 8-12 mesh, 740 m2/g
Ref.: Sircar, S. and A. L. Myers, J. Phys. Chem., 74, 2828 (1970)

T = 303.15 K

x_1	Δn_1^{ex}
0.041	0.036
0.055	0.000
0.157	0.014
0.190	0.054
0.025	0.000
0.349	-0.021
0.484	0.021
0.513	0.000
0.633	0.020
0.780	0.020
0.831	0.000
0.862	0.000
0.911	0.000

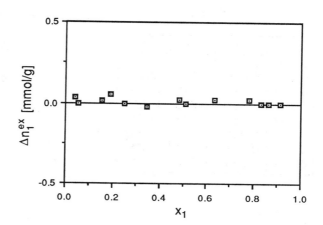

(1) Cyclohexane   (2) Nitrobenzene
Alumina, 65.2 m2/g
Ref.: Suri, S. K. and V. Ramakrishna, Acta Chim. Acad. Sci. Hung.,
       63, 301 (1970)

T = 303.15 K

x_1	Δn_1^{ex}
0.1667	0.5663
0.3212	0.2802
0.4890	-0.4341
0.5304	-0.5860
0.6616	-1.0570
0.8485	-1.0730

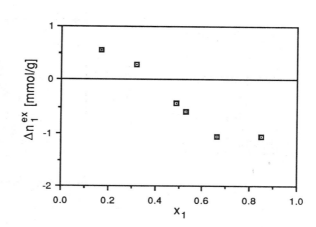

(1) 1,2-Dichloroethane   2) Benzene
Silica gel, PA 400, Davison Chemical Co., 8-12 mesh, 740 m2/g
Ref.: Sircar, S. and A. L. Myers, J. Phys. Chem., 74, 2828 (1970)

T = 303.15 K

x_1	Δn_1^{ex}	x_1	Δn_1^{ex}
0.046	0.039	0.614	0.246
0.069	0.095	0.689	0.214
0.082	0.128	0.810	0.136
0.128	0.167	0.836	0.120
0.162	0.156	0.847	0.071
0.273	0.241	0.871	0.080
0.317	0.268	0.934	0.025
0.524	0.233	0.963	0.011

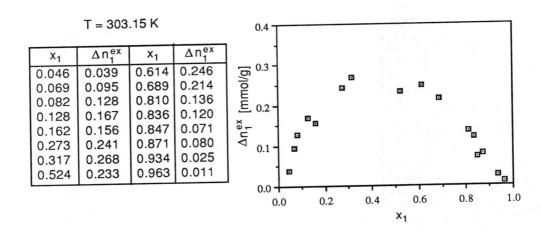

(1) Dioxane   (2) Nitrobenzene
Alumina, 65.2 sq. m./g
Ref.: Suri, S. K. and V. Ramakrishna, Acta Chim. Acad. Sci. Hung.,
      63, 301 (1970)

T = 303.15 K

x_1	Δn_1^{ex}
0.1966	0.1873
0.3790	0.2184
0.5756	0.1615
0.7069	0.1180
0.8602	0.0540

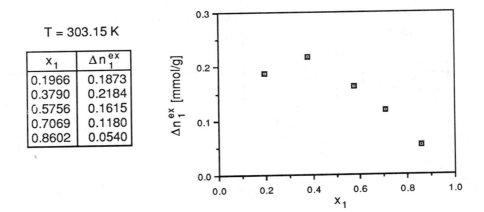

(1) Dioxane   (2) Nitrobenzene
Silica gel, 412 m2/g
Ref.: Suri, S. K. and V. Ramakrishna, Acta Chim. Acad. Sci. Hung.,
    63, 301 (1970)

T = 293.15 K

x_1	Δn_1^{ex}
0.1126	0.9893
0.2181	1.0139
0.3316	0.9464
0.4313	0.8501
0.5322	0.7588
0.6378	0.6105
0.7419	0.4398
0.8210	0.3235
0.9098	0.1707

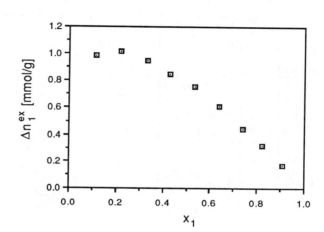

(1) Dioxane   (2) Nitromethane
Silica gel, 412 m2/g
Ref.: Suri, S. K. and V. Ramakrishna, Acta Chim. Acad. Sci. Hung.,
    63, 301 (1970)

T = 293.15 K

x_1	Δn_1^{ex}
0.0668	0.6356
0.1333	0.8464
0.2104	0.7622
0.2834	0.6605
0.3743	0.5302
0.5942	0.2425
0.7007	0.0716
0.8435	-0.0548

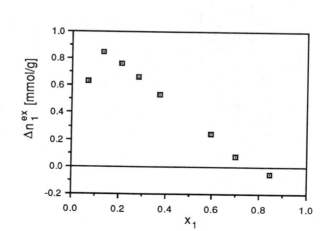

(1) Ethanol   (2) Benzene
Titania gel
Ref.: Kipling, J. J. and D. B. Peakall, J. Chem. Soc., 4054 (1957)

T=293.15K   T=333.15K

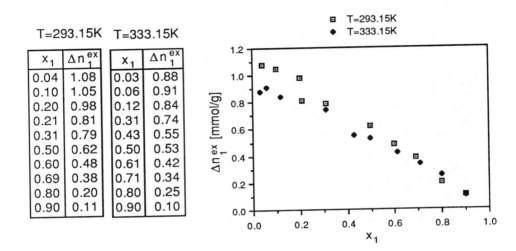

x_1	Δn_1^{ex}	x_1	Δn_1^{ex}
0.04	1.08	0.03	0.88
0.10	1.05	0.06	0.91
0.20	0.98	0.12	0.84
0.21	0.81	0.31	0.74
0.31	0.79	0.43	0.55
0.50	0.62	0.50	0.53
0.60	0.48	0.61	0.42
0.69	0.38	0.71	0.34
0.80	0.20	0.80	0.25
0.90	0.11	0.90	0.10

(1) Ethyl acetate   (2) Cyclohexane
Activated carbon, BPL, Pittsburgh Activated Carbon Co.,4x10, 1100 m2/g
Ref.: Minka, C. and A. L. Myers, AIChE J., 19, 453 (1973)

T= 303.15 K

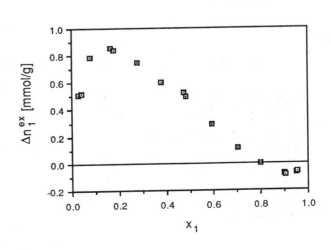

x_1	Δn_1^{ex}
0.028	0.510
0.039	0.520
0.079	0.787
0.163	0.852
0.179	0.840
0.278	0.747
0.378	0.599
0.477	0.525
0.485	0.499
0.593	0.288
0.702	0.116
0.801	0.000
0.898	-0.075
0.904	-0.087
0.951	-0.068
0.953	-0.065

(1) n-Heptane   (2) Ethanol
Graphitized Carbon Black, Graphon, Cabot Corp., 86 m2/g
Ref.: Brown, C. E., D. H. Everett and C. J. Morgan, Trans. Faraday Soc.,
71, 883(1975)

x_1	Δn_1^{ex} [mmol/g]								
	T [K]								
	282	288	293	298	303	308	313	323	333
0.049	0.112	0.105	0.098	0.091	0.085	0.079	0.074	0.065	0.056
0.090	0.201	0.190	0.178	0.167	0.158	0.149	0.141	0.123	0.108
0.171	0.316	0.301	0.287	0.273	0.256	0.244	0.234	0.211	0.191
0.243	0.370	0.352	0.334	0.319	0.306		0.280	0.254	0.232
0.286	0.389	0.367	0.350	0.334	0.319		0.290	0.261	0.236
0.346	0.401	0.381	0.360	0.342	0.323	0.307	0.294	0.263	0.235
0.409	0.393	0.371	0.350	0.330	0.312		0.279	0.245	0.219
0.416	0.390			0.329					0.222
0.491	0.353	0.332	0.309	0.289	0.277	0.261	0.245	0.217	0.192
0.553	0.315	0.294	0.276	0.261	0.248	0.237	0.223	0.200	0.177
0.635	0.252	0.233	0.215	0.200	0.191	0.180	0.172	0.150	0.134
0.698	0.163	0.153	0.144	0.134	0.130		0.112	0.103	0.092
0.745	0.121	0.114	0.107	0.102	0.096	0.091	0.085	0.075	0.067
0.824	0.067			0.059	0.057		0.046		0.041
0.952	0.007			0.006	0.005		0.002		0.000

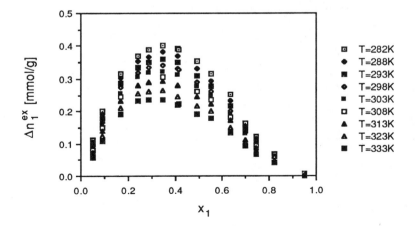

(1) Methyl acetate   (2) Benzene
Silica gel
Ref.: Kipling, J. J. and D. B. Peakall, J. Chem. Soc., 4054 (1957)

T=293.15 K   T=333.15K

x_1	Δn_1^{ex}	x_1	Δn_1^{ex}
0.03	1.27	0.04	1.08
0.07	1.42	0.07	1.22
0.11	1.48	0.12	1.34
0.17	1.50	0.17	1.41
0.22	1.49	0.22	1.43
0.27	1.48	0.28	1.43
0.32	1.48	0.32	1.38
0.38	1.45	0.38	1.34
0.42	1.37	0.42	1.31
0.46	1.35	0.46	1.25
0.53	1.19	0.53	1.03
0.57	1.17	0.62	0.86
0.62	0.99	0.66	0.88
0.66	0.88	0.75	0.53
0.75	0.66	0.78	0.48
0.78	0.58	0.83	0.38
0.83	0.47	0.88	0.27
0.88	0.37	0.95	0.11
0.95	0.17		

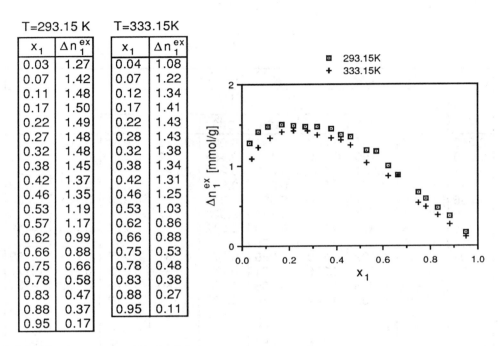

(1) Nitrobenzene   (2) Cyclohexane
Silica gel, 412 m2/g
Ref.: Suri, S. K. and V. Ramakrishna, Acta Chim. Acad. Sci. Hung.,
        63, 301 (1970)

T = 293.15 K

x_1	Δn_1^{ex}
0.0776	2.2460
0.1647	2.8770
0.2690	2.6860
0.3903	2.4950
0.4887	1.8431
0.6137	1.1290
0.7113	0.9824
0.8120	0.5464
0.8867	0.3874

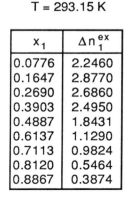
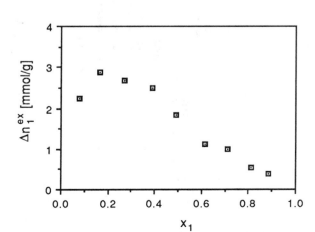

(1) Nitromethane   (2) Benzene
Alumina, 65.2 m2/g
Ref.: Suri, S. K. and V. Ramakrishna, Acta Chim. Acad. Sci. Hung.,
    63, 301 (1970)

T = 303.15 K

x_1	Δn_1^{ex}
0.2642	0.8167
0.4697	0.7161
0.5947	0.5780
0.6085	0.5575
0.7657	0.3505
0.8916	0.1684

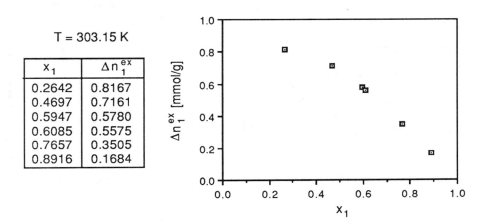

(1) Nitromethane   (2) Benzene
Silica gel, 412 m2/g
Ref.: Suri, S. K. and V. Ramakrishna, Acta Chim. Acad. Sci. Hung.,
    63, 301 (1970)

T = 293.15 K

x_1	Δn_1^{ex}
0.2841	1.6184
0.3829	1.7113
0.5013	1.6240
0.6117	1.3066
0.7018	1.0285
0.7874	0.7577
0.8595	0.4712
0.9332	0.1936

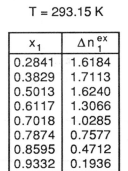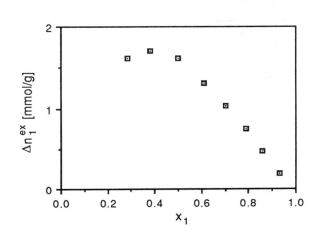

(1) Nitromethane   (2) Dioxane
Alumina, 65.2 m2/g
Ref.: Suri, S. K. and V. Ramakrishna, Acta Chim. Acad. Sci. Hung.,
    63, 301 (1970)

T = 303.15 K

x_1	Δn_1^{ex}
0.2463	0.0542
0.4361	0.0697
0.6114	0.0678
0.7903	0.0399
0.8860	0.0237

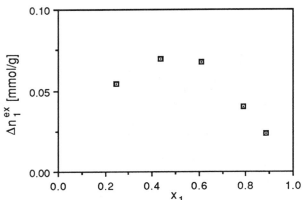

(1) Toluene   (2) Bromobenzene
Silica gel, 420 m2/g
Ref.: Peisen, L. and T. Gu, Scientia Sinica, 22, 1384 (1979)

T = 293.15 K

x_1	Δn_1^{ex}
0.0273	0.0539
0.0688	0.1004
0.2255	0.1947
0.2608	0.2153
0.3439	0.2443
0.4146	0.2333
0.5953	0.1744
0.6620	0.1461
0.8081	0.0954
0.9153	0.0227
0.9258	0.0141
0.9518	0.0124
0.9705	-0.0032
0.9737	-0.0050

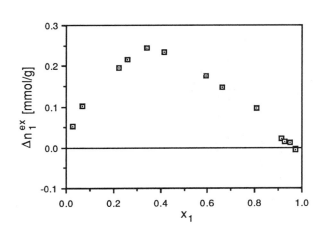

(1) Toluene　(2) Chlorobenzene
Silica gel, 420 m2/g
Ref.: Peisen, L. and T. Gu, Scientia Sinica, 22, 1384 (1979)

T = 293.15 K

x_1	Δn_1^{ex}
0.0631	0.0923
0.1917	0.1587
0.2890	0.1856
0.3222	0.1839
0.3872	0.1752
0.5878	0.1497
0.6896	0.1035
0.7922	0.0512
0.8963	0.0109
0.9592	-0.0172
0.9799	-0.0216

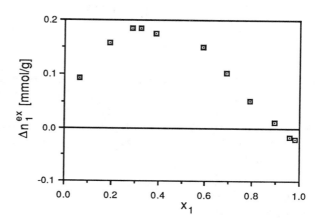

APPENDIX A

INDEX OF SINGLE-GAS ISOTHERMS
IN CHAPTER 2

Adsorption isotherms tabulated in Chapter 2 are indexed alphabetically by adsorbates. For each entry, the adsorbent is on the second line, the temperatures studied are on the third line, and the abbreviated reference is on the fourth line. The abbreviated reference is the first two letters of the name of the first author, the page number, and the year of publication. Complete references are in the bibliography, Appendix G.

APPENDIX B

INDEX OF BINARY GAS ISOTHERMS
IN CHAPTER 3

Binary adsorption isotherms tabulated in Chapter 3 are indexed alphabetically by the more strongly adsorbed gas. For each entry, the gas mixture is on the first line, the adsorbent is on the second line, the experimental conditions are on the third line, and the abbreviated reference is on the fourth line. The abbreviated reference is the first two letters of the name of the first author, the page number, and the year of publication. Complete references are in the bibliography, Appendix G.

APPENDIX C

INDEX OF BINARY LIQUID ISOTHERMS
IN CHAPTER 4

Binary liquid isotherms tabulated in Chapter 4 are indexed alphabetically by the more strongly adsorbed component. For each entry, the components are on the first line, the adsorbent is on the second line, the temperature is on the third line, and the abbreviated reference is on the fourth line. The abbreviated reference is the first two letters of the name of the first author, the page number, and the year of publication. Complete references are in the bibliography, Appendix G.

APPENDIX D

COMPILATION OF REFERENCES FOR
SINGLE-GAS ADSORPTION ISOTHERMS

Deitz [De(44)] published an annotative bibliographical survey of the literature on solid adsorbents for the period 1900–1942.

Explanation of Table

Adsorbate   Names of adsorbate gases, listed alphabetically.

Adsorbent   Name of adsorbent.

AC	Activated carbon
AL	Alumina
CB	Carbon black
CMS	Carbon molecular sieve
GCB	Graphitized carbon black
SG	Silica gel
SI	Silica
Z	Zeolite

A, m²/g   Surface area of adsorbent, m^2/g.

Temp., K   Temperature or range of temperatures studied, K.

     B.P.   Normal boiling point.

Max. P, kPa   Maximum pressure of adsorption isotherm, kPa.

     P_{sat} Saturated vapor pressure of liquid adsorbate

Method   Experimental technique.

     D   Dynamic
     G   Gravimetric
     V   Volumetric

Data   Method of presenting experimental data.

     G   Graphical
     T   Tabular

Reference   Abbreviation of reference. For example, Ka928(87) refers to B. K. Kaul, *Ind. Eng. Chem. Res.* **26**, 928 (1987).

Adsorbate	Adsorbent	A, m2/g	Temp., K	Max. P, kPa	Method	Data	Reference
Acetone	AC (G-2X, Takeda)		303	14	V	G	Ko398(85)
Acetone	AC (HGI-780)	724	303	0.8	G	G	Ok209(78)
Acetone	AC (Shirasagi S)	972	303	0.8	G	G	Ok209(78)
Acetone	CMS (5A, Takeda)		303	14	V	G	Ko398(85)
Acetonitrile	SI (Aerosilogel)	73, 91		0.7 Psat		G	Cu58(74)
Acetylene	AC (4LXC, Columbia)	1130	298	4	D	G	Le1(84)
Acetylene	AC (BPL, Pittsburgh)	1050-1150	303-323	103	G	T	La48(69)
Acetylene	AC (Columbia L)	1152	311-394	98	V	T	Ra1315(50)
Acetylene	AC (EY-51-C, Pittsburgh)	805	298	101	V	T	Le1157(50)
Acetylene	AC (Nuxit-AL)		293-363	118	V	T	Sz37(63)
Acetylene	AC (Pittsburgh coke)		298	101	V	G	Le1326(50)
Acetylene	SG (Davison)	751	298	91	V	T	Le1157(50)
Acetylene	SG (Davison)		298	101	V	G	Le1326(50)
Ammonia	AC (Nitrogen-containing)	823	303	80	G	G	Bo286(87)
Ammonia	AC (Wako Pure Chemicals)	957	303	80	G	G	Bo286(87)
Ammonia	GCB (Vulcan III)	71	163-211	Psat	G	T	Bo1535(79)
Ammonia	SG (Davison 59)	340	298-333		D	T	Ku330(85)
Ammonia	Z (CaY)		333-483	53	G	G	Sh1(85)
Ammonia	Z (HY)		333-483	53	G	G	Sh1(85)
Ammonia	Z (KY)		341-382	0.7	G	G	Mo2525(85)
Ammonia	Z (La-mordenite)		303-473	80	V	G	Co1969(79)
Ammonia	Z (LaCaY)		333-483	53	G	G	Sh1(85)
Ammonia	Z (LaHY)		333-483	53	G	G	Sh1(85)
Ammonia	Z (LaY)		333-483	53	G	G	Sh1(85)
Ammonia	Z (NaX)		432-463	0.7	G	G	Mo2525(85)
Ammonia	Z (NaY)		341-382	0.7	G	G	Mo2525(85)
Ammonia	Z (NaY, Linde)		333-483	53	G	G	Sh1(85)
Ammonia	Z (Ru-mordenite)		303-473	80	V	G	Co1969(79)

Adsorbate	Adsorbent	A, m2/g	Temp., K	Max. P, kPa	Method	Data	Reference
Argon	AC (Nitrogen-containing)	823	77	33	V	G	Bo286(87)
Argon	AC (Wako Pure Chemicals)	957	77	33	V	G	Bo286(87)
Argon	GCB (Graphon)		77	0.9 Psat	V	G	Ba649(68)
Argon	GCB (P-33, Cabot)	13	78, 90	2	V	G	Ro319(55)
Argon	GCB (Sterling FTG-D5, Cabot)	13	298	93	G	G	Fr439(72)
Argon	SI (G-200-0, Unilever)		77	0.8 Psat	V	G	Pe439(84)
Argon	SI (HiSil 233)		77	0.9 Psat	V	G	Ba649(68)
Argon	Z (13X, Linde)		273-348	9693		T	Wa399(81)
Argon	Z (4A, Linde)		173-195	101	V	G	Ea419(75)
Argon	Z (5A)		203-297	450	V	G	Mi194(87)
Argon	Z (5A, Linde)		195-348	11039		T	Wa399(81)
Argon	Z (5A, Sillipore NK20 Ceca)		274-348	444	V	T	Ve66(85)
Argon	AC (208C Sutcliffe Speakman)	1038	423		D	G	Th240(71)
Benzene	AC (Cloth CDE Porton Down)	1350	303	Psat	G	G	Ha301(86)
Benzene	AC (Filtrasorb 400, Calgon)	1100	303	15	G	G	My97(82)
Benzene	AC (G-2X, Takeda)		303	14	V	G	Ko398(85)
Benzene	AC (Shirasagi S)	972	303	0.8	G	G	Ok209(78)
Benzene	AL (Alpha)	3	273, 259	Psat	G	T	Hs490(74)
Benzene	CMS (5A)	650	279, 303	12	G	G	Na310(74)
Benzene	GCB (Graphon)	89	293	0.8 Psat	V	G	Pe473(75)
Benzene	GCB (Sterling FT-FF Cabot)	12	297-383	3		G	De1685(85)
Benzene	GCB (Sterling MT-D5, Cabot)	10	303	15	G	G	My97(82)
Benzene	GCB (Sterling MTFF-D-7)	10	273-298	3	G	G	Sl347(75)
Benzene	SG	320	299	Psat	G	G	Ba80(52)
Benzene	SG	39-399	308	0.4 Psat	V	G	Mi260(75)
Benzene	SG (Davison 3)	793	308	Psat		G	Mi505(70)
Benzene	SG (Davison 8)	618	339	64	G	G	Ci710(53)
Benzene	SG (Davison)	823	343-403	2	D	G	Sh100(68)

Adsorbate	Adsorbent	A, m2/g	Temp., K	Max. P, kPa	Method	Data	Reference
Benzene	SG (Davison)		303	14	G	T	Si159(73)
Benzene	SG (Mallinckrodt)	637	339	64	G	G	Ci710(53)
Benzene	SG (PA 400, Davison)	660	303	15	G	G	My3412(72)
Benzene	SI (Aerosilogel)	73, 91		0.7 Psat	G	G	Cu58(74)
Benzene	SI (Cab-O-Sil)		303	0.3 Psat	V	G	Pe313(74)
Benzene	SI (Cab-O-Sil, Cabot)	214	293-313	21	V	G	Pe252(76)
Benzene	SI (G-200-0, Unilever)		298	0.8 Psat	G	G	Pe439(84)
Benzene	Z (10X, Linde)		437-517	5	G	G	Ru654(81)
Benzene	Z (H-ZSM-5)		373-673	8	D	G	Ch382(87)
Benzene	Z (Na-mordenite, Norton)		496-533	0.1	G	G	Eb2404(63)
Benzene	Z (NaHX)		445	2	D	G	Ba2147(73)
Benzene	Z (NaHY)		445	2	D	G	Ba2147(73)
1,3-Butadiene	SI (Aerosil 200, Degussa)	186	298-318	40	V	G	Co2127(84)
Butane	AC (Nitrogen-containing)	823	B.P.	93	V	G	Bo286(87)
Butane	AC (Nuxit-Al)		293-363	130	V	T	Sz37(63)
Butane	AC (Wako Pure Chemicals)	957	B.P.	80	V	G	Bo286(87)
Butane	Z (5A, Davison)		298-348	20	G	G	Yo308(71)
iso-Butane	AC (Nitrogen-containing)	823	B.P.	87	V	G	Bo286(87)
iso-Butane	AC (Pittsburgh coke)		298	101	V	G	Le1326(50)
iso-Butane	AC (Wako Pure Chemicals)	957	B.P.	87	V	G	Bo286(87)
iso-Butane	AL	344	283, 303	100	V	G	Ta136(81)
iso-Butane	Glass (Vycor)	117	273-323	100	V	G	Ok262(81)
iso-Butane	SG (Davison 3)	807-830	311	186	V	G	Ma1097(66)
iso-Butane	SG (Davison)		298	101	V	G	Le1326(50)
iso-Butane	Z (13X, Linde)	525	298-373	138	V	T	Hy196(82)
n-Butane	AC (BPL, Pittsburgh)	1004	250-349	101	G	G	Gr403(62)
n-Butane	AC (Columbia G)	1157	313-343	755	V	T	Pa363(68)
n-Butane	AC (Columbia L)	1152	311-478	99	V	T	Ra1315(50)

Adsorbate	Adsorbent	A, m2/g	Temp., K	Max. P, kPa	Method	Data	Reference
n-Butane	CB (H, Acheson)	542	303	12	V,G	G	Ha2265(78)
n-Butane	CMS (5A)	650	279-324	86	G	T	Na310(74)
n-Butane	GCB (Sterling FT, Cabot)	12	195	0.8 Psat	V	G	Da353(69)
n-Butane	SG (Davison 3)	807-830	311	186	V	G	Ma1097(66)
n-Butane	SG (PA 400, Davison)	670	283-303	267	G	G	Al658(81)
n-Butane	SI (Aerosil 200, Degussa)	186	298-318	50	V	G	Co2127(84)
n-Butane	Z (5A)		196-313	53	G	G	Ve1071(75)
n-Butane	Z (5A, Linde)		308	81	V,G	T	Gl73(69)
n-Butane	Z (5A, Linde)		358-498	50	G	G	Ru1145(71)
n-Butane	Z (5A, Linde)		273-498	40	G	G	Lo331(72)
n-Butane	Z (5A, Linde)		273-498	47	G	G	Ru696(72)
n-Butane	Z (Na-mordenite, Norton)		298-413	27	G	G	Sa731(67)
1-Butanol	AL (Alpha)	3, 104	293	Psat	G	G	Ba491(66)
1-Butanol	Glass (Vycor, Corning)	126	293	Psat	G	G	Jo415(68)
1-Butene	AC (Pittsburgh coke)		298	101	V	G	Le1326(50)
1-Butene	SG (Davison)		298	101	V	G	Le1326(50)
iso-Butene	SI (Aerosil 200, Degussa)	110-166	291-315	80	V	G	Co2127(84)
Butylene	Z (5A, Davison)		298-348	20	G	G	Yo308(71)
Carbon dioxide	AC (BPL, Calgon)		303, 342	202	V	G	Si919(86)
Carbon dioxide	AC (BPL, Pittsburgh)	1050-1150	303-323	102	G	T	Me259(67)
Carbon dioxide	AC (BPL, Pittsburgh)	988	213-301	3840	V	T	Re336(80)
Carbon dioxide	AC (BPL, Pittsburgh)	1053	298	700	V	G	Wi14(83)
Carbon dioxide	AC (Columbia L)	1152	311-394	99	V	T	Ra1315(50)
Carbon dioxide	AC (Fiber, KF-1500, Toyobo)	1440	273-323	97	G	T	Ku588(84)
Carbon dioxide	AC (Metal-impregnated BPL)	800-850	303-323	94	G	T	Me259(67)
Carbon dioxide	AC (Nuxit-AL)		293-363	133	V	T	Sz37(63)
Carbon dioxide	AC (Nuxit-AL)		293-363	744	V	T	Sz53(63)
Carbon dioxide	AC (PCB, Calgon)		296-480	3674	V	T	Ri1679(87)

Adsorbate	Adsorbent	A, m2/g	Temp., K	Max. P, kPa	Method	Data	Reference
Carbon dioxide	SG (E, Unilever)	150, 572	195	0.5 Psat	V	G	Le227(77)
Carbon dioxide	SG (J, Unilever)	298, 327	195	0.5 Psat	V	G	Le227(77)
Carbon dioxide	SI (TK 800, Degussa)	143, 163	195	0.5 Psat	V	G	Le227(77)
Carbon dioxide	Z (13X, Linde)	525	298, 323	138	V	T	Hy196(82)
Carbon dioxide	Z (4A, Linde)		203-263	101	V	G	Ea419(75)
Carbon dioxide	Z (5A, Linde)		273-323	4	G	G	Fu269(68)
Carbon dioxide	Z (5A, Linde)		308	79	V,G	T	Gl73(69)
Carbon dioxide	Z (5A, Linde)		298-348	11191		T	Wa399(81)
Carbon dioxide	Z (Ca-mordenite)		243-273	27	V	G	Va1371(81)
Carbon dioxide	Z (Cr-mordenite)		273-423	70	V	G	Co1969(79)
Carbon dioxide	Z (CrA)		303-423	5	V	G	Co1809(75)
Carbon dioxide	Z (CrX)		303-423	5	V	G	Co1809(75)
Carbon dioxide	Z (CrY)		303-423	5	V	G	Co1809(75)
Carbon dioxide	Z (Fe-mordenite)		273-423	70	V	G	Co1969(79)
Carbon dioxide	Z (FeY)		303-423	5	V	G	Co1809(75)
Carbon dioxide	Z (H-mordenite)		303, 423	60	V	G	Co1969(79)
Carbon dioxide	Z (H-mordenite)		273	27	V	G	Va1371(81)
Carbon dioxide	Z (H-mordenite, Norton)		283-323	293	V	T	Ta1263(86)
Carbon dioxide	Z (La-mordenite)		273-423	70	V	G	Co1969(79)
Carbon dioxide	Z (Na-mordenite)		303-423	60	V	G	Co1969(79)
Carbon dioxide	Z (Na-mordenite, Norton)		273	27	V	G	Va1371(81)
Carbon dioxide	Z (Ru-mordenite)		273-423	70	V	G	Co1969(79)
Carbon dioxide	Z (RuA)		273-423	70	V	G	Co1969(79)
Carbon dioxide	Z (RuL)		273-423	70	V	G	Co1969(79)
Carbon dioxide	Z (RuX)		273-423	70	V	G	Co1969(79)
Carbon dioxide	Z (RuY)		273-423	70	V	G	Co1969(79)
Carbon dioxide	Z (TlY)		303-423	5	V	G	Co1809(75)
Carbon dioxide	Z (YA)		303-423	5	V	G	Co1809(75)

Adsorbate	Adsorbent	A, m2/g	Temp., K	Max. P, kPa	Method	Data	Reference
Carbon dioxide	Z (YX)		303-423	5	V	G	Co1809(75)
Carbon dioxide	Z (YY)		303-423	5	V	G	Co1809(75)
Carbon disulfide	AC (BPL, Pittsburgh)	1040	252-306	1.5	G	G	Gr403(62)
Carbon monoxide	AC (BPL, Pittsburgh)	1053	298	700	V	G	Wi14(83)
Carbon monoxide	AC (Columbia L)	1152	311-478	1786	V	T	Ra1315(50)
Carbon monoxide	AC (PCB, Calgon)		296-473	5812	V	T	Ri1679(87)
Carbon monoxide	CMS (Cecalite, Ceca)		298	1500	V	G	Ve1417(81)
Carbon monoxide	Z (10A, Linde)	676	144	150	V	T	Da515(69)
Carbon monoxide	Z (10X, Linde)	676	144	293	V	G	Do30(75)
Carbon monoxide	Z (10X, Linde)	672	172-273	212	V	T	No112(81)
Carbon monoxide	Z (4A, Linde)		193-213	101	V	G	Ea419(75)
Carbon monoxide	Z (4A, Linde)		305-366	101	D	G	Ru27(80)
Carbon monoxide	Z (5A)		298	1500	V	G	Ve1417(81)
Carbon monoxide	Z (5A, Linde)		298	400	D	G	Da2129(80)
Carbon monoxide	Z (5A, Linde)	572	144	110	V	T	Da515(69)
Carbon monoxide	Z (5A, Linde)		77-273	133	V	G	Fl205(74)
Carbon monoxide	Z (5A, Linde)		298-348	9231		T	Wa399(81)
Carbon monoxide	Z (BaY)		273-302	0.7	V	G	Eg22(73)
Carbon monoxide	Z (Ca-mordenite)		243-273	27	V	G	Va1371(81)
Carbon monoxide	Z (CaY)		273	1	V	G	Eg22(73)
Carbon monoxide	Z (CeY)		273	1	V	G	Eg22(73)
Carbon monoxide	Z (CuY)		273	1	V	G	Eg22(73)
Carbon monoxide	Z (H-mordenite)		273	27	V	G	Va1371(81)
Carbon monoxide	Z (MnY)		273-315	2	V	G	Eg22(73)
Carbon monoxide	Z (Na-mordenite, Norton)		273	27	V	G	Va1371(81)
Carbon monoxide	Z (NaY)		273	1	V	G	Eg22(73)
Carbon monoxide	Z (NiY)		273-315	2	V	G	Eg22(73)
Carbon monoxide	Z (NK20B, Ceca)		298	1500	V	G	Ve1417(81)

Adsorbate	Adsorbent	A, m2/g	Temp., K	Max. P, kPa	Method	Data	Reference
Carbon monoxide	Z (UO2Y)		273	0.7	V	G	Eg22(73)
Carbon monoxide	Z (ZnY)		273-325	1	V	G	Eg22(73)
Carbon tetrachloride	SG	320, 355	299, 313	Psat	G	G	Ba80(52)
Carbon tetrachloride	SG	39-399	308	0.4 Psat	V	G	Mi260(75)
Carbon tetrachloride	SG (Davison 3)		283	Psat	G	G	Ma46(85)
Carbon tetrachloride	SG (Davison 3)	793	308	Psat		G	Mi505(70)
Carbon tetrachloride	SI (Aerosilogel)	73, 91		0.7 Psat		G	Cu58(74)
Carbon tetrafluoride	Z (5A, Linde)		300	101	D	G	Ru27(80)
Carbonyl sulfide	AC (BPL, Pittsburgh)	1040	252-373	7	G	G	Gr403(62)
Cyclohexane	AC (Filtrasorb 400, Calgon)	1100	303	15	G	G	My97(82)
Cyclohexane	AL (Alpha)	3	259-294	Psat	G	G	Hs490(74)
Cyclohexane	CB (Spheron 9)	120	349-365	12	D	G	Do832(72)
Cyclohexane	CMS (5A)	650	303	12	G	T	Na310(74)
Cyclohexane	GCB (Graphon)	89	293	0.8 Psat	V	G	Pe473(75)
Cyclohexane	GCB (Sterling FT-FF, Cabot)	12	296, 365	0.4	G	G	De1685(85)
Cyclohexane	GCB (Sterling MT-D5, Cabot)	10	303	15	G	G	My97(82)
Cyclohexane	SG	39-399	308	0.4 Psat	V	G	Mi260(75)
Cyclohexane	SG (Davison 3)	793	308	Psat		G	Mi505(70)
Cyclohexane	SG (Davison)		303	15	G	T	Si159(73)
Cyclohexane	SG (PA 400, Davison)	660	303	16	G	G	My3412(72)
Cyclohexane	SI (Aerosilogel)	73, 91		0.7 Psat		G	Cu58(74)
Cyclohexane	SI (Cab-O-Sil)		303	0.3 Psat	V	G	Pe313(74)
Cyclohexane	SI (Cab-O-Sil, Cabot)	214	293-313	23	V	G	Pe252(76)
Cyclohexane	Z (10X, Linde)		409-488	9	G	G	Ru654(81)
Cyclohexane	Z (13X, Linde)		409, 458	13	V	G	Ru27(85)
Cyclohexane	Z (NaHX)		445	2	D	G	Ba2147(73)
Cyclohexane	Z (NaHY)		445	2	D	G	Ba2147(73)
Cyclopropane	Z (5A, Linde)		273-458	34	G	G	De1947(72)

Adsorbate	Adsorbent	A, m2/g	Temp., K	Max. P., kPa	Method	Data	Reference
n-Decane	Z (5A, Linde)	660	523-596	0.2	G	G	Va526(81)
1,2-Dichloroethane	SG (PA 400, Davison)	823	303	14	G	G	My3412(72)
Dimethyl sulfide	AC (Nitrogen-containing)	957	303	40	G	G	Bo286(87)
Dimethyl sulfide	AC (Wako Pure Chemicals)	637	303	40	G	G	Bo286(87)
Dimethylpentane	SG (Mallinckrodt)	618	339	64	G	G	Ci710(53)
2,4-Dimethylpentane	SG (Davison 8)	542	339	64	G	G	Ci710(53)
2,2-Dimethylpropane	CB (H, Acheson)		303	12	V,G	G	Ha2265(78)
Ethane	AC (40, Comp. Esp. C. A.)	700	293	100	V	T	Co5(81)
Ethane	AC (4LXC, Columbia)	1130	298	2	D	G	Le1(84)
Ethane	AC (4LXC, Columbia)	1130	298	2	D	G	Le461(85)
Ethane	AC (BPL, Pittsburgh)	1050-1150	303-323	104	G	T	La48(69)
Ethane	AC (BPL, Pittsburgh)	988	213-301	1711	V	T	Re336(80)
Ethane	AC (Columbia G)		298	101	V	G	Le1326(50)
Ethane	AC (Columbia L)	1152	311-478	1493	V	T	Ra1315(50)
Ethane	AC (Fiber, KF-1500, Toyobo)	1440	273-323	100	G	T	Ku588(84)
Ethane	AC (Kureha beads)	800-1200	311-422	444	V	T	Ka928(87)
Ethane	AC (Nuxit-AL)		293-363	119	V	T	Sz37(63)
Ethane	AC (Nuxit-AL)		293-363	670	V	T	Sz53(63)
Ethane	AC (Pittsburgh coke)		298	101	V	G	Le1326(50)
Ethane	CMS (5A)	650	279-323	90	G	T	Na310(74)
Ethane	CMS (5A, Takeda)		292	13	V	G	Su279(82)
Ethane	CMS (Cecalite, Ceca)		298	1500	V	G	Ve1417(81)
Ethane	SG (Davison 15)	804	298	2700	D	G	Ma197(68)
Ethane	SG (Davison 3)	807-830	311	3447	V	G	Ma1097(66)
Ethane	SG (Davison)		298	2200	V	G	Le1326(50)
Ethane	Z (13X)		323-423	362	V	T	Ka928(87)
Ethane	Z (13X, Linde)		298	200	D	G	Da2129(80)
Ethane	Z (13X, Linde)	525	298, 323	142	V	T	Da248(78)

Adsorbate	Adsorbent	A, m2/g	Temp., K	Max. P, kPa	Method	Data	Reference
Ethane	Z (13X, Linde)	525	273-373	138	V	T	Hy196(82)
Ethane	Z (13X, Linde)	420	323	140	D	G	Hy95(85)
Ethane	Z (4A, Linde)		273-293	101	V	G	Ea419(75)
Ethane	Z (5A)		345	22	G	G	Ru397(75)
Ethane	Z (5A)		196-313	53	G	G	Ve1071(75)
Ethane	Z (5A)		298	1500	V	G	Ve1417(81)
Ethane	Z (5A, Davison)		298	20	G	G	Yo308(71)
Ethane	Z (5A, Linde)		308	80	V,G	T	Gl73(69)
Ethane	Z (5A, Linde)		195-300	67	V	G	Ri1759(76)
Ethane	Z (5A, Linde)		298-348	101	D	G	Ru27(80)
Ethane	Z (5A, Linde)		298-348	5656		T	Wa399(81)
Ethane	Z (5A, Linde)		188-273	40	G	G	Lo331(72)
Ethane	Z (5A, Linde)		230-345	53	G	G	Ru696(72)
Ethane	Z (CaNaX)		323	15	V	G	Am306(78)
Ethane	Z (CaNaX)		368	9	V	G	Be223(72)
Ethane	Z (Erionite)		230-349	13	G	G	Ru397(75)
Ethane	Z (MgNaX)		323	15	V	G	Am306(78)
Ethane	Z (Na-mordenite, Norton)		298-413	27	G	G	Sa731(67)
Ethane	Z (NaX)		323	15	V	G	Am306(78)
Ethane	Z (NaX)		323	6	V	G	Be223(72)
Ethane	Z (NaY)		273-368	107	V	G	Be223(72)
Ethane	Z (NK20B, Ceca)		298	1500	V	G	Ve1417(81)
Ethane	Z (SrNaX)		298-368	107	V	G	Be223(72)
Ethanol	AL (Alpha)	3, 104	293	Psat	G	G	Ba491(66)
Ethanol	GCB (Graphon)	89	293	0.8 Psat	V	G	Pe473(75)
Ethanol	Glass (Vycor, Corning)	126	293	Psat	G	G	Jo415(68)
Ethanol	SG	320, 455	298, 318	Psat	G	G	Ba80(52)
Ethanol	SI (Cab-O-Sil)		303	0.3 Psat	V	G	Pe313(74)

Adsorbate	Adsorbent	A, m2/g	Temp., K	Max. P, kPa	Method	Data	Reference
Ethanol	SI (Cab-O-Sil, Cabot)	214	293-313	17	V	G	Pe252(76)
Ethyl acetate	SI (Aerosilogel)	73, 91		0.4 Psat		G	Cu58(74)
Ethyl formate	SI (Aerosil, Degussa)	200	293-313	0.6 Psat	V	G	Ro541(74)
Ethyl mercaptan	AC (BPL, Pittsburgh)	1040	252-373	6	G	G	Gr403(62)
Ethylene	AC (40, Comp. Esp. C. A.)	700	293	100	V	T	Co5(81)
Ethylene	AC (BPL, Pittsburgh)	988	213-301	1696	V	T	Re336(80)
Ethylene	AC (Columbia G)		298	2200	V	G	Le1326(50)
Ethylene	AC (Columbia L)	1152	311-478	1485	V	T	Ra1315(50)
Ethylene	AC (EY-51-C, Pittsburgh)	805	298	101	V	T	Le1157(50)
Ethylene	AC (Kureha beads)	800-1200	311-422	573	V	T	Ka928(87)
Ethylene	AC (Nuxit-AL)		293-363	125	V	T	Sz37(63)
Ethylene	AC (Nuxit-AL)		293-363	648	V	T	Sz53(63)
Ethylene	AC (Pittsburgh coke)		298	101	V	G	Le1326(50)
Ethylene	AL	344	283, 303	100	V	G	Ta136(81)
Ethylene	CMS (5A)	650	279-323	91	G	T	Na310(74)
Ethylene	CMS (5A)	650	275-323	33	V,G	T	Na317(82)
Ethylene	Glass (Vycor)	117	273-323	100	V	G	Ok262(81)
Ethylene	SG (Davison)	751	298	102	V	T	Le1157(50)
Ethylene	SG (Davison)	751	273-313	102	V	T	Le1160(50)
Ethylene	SG (Davison)		273-499	2200	V	G	Le1326(50)
Ethylene	Z (13X)		323-423	545	V	T	Ka928(87)
Ethylene	Z (13X, Linde)	525	298	213	D	G	Da2129(80)
Ethylene	Z (13X, Linde)	525	298, 323	157	V	T	Da248(78)
Ethylene	Z (13X, Linde)	525	298-373	138	V	T	Hy196(82)
Ethylene	Z (13X, Linde)	420	323	140	D	G	Hy95(85)
Ethylene	Z (4A)		306-394	21	G	G	Ru397(75)
Ethylene	Z (5A, Davison)		298-348	20	G	G	Yo308(71)
Ethylene	Z (5A, Linde)		230-358	34	G	G	De1947(72)

Adsorbate	Adsorbent	A, m2/g	Temp., K	Max. P, kPa	Method	Data	Reference
Ethylene	Z (5A, Linde)		398	101	D	G	Ru27(80)
Ethylene	Z (5A, Linde)		298-348	7979		T	Wa399(81)
Ethylene	Z (CaNaX)		393	2	V	G	Am306(78)
Ethylene	Z (CaNaX)		443	5	V	G	Be223(72)
Ethylene	Z (Erionite)		273, 358	10	G	G	Ru397(75)
Ethylene	Z (MgNaX)		368	2	V	G	Am306(78)
Ethylene	Z (NaX)		368, 393	2	V	G	Am306(78)
Ethylene	Z (NaX)		323	4	V	G	Be223(72)
Ethylene	Z (NaY)		323	5	V	G	Be223(72)
Ethylene	Z (SrNaX)		443	5	V	G	Be223(72)
Helium	AC (BPL, Pittsburgh)	1123	100, 150	11244	V	T	Fe457(72)
Helium	AC (IG-1, Barneby-Cheney)	904	76	9099	V	G	Ki949(70)
Helium	AC (KH-1, Barneby-Cheney)		77-303	101	G	T	Sp78(69)
Helium	AC (SXC, Pittsburgh)		77-294	101	G	T	Sp78(69)
Helium	Z (13X, Linde)		77-296	101	G	T	Sp78(69)
Helium	Z (4A, Linde)		77-195	101	G	T	Sp78(69)
Helium	Z (5A, Linde)		77-295	101	G	T	Sp78(69)
n-Heptane	SG (Davison)		303	7	G	T	Si159(73)
n-Heptane	SG (KSK)	380	348	Psat	V,G	G	Be1817(61)
n-Heptane	SG (PA 400, Davison)	660	303	7	G	G	My3412(72)
n-Heptane	SI (Aerosil, Degussa)	200	302-328	6	V	G	Al513(85)
n-Heptane	Z (13X)		409-491	4	G	G	Ru654(81)
n-Heptane	Z (13X, Linde)		409, 458	13	V	G	Ru27(85)
n-Heptane	Z (5A)		423, 673	5	V	G	Ha252(73)
1-Heptene	SI (Aerosil, Degussa)	200	302-328	6	V	G	Al513(85)
1,5-Hexadiene	SI (Aerosil, Degussa)	200	288-314	22	V	G	Al420(84)
n-Hexane	AC (BPL, Pittsburgh)	1004	250-350	101	G	G	Gr403(62)
n-Hexane	CMS (5A, Takeda)		303	14	V	G	Ko398(85)

Adsorbate	Adsorbent	A, m2/g	Temp., K	Max. P, kPa	Method	Data	Reference
n-Hexane	SG	320, 355	299, 313	Psat	G	G	Ba80(52)
n-Hexane	SG (Davison)	823	343-403	2	D	G	Sh100(68)
n-Hexane	SG (PA 400, Davison)	670	283-303	24	G	G	Al658(81)
n-Hexane	SI (Aerosil, Degussa)	200	288-314	18	V	G	Al420(84)
n-Hexane	SI (Aerosilogel)	73, 91		0.7 Psat		G	Cu58(74)
n-Hexane	Z (NaX)		293	12	G	G	He559(84)
1-Hexene	SI (Aerosil, Degussa)	200	288-314	20	V	G	Al420(84)
1-Hexene	Z (NaX)		293	13	G	G	He559(84)
Hydrogen	AC (BPL, Pittsburgh)	1053	298	700	V	G	Wi14(83)
Hydrogen	AC (Columbia L)	1152	311-366	1772	V	T	Ra1315(50)
Hydrogen	AC (PCB, Calgon)		296-480	6667	V	T	Ri1679(87)
Hydrogen	CB (DAG 621, Acheson)	70	296	16	V	G	Sa3121(82)
Hydrogen	Z (5A, Linde)		76	7280	V	G	Ki58(66)
Hydrogen sulfide	AC (BPL, Pittsburgh)	1040	252-373	101	G	G	Gr403(62)
Hydrogen sulfide	AC (Nitrogen-containing)	823	303	80	G	G	Bo286(87)
Hydrogen sulfide	AC (PCB, Calgon)		296-480	1296	V	T	Ri1679(87)
Hydrogen sulfide	AC (Wako Pure Chemicals)	957	303	80	G	G	Bo286(87)
Hydrogen sulfide	Z (H-mordenite, Norton)		283-368	102	V	T	Ta1263(86)
Hydrogen sulfide	Z (NaX)		298, 313	7	V,G	G	Be2480(67)
Krypton	AC (Barneby-Cheney)		195, 298	67	G	G	Ma129(69)
Krypton	CMS (4A)		298	67	G	G	Ma129(69)
Krypton	CMS (5A)		195, 298	67	G	G	Ma129(69)
Krypton	CMS (6A)		195, 298	67	G	G	Ma129(69)
Krypton	GCB (P-33, Cabot)	13	78, 90	0.2	V	G	Ro330(55)
Krypton	GCB (Sterling FT-D5, Cabot)	12	95-105	13	V	G	Pu459(75)
Krypton	SG (Gasil I, xerogel)	179-183	78	0.1	V	G	Le3163(82)
Krypton	Z (5A, Linde)		195, 298	67	G	G	Ma129(69)
Methane	AC (40, Comp. Esp. C. A.)	700	293	100	V	T	Co5(81)

Adsorbate	Adsorbent	A, m2/g	Temp., K	Max. P, kPa	Method	Data	Reference
Methane	AC (BPL, Calgon)		303, 342	121	V	G	Si919(86)
Methane	AC (BPL, Pittsburgh)	1004	250-350	101	G	G	Gr403(62)
Methane	AC (BPL, Pittsburgh)	988	213-301	3829	V	T	Re336(80)
Methane	AC (BPL, Pittsburgh)	1053	298	700	V	G	Wi14(83)
Methane	AC (Columbia G)		298	101	V	G	Le1326(50)
Methane	AC (Columbia G)	1157	283-323	14000	V	T	Pa363(68)
Methane	AC (Columbia L)	1152	311-422	1479	V	T	Ra1315(50)
Methane	AC (Fiber, KF-1500, Toyobo)	1440	273-323	100	G	T	Ku588(84)
Methane	AC (IG-1, Barneby-Cheney)	904	76	1	V	G	Ki949(70)
Methane	AC (Kureha beads)	800-1200	311-422	569	V	T	Ka928(87)
Methane	AC (Nuxit-AL)		293-363	129	V	T	Sz37(63)
Methane	AC (Nuxit-AL)		293-363	681	V	T	Sz53(63)
Methane	AC (PCB, Calgon)		296-480	6688	V	T	Ri1679(87)
Methane	CB (Spheron 6, Cabot)	123	311-394	60000	V	T	St74(68)
Methane	CMS (5A)	650	279, 303	87	G	T	Na310(74)
Methane	CMS (Cecalite, Ceca)		298	1000	V	G	Ve1417(81)
Methane	SG		293	6870	D	G	Gi702(65)
Methane	SG	532	233-313	14140	G	G	Gi797(64)
Methane	SG (Davison 15)	804	273-313	6875	D	T	Ha546(67)
Methane	SG (Davison 15)	804	298	9600	D	G	Ma197(68)
Methane	SG (Davison 3)	807-830	311	14000	V	G	Ma1097(66)
Methane	SG (Davison)		298	101	V	G	Le1326(50)
Methane	SI (Linde)	145	311-394	69000	V	G	St1(70)
Methane	Z (13X)		288-308	7000	D	G	Ro616(80)
Methane	Z (4A, Linde)		253-273	101	V	G	Ea419(75)
Methane	Z (4A, Linde)		306, 366	101	D	G	Ru27(80)
Methane	Z (5A)		288-308	9000	D	G	Ro616(80)
Methane	Z (5A)		190-273	80	D	G	Ru753(76)

Adsorbate	Adsorbent	A, m2/g	Temp., K	Max. P, kPa	Method	Data	Reference
Methane	Z (5A)		298	1000	V	G	Ve1417(81)
Methane	Z (5A, Linde)		76	Psat	V	G	Ki58(66)
Methane	Z (5A, Linde)		299, 348	101	D	G	Ru27(80)
Methane	Z (5A, Linde)		195-300	53		G	St942(76)
Methane	Z (5A, Linde)		298-348	9151		T	Wa399(81)
Methane	Z (5A, Linde)		185-273	40	G	G	Lo331(72)
Methane	Z (5A, Linde)		185-273	43	G	G	Ru696(72)
Methane	Z (Ca-mordenite)		273	27	V	G	Va1371(81)
Methane	Z (Na-mordenite, Norton)		298-413	27	G	G	Sa731(67)
Methane	Z (NK20B, Ceca)		298	1000	V	G	Ve1417(81)
Methanol	AC (HGI-780)	724	303	0.8	G	G	Ok209(78)
Methanol	AC (Shirasagi S)	972	303	0.8	G	G	Ok209(78)
Methanol	AL (Alpha)	3, 104	293	Psat	G	G	Ba491(66)
Methanol	Glass (Vycor, Corning)	126	293	Psat	G	G	Jo415(68)
Methanol	SG	320	298	Psat	G	G	Ba80(52)
Methanol	SG	39-399	308	0.4 Psat	V	G	Mi260(75)
Methanol	SI (Aerosil 200, Degussa)	186	298-308	12	V	G	Co1351(86)
Methanol	SI (Aerosil R972, Degussa)	115	298-308	12	V	G	Co1351(86)
Methanol	SI (G-200-0, Unilever)		298	0.8 Psat	G	G	Pe439(84)
Methyl mercaptan	AC (BPL, Pittsburgh)	1040	252-373	5	G	G	Gr403(62)
Neon	CB (DAG 621, Acheson)	70	296	16	V	G	Sa3121(82)
Nitrogen	AC	1-1880	77-90	Psat	V	G	De191(42)
Nitrogen	AC	83-949		0.8 Psat	V	G	La104(69)
Nitrogen	AC	2-120	70-90	Psat	V	T	Lo11(55)
Nitrogen	AC	775-1210	77	Psat	V	G	Sk476(71)
Nitrogen	AC (BPL)	1050-1150	303-313	99	G	T	Me259(67)
Nitrogen	AC (BPL, Calgon)		303, 343	7091		G	Si1101(84)
Nitrogen	AC (BPL, Calgon)		302, 342	150	V	G	Si919(86)

Adsorbate	Adsorbent	A, m2/g	Temp., K	Max. P, kPa	Method	Data	Reference
Nitrogen	AC (BPL, Pittsburgh)	1123	100, 150	88	V	T	Fe457(72)
Nitrogen	AC (Cloth CDE Porton Down)	1350	77	Psat	G	G	Ha301(86)
Nitrogen	AC (Columbia L)	1152	311-422	1500	V	T	Ra1315(50)
Nitrogen	AC (IG-1, Barneby-Cheney)	904	76	2	V	G	Ki949(70)
Nitrogen	AC (Metal-impregnated BPL)	800-850	303-313	97	G	T	Me259(67)
Nitrogen	AC (Nitrogen-containing)	823	B.P.	93	V	G	Bo286(87)
Nitrogen	AC (Wako Pure Chemicals)	957	B.P.	93	V	G	Bo286(87)
Nitrogen	CB (DAG 621, Acheson)	70	296-348	16	V	G	Sa3121(82)
Nitrogen	CB (Spheron 6, Cabot)	128	77	Psat	V	G	Sk476(71)
Nitrogen	CMS (5A, HGM-366, Takeda)		77	93	G	T	Ho470(83)
Nitrogen	CMS (5A, HGS-638, Takeda)		77	98	G	T	Ho470(83)
Nitrogen	CMS (5A, Takeda)		77-323	90	G	G	Ka158(74)
Nitrogen	CMS (Bergbau-Forschung)		273-333	90	G	G	Ru1325(86)
Nitrogen	GCB (3-G, Vulcan)	70	77	Psat	V	G	Ma1591(82)
Nitrogen	GCB (FT-G, Sterling)	12	77	Psat	V	G	Ma1591(82)
Nitrogen	GCB (Graphon)		77	0.9 Psat	V	G	Ba649(68)
Nitrogen	GCB (P-33, Cabot)	13	78, 90	2	V	G	Ro319(55)
Nitrogen	SG	300-767	77	0.9 Psat	V	G	Bh109(72)
Nitrogen	SG	160-830	77		V	G	Ta347(69)
Nitrogen	SG (Davison 70)	253	77-110	Psat	V	G	Ne999(70)
Nitrogen	SG (Davison)	273, 399	77	Psat	V	G	Mi54(68)
Nitrogen	SG (Davison 3)	793	77	Psat		G	Mi45(68)
Nitrogen	SG (Mallinckrodt)	396, 658	77	Psat	V	G	Mi54(68)
Nitrogen	SI	219-630	77	Psat	V	G	Av597(73)
Nitrogen	SI	39-313	77	0.9 Psat	V	T	Bh109(72)
Nitrogen	SI	213-739	77	Psat		G	Ra205(75)
Nitrogen	SI		77		V	G	Ta347(69)
Nitrogen	SI (HiSil 233)	685	77	0.9 Psat	V	G	Ba649(68)

Adsorbate	Adsorbent	A, m2/g	Temp., K	Max. P, kPa	Method	Data	Reference
Nitrogen	Z (10A, Linde)	676	144	120	V	T	Da515(69)
Nitrogen	Z (10X, Linde)	676	144	293	V	G	Do30(75)
Nitrogen	Z (10X, Linde)	672	78-273	212	V	T	No112(81)
Nitrogen	Z (13X, Linde)		77-348	8201		T	Wa399(81)
Nitrogen	Z (4A)		235	133		G	Ru753(76)
Nitrogen	Z (4A, Linde)		195-223	101	V	G	Ea419(75)
Nitrogen	Z (4A, Linde)		304	101	D	G	Ru27(80)
Nitrogen	Z (5A)		203-297	450	V	G	Mi194(87)
Nitrogen	Z (5A)		205-252	133		G	Ru753(76)
Nitrogen	Z (5A, Laporte)		278-303	800	V	G	So1517(83)
Nitrogen	Z (5A, Linde)	572	144	104	V	T	Da515(69)
Nitrogen	Z (5A, Linde)		76	Psat	V	G	Ki58(66)
Nitrogen	Z (5A, Linde)		298, 304	101	D	G	Ru27(80)
Nitrogen	Z (5A, Linde)		77-348	17554		T	Wa399(81)
Nitrogen	Z (5A, Sillipore NK20, Ceca)		274-348	428	V	T	Ve66(85)
Nitrogen	Z (Ca-mordenite)		273	27	V	G	Va1371(81)
Nitrogen	Z (HY)		78	93	V	G	Ch1679(83)
Nitrogen	Z (Na-mordenite)		298-348	8080	V	G	Ta671(74)
Nitrogen	Z (NaY)		78	93	V	G	Ch1679(83)
Nitromethane	SI (Aerosilogel)	73, 91		0.8 Psat		G	Cu58(74)
Nitrous oxide	Z (5A, Linde)		298-348	7979		T	Wa399(81)
iso-Octane	GCB (Sterling FT-FF, Cabot)	12	296-366	0.2		G	De1685(85)
iso-Octane	Z (13X)		373-459	2	G	G	Ru654(81)
n-Octane	Z (5A, Linde)		523-668	2	G	G	Va526(81)
Oxygen	CMS (5A, Takeda)		195-323	1100	G	G	Ka158(74)
Oxygen	CMS (Bergbau-Forschung)		303	90	G	G	Ru1325(86)
Oxygen	SG	241-805	77, 90	Psat	V	G	Ha38(73)
Oxygen	Z (10A, Linde)	676	144	178	V	T	Da515(69)

Adsorbate	Adsorbent	A, m2/g	Temp., K	Max. P, kPa	Method	Data	Reference
Oxygen	Z (10X, Linde)	676	144	293	V	G	Do30(75)
Oxygen	Z (10X, Linde)	672	172-273	212	V	T	No112(81)
Oxygen	Z (4A, Davison)		77	Psat	G	G	St112(85)
Oxygen	Z (4A, Linde)		123-173	101	V	G	Ea419(75)
Oxygen	Z (5A)		203-297	450	V	G	Mi194(87)
Oxygen	Z (5A)		201, 298	80		G	Ru753(76)
Oxygen	Z (5A, Laporte)		278-303	800	V	G	Sol517(83)
Oxygen	Z (5A, Linde)	572	144	213	V	T	Da515(69)
Oxygen	Z (5A, Linde)		77	Psat	G	G	St112(85)
Oxygen	Z (5A, Sillipore NK20 Ceca)		274-348	410	V	T	Ve66(85)
Ozone	SG (KSS)	365	156-273	17	V	G	Ko869(74)
n-Pentane	AC (Fiber, KF-1500, Toyobo)	1440	273-323	44	G	T	Ku588(84)
n-Pentane	AC (BPL, Pittsburgh)	1004	250-350	101	G	G	Gr403(62)
n-Pentane	CMS (5A)	650	303	72	G	T	Na310(74)
n-Pentane	SG (Davison 3)	807-830	311	48	V	G	Ma1097(66)
n-Pentane	SG (PA 400, Davison)	670	283-303	80	G	G	Al658(81)
n-Pentane	Z (5A)		273-477	15	G	G	Ru397(75)
n-Pentane	Z (5A, Linde)		523-674	12	G	G	Va526(81)
n-Pentane	Z (Erionite)		423, 473	8	G	G	Ru397(75)
Propane	AC (4LXC, Columbia)	1130	298	1	D	G	Le461(85)
Propane	AC (Black pearls I, Cabot)	705	298	91	V	T	Le1153(50)
Propane	AC (BPL, Pittsburgh)	1004	250-350	101	G	G	Gr403(62)
Propane	AC (Columbia G)		298	820	V	G	Le1326(50)
Propane	AC (Columbia G)	1157	293-333	1370	V	T	Pa363(68)
Propane	AC (Columbia L)	1152	311-505	803	V	T	Ra1315(50)
Propane	AC (Kureha beads)	800-1200	311-422	296	V	T	Ka928(87)
Propane	AC (Nuxit-AL)		293-363	131	V	T	Sz37(63)
Propane	AC (Nuxit-AL)		293-363	692	V	T	Sz53(63)

Adsorbate	Adsorbent	A, m2/g	Temp., K	Max. P, kPa	Method	Data	Reference
Propane	AC (Pittsburgh coke)		298	101	V	G	Le1326(50)
Propane	CMS (5A)	650	279-323	85	G	T	Na310(74)
Propane	SG (Davison 15)	804	273-313	832	D	T	Ha546(67)
Propane	SG (Davison 3)	807-830	311	172	V	G	Ma1097(66)
Propane	SG (Davison)	751	298	103	V	T	Le1153(50)
Propane	SG (Davison)	751	273-373	102	V	T	Le1160(50)
Propane	SG (Davison)		298	810	V	G	Le1326(50)
Propane	SG (PA-100, Davison)	751	238-273	102	G	T	Ma286(72)
Propane	Z (5A)		196-313	53	G	G	Ve1071(75)
Propane	Z (5A, Davison)		298-348	20	G	G	Yo308(71)
Propane	Z (5A, Linde)		398	101	D	G	Ru27(80)
Propane	Z (5A, Linde)		190-398	40	G	G	Lo331(72)
Propane	Z (5A, Linde)		273-398	53	G	G	Ru696(72)
Propane	Z (H-mordenite, Norton)		283-324	208	V	T	Ta1263(86)
Propane	Z (HNaY)		297	44	G	T	Dz2662(79)
Propane	Z (KNaX)		297	23	G	T	Dz2662(79)
Propane	Z (KNaX)		297-343	23	G	T	Dz2662(79)
Propane	Z (KNaY)		297-323	6	G	T	Dz2662(79)
Propane	Z (LiNaX)		297-403	33	G	T	Dz2662(79)
Propane	Z (LiNaY)		297-373	37	G	T	Dz2662(79)
Propane	Z (Na-mordenite, Norton)		298-413	27	G	G	Sa731(67)
Propane	Z (NaX)		298, 313	7	V,G	G	Be2480(67)
Propane	Z (NaX)		297	47	G	T	Dz2662(79)
Propane	Z (NaY)		297	29	G	T	Dz2662(79)
1-Propanol	AL (Alpha)	3, 104	293	Psat	G	G	Ba491(66)
1-Propanol	Glass (Vycor, Corning)	126	293	Psat	G	G	Jo415(68)
1-Propanol	SG	320	298	Psat	G	G	Ba80(52)
2-Propanol	SG	39-399	308	0.4 Psat	V	G	Mi260(75)

Adsorbate	Adsorbent	A, m2/g	Temp., K	Max. P, kPa	Method	Data	Reference
2-Propanol	SG (Davison 3)	793	308	Psat		G	Mi505(70)
2-Propanol	SI (G-200-0, Unilever)		298	0.8 Psat	G	G	Pe439(84)
n-Propyl mercaptan	AC (BPL, Pittsburgh)	1040	252-373	2	G	G	Gr403(62)
Propylene	AC (40, Comp. Esp. C. A.)	700	293	100	V	T	Co5(81)
Propylene	AC (Black pearls I, Cabot)	705	298	87	V	T	Le1153(50)
Propylene	AC (BPL, Pittsburgh)	1050-1150	303-323	114	G	T	La48(69)
Propylene	AC (Columbia G)		298	101	V	G	Le1326(50)
Propylene	AC (Columbia L)	1152	311-450	99	V	T	Ra1315(50)
Propylene	AC (Fiber, KF-1500, Toyobo)	1440	273-323	100	G	T	Ku588(84)
Propylene	AC (Nuxit-AL)		293-363	131	V	T	Sz37(63)
Propylene	AC (Nuxit-AL)		293-363	804	V	T	Sz53(63)
Propylene	AL	344	283, 303	100	V	G	Ta136(81)
Propylene	CMS (5A)	650	279-323	91	G	T	Na310(74)
Propylene	CMS (5A)	650	275-323	38	V,G	T	Na317(82)
Propylene	Glass (Vycor)	117	273-323	100	V	G	Ok262(81)
Propylene	SG (Davison)	751	298	102	V	T	Le1153(50)
Propylene	SG (Davison)	751	273-313	102	V	T	Le1160(50)
Propylene	SG (Davison)		298	101	V	G	Le1326(50)
Propylene	SG (PA-100, Davison)	751	238-273	101	G	T	Ma286(72)
Propylene	Z (5A, Davison)		298-348	27	G	G	Yo308(71)
Propylene	Z (5A, Linde)		273-458	34	G	G	De1947(72)
Pyridine	SI (Aerosilogel)	73, 91		0.4 Psat		G	Cu58(74)
Sulfur dioxide	Glass (Vycor)	117	263-303	100	V	G	Ok262(81)
Sulfur hexafluoride	Z (13X)		224-323	40	G	G	Ru1043(76)
Sulfur hexafluoride	Z (13X, Linde)		195-297	101	G	G	Th279(68)
Tetrahydrofuran	SI (Aerosilogel)	73, 91		0.4 Psat		G	Cu58(74)
Toluene	AC (208C Sutcliffe Speakman)	1038	423		D	G	Th240(71)
Toluene	AC (Shirasagi S)	972	303	0.8	G	G	Ok209(78)

Adsorbate	Adsorbent	A, m2/g	Temp., K	Max. P, kPa	Method	Data	Reference
Trichlorofluoromethane	GCB (Sterling MTFF-D-7)	10	273, 298	17	G	G	Sl347(75)
Triethylamine	GCB (Sterling FT-FF Cabot)	12	300-383	0.5		G	De1685(85)
Triethylamine	SI (Aerosilogel)	73, 91		0.4 Psat		G	Cu58(74)
Trimethylamine	AC (Nitrogen-containing)	823	303	80	G	G	Bo286(87)
Trimethylamine	AC (Wako Pure Chemicals)	957	303	80	G	G	Bo286(87)
Water	AC (Cloth CDE Porton Down)	1350	303	0.8 Psat	G	G	Ha301(86)
Water	AC (HGI-780)	724	303	0.9 Psat	G	G	Ok209(78)
Water	AC (Shirasagi S)	972	303	0.9 Psat	G	G	Ok209(78)
Water	CMS (Cecalite, Ceca)		303	3	V	G	Ve1417(81)
Water	SG	355-455	299	Psat	G	G	Ba80(52)
Water	SG (Davison 3)	299	298, 274	Psat	G	G	Ma597(85)
Water	SG (KSK)	380	348	Psat	V,G	G	Be1817(61)
Water	SG (Mobil Sorbead R)	650	301-326	12	G	T	Pe11(83)
Water	SI (G-200-0, Unilever)		298	0.8 Psat	G	G	Pe439(84)
Water	SI (HiSil 233)		298	0.6 Psat	V	G	Ba649(68)
Water	Z (KNaX)		296-373	76		G	Ch1345(76)
Water	Z (NaX)		296-373	76		G	Ch1345(76)
Water	Z (NK20B, Ceca)		303	3	V	G	Ve1417(81)
Water	Z(5A)		303	3	V	G	Ve1417(81)
Xenon	CMS (5A, HGS-638)		273, 292	0.2	V	G	Sa156(83)
Xenon	GCB (Graphon)	90	162-195	33	V, D	G	Co405(67)
Xenon	GCB (Sterling, MT3100)	8	162-195	33	V, D	G	Co405(67)

APPENDIX E

COMPILATION OF REFERENCES
FOR GAS MIXTURES

Young and Crowell published [Yo(62)] a comprehensive review of experimental data on adsorption of gas mixtures for the period up to 1959. Hall and Müller [Ha165(78)] covered the period from 1959 to 1976.

Explanation of Table

Adsorbate   Names of gaseous adsorbates, listed alphabetically
                   by adsorbate no. 1. Multicomponent mixtures
                   (ternary and higher) listed alphabetically by adsorbent.

Adsorbent   Name of adsorbent.
                   AC   Activated carbon
                   AL   Alumina
                   CB   Carbon black
                   CMS   Carbon molecular sieve
                   GCB   Graphitized carbon black
                   SG   Silica gel
                   SI   Silica
                   Z   Zeolite

A, m^2/g   Surface area of adsorbent, m^2/g.

Temp., K   Temperature or range of temperatures studied, K.

P, kPa   Pressure or range of pressures studied, kPa.

Method   Experimental technique.
                   D   Dynamic
                   G   Gravimetric
                   V   Volumetric

Data   Method of presenting experimental data.
                   G   Graphical
                   T   Tabular

Reference   Abbreviation of reference. For example, Ri1679(87) refers to
                   J. A. Ritter and R.T Yang, *Ind. Eng. Chem. Res.* **26**, 1679 (1987).

Adsorbate No. 1	Adsorbate No. 2	Adsorbent	A, m2/g	Temp., K	P, kPa	Method	Data	Reference
Acetic acid	Benzene	AC		293			T	Ki4123(52)
Acetone	Benzene	AC (G-2X, Takeda)		303	4	V	G	Ko398(85)
Acetone	Benzene	CMS (5A, Takeda)		303	4	V	G	Ko398(85)
Acetone	Benzene	Z (13X, Linde)		303	4	V	G	Ko398(85)
Acetone	n-Hexane	AC (G-2X, Takeda)		303	4	V	G	Ko398(85)
Acetone	n-Hexane	CMS (5A, Takeda)		303	4	V	G	Ko398(85)
Acetone	n-Hexane	Z (13X, Linde)		303	4	V	G	Ko398(85)
Acetone	Water	AC (HGI-780)	724	303	3-5	V	T	Ok209(78)
Acetone	Water	AC (Shirasagi-S)	972	303	3-5	V	T	Ok209(78)
Acetylene	Ethylene	AC (EY-51-C, Pittsburgh)	805	298	101	V	T	Le157(50)
Acetylene	Ethylene	SG (Davison)	751	298	101	V	T	Le157(50)
Argon	Nitrogen	SG (Davison 8)		195	101	D	G	Ca339(66)
Argon	Nitrogen	Z (NaX)		140-160	0.5-65	V	G	Va2547(75)
Benzene	Acetone	GCB (Carbopack C, Supelco)	9	293	3-10	D	T	Li420(84)
Benzene	Cyclohexane	AC (Filtrasorb 400, Calgon)	1100	303	0.01-15	G	G	My97(82)
Benzene	Cyclohexane	GCB (Graphon)	89	293-313	4	V	G	Pe473(75)
Benzene	Cyclohexane	GCB (Sterling MT-D5, Cabot)	10	303	0.3-15	G	G	My97(82)
Benzene	Cyclohexane	SI (Cab-O-Sil)		303	4-6		T	Pe313(74)
Benzene	Cyclohexane	SI (Cab-O-Sil, Cabot)	214	293	4	V	G	Pe252(76)
Benzene	2,4-Dimethylpentane	SG (Davison 8)	619	339	7-67	G	G	Ci710(53)
Benzene	2,4-Dimethylpentane	SG (Mallinckrodt)	637	339	7-67	G	G	Ci710(53)
Benzene	Ethanol	GCB (Graphon)	89	293-313	4	V	G	Pe473(75)
Benzene	n-Hexane	SG (Davison)	832	363		D	G	Sh100(68)
Benzene	2-Propanol	AC	1230	303	2-4	D	T	Ho1169(82)
Benzene	Toluene	AC (208C, Sutcliffe Speakman)	1038	423		D	G	Th240(71)
Benzene	Trichlorofluoromethane	GCB (Sterling MTFF-D-7)	10	298	0.5-1	G	G	Sl347(75)
Benzene	Water	AC (Shirasagi-S)	972	303	3-4	V	T	Ok209(78)
iso-Butane	Ethylene	Z (13X, Linde)	420	298	138	D	G	Hy95(85)
n-Butane	Hydrogen sulfide	Z (CaA)		273-348			T	Zh1172(70)
1-Butene	iso-Butane	AC (Pittsburgh)	805	298	101	V	G	Le1319(50)
1-Butene	iso-Butane	SG (Davison)	751	298	101	V	G	Le1319(50)
Carbon dioxide	Helium	SG (KSM)		140-160	150-8000		T	Mo568(74)

Adsorbate No. 1	Adsorbate No. 2	Adsorbent	A, m2/g	Temp., K	P, kPa	Method	Data	Reference
Carbon dioxide	Hydrogen	SG (KSM)	510	120-180	253-10000		T	Mo1662(72)
Carbon dioxide	Hydrogen sulfide	AC (PCB, Calgon)		293	698-844	V	T	Ri1679(87)
Carbon dioxide	Hydrogen sulfide	Z (H-mordenite, Norton)		303	0.6-16	V	T	Ta1263(86)
Carbon dioxide	Methane	AC (BPL, Calgon)		304	101	D	G	Si919(86)
Carbon dioxide	Methane	Z (Ca-mordenite)		273		V	G	Va1371(81)
Carbon dioxide	Nitrogen	AC (BPL, Calgon)		304	101	D	G	Si919(86)
Carbon dioxide	Nitrogen	SG (KSM)		140-160	150-8000		T	Mo568(74)
Carbon dioxide	Nitrogen	SG (KSM)		100-180	250-9000		T	Mo599(76)
Carbon dioxide	Nitrogen	Z (Ca-mordenite)		273		V	G	Va1371(81)
Carbon monoxide	Carbon dioxide	AC (BPL, Pittsburgh)	1053	298	345	V	T	Wi14(83)
Carbon monoxide	Carbon dioxide	AC (PCB, Calgon)		293-297	798-2430	V	T	Ri1679(87)
Carbon monoxide	Carbon dioxide	Z (Ca-mordenite)		243-273		V	G	Va1371(81)
Carbon monoxide	Carbon dioxide	Z (H-mordenite)		273		V	G	Va1371(81)
Carbon monoxide	Carbon dioxide	Z (Na-mordenite, Norton)		273		V	G	Va1371(81)
Carbon monoxide	Hydrogen	AC (PCB, Calgon)		290-295	1139-1794	V	T	Ri1679(87)
Carbon monoxide	Nitrogen	Z (10A, Linde)	676	144	101	V	T	Da515(69)
Carbon monoxide	Nitrogen	Z (5A, Linde)	572	144	101	V	T	Da515(69)
Carbon monoxide	Oxygen	Z (10A, Linde)	676	144	101	V	T	Da515(69)
Carbon monoxide	Oxygen	Z (5A, Linde)	572	144	101	V	T	Da515(69)
Chloroform	Diethyl ether	AC		333		V,G	G	Be753(65)
Cyclohexane	Ethanol	GCB (Graphon)	89	293-313	4	V	G	Pe473(75)
Cyclohexane	n-Heptane	Z (13X, Linde)		409, 458	2-13	V	G	Ru27(85)
Deuterium	Hydrogen	AC (Columbia G)	1250	75-90	27-100	V	T	Ba141(60)
Deuterium	Hydrogen	AC (Fischer)	1100	75-90	27-100	V	T	Ba141(60)
Deuterium	Hydrogen	SG (Fischer)	750	75-90	27-100	V	T	Ba141(60)
Deuterium	Hydrogen	Z (4A, Linde)		55-112	1-40	V	T	Pa1515(74)
Deuterium	Hydrogen	Z (5A, Linde)		50-112	0.3-40	V	T	Pa1515(74)
Deuterium	Hydrogen	Z (Linde, 4A, 5A, 13X)	700-800	75-90	27-100	V	T	Ba141(60)
Deuterium	Hydrogen	Z (NaA, GROZNII)		48-90	0.2-15	V	T	Pa1515(74)
Deuterium	Hydrogen	Z (NaX, VNIINP)		35-90	1-6	V	T	Pa1515(74)
2,2-Dimethylpropane	n-Butane	CB (H, Acheson)	542	303	12	V,G	T	Ha2265(78)
Ethane	Acetylene	AC (4LXC, Columbia)	1130	298	1	D	T	Le1(84)

Adsorbate No. 1	Adsorbate No. 2	Adsorbent	A, m2/g	Temp., K	P, kPa	Method	Data	Reference
Ethane	iso-Butane	Z (13X, Linde)	525	298, 323	138	V	T	Hy196(82)
Ethane	n-Butane	Z (5A)		273	1-53	G	T	Ve1071(75)
Ethane	n-Butane	Z (5A, Linde)		308	7	V,G	T	Gl73(69)
Ethane	Carbon dioxide	Z (5A, Linde)		308	13	V,G	T	Gl73(69)
Ethane	Ethylene	Z (13X)		323-423	138	V	T	Ka928(87)
Ethane	Ethylene	Z (13X, Linde)		298	138	D	G	Da2129(80)
Ethane	Nitrogen	Z (5A, Linde)		300-323	101	D	G	Ru27(80)
Ethane	Propane	AC (Nuxit-AL)		293-333	100	V	T	Sz245(63)
Ethane	Propane	CMS (5A)	650	278-323	13-40	V,G	T	Na161(81)
Ethane	Propane	CMS (5A, Takeda)		303	13, 40	V	T	Ko394(85)
Ethane	Propane	GCB (Sterling FTG-D5, Cabot)	13	298	93	G	G	Fr439(72)
Ethane	Propane	Z (5A)		273	1-53	G	T	Ve1071(75)
Ethane	Propylene	AC (40, Comp. Esp. C. A.)	700	293	1-107	V	T	Co5(81)
Ethanol	Benzene	AC		293			T	Ki4123(52)
Ethanol	Benzene	SI (Cab-O-Sil)		303	3-5		T	Pe313(74)
Ethanol	Benzene	SI (Cab-O-Sil, Cabot)	214	293	4	V	G	Pe252(76)
Ethanol	Cyclohexane	SI (Cab-O-Sil)		303	3-5		T	Pe313(74)
Ethanol	Cyclohexane	SI (Cab-O-Sil, Cabot)	214	293	4	V	G	Pe252(76)
Ethanol	1-Propanol	Z (CaA)		358-413	40	V	G	De1050(75)
Ethanol	Diethyl ether	AC		323-344		V,G	G	Be753(65)
Ethyl chloride	Benzene	AC		293			T	Ki4123(52)
Ethyl dichloride	iso-Butane	Z (13X, Linde)	525	298-373	138	V	T	Hy196(82)
Ethylene	Carbon dioxide	AC (Nuxit-AL)		293	100	V	T	Sz245(63)
Ethylene	Carbon dioxide	Z (13X, Linde)	525	298, 323	138	V	T	Hy196(82)
Ethylene	Ethane	AC (40, Comp. Esp. C. A.)	700	293	1-107	V	T	Co5(81)
Ethylene	Ethane	AC (BPL, Pittsburgh)	988	213-301	140-2100	V	T	Re336(80)
Ethylene	Ethane	AC (Columbia G)		298	101	V	G	Le1319(50)
Ethylene	Ethane	AC (Nuxit-AL)		293-333	100	V	T	Sz245(63)
Ethylene	Ethane	CMS (5A)	650	283-313	0.7-23	V,G	T	Na114(87)
Ethylene	Ethane	CMS (5A)	650	303	3-13	V,G	T	Na202(84)
Ethylene	Ethane	GCB (Sterling FTG-D5, Cabot)	13	298	93	G	G	Fr439(72)
Ethylene	Ethane	SG (Davison)	751	298	33-1940	V	G	Le1319(50)

Adsorbate No. 1	Adsorbate No. 2	Adsorbent	A, m2/g	Temp,, K	P, kPa	Method	Data	Reference
Ethylene	Ethane	Z (13X, Linde)	525	298-323	138	V	T	Da248(78)
Ethylene	Propane	AC (Columbia G)		298	101-750	V	G	Le1319(50)
Ethylene	Propane	CMS (5A, Takeda)		303	13	V	G	Ko394(85)
Ethylene	Propane	SG (Davison)	751	273-313	101	V	T	Le1160(50)
Ethylene	Propane	SG (Davison)		273-313	101	V	G	Le1319(50)
Ethylene	Propane	Z (5A, Linde)		399	101	D	G	Ru27(80)
Ethylene	Propylene	AC (40, Comp. Esp. C. A.)	700	293	1-107	V	T	Co5(81)
Ethylene	Propylene	AC (Columbia G)		298	101	V	G	Le1319(50)
Ethylene	Propylene	AC (Nuxit-AL)		293	100	V	T	Sz245(63)
Ethylene	Propylene	CMS (5A)	650	275-323	1-10	V,G	T	Na202(84)
Ethylene	Propylene	CMS (5A)	650	275-323	13	V,G	T	Na317(82)
Ethylene	Propylene	SG (Davison)	751	273-313	101	V	T	Le1160(50)
Ethylene	Propylene	SG (Davison)	751	273-313	101	V	G	Le1319(50)
Helium	Nitrogen	AC (BPL, Pittsburgh)	1123	100, 150	105-10500	V	T	Fe457(72)
n-Heptane	Water	SG (KSK)	380	348		V,G	G	Be1817(61)
n-Hexane	n-Heptane	Z (CaA)		358, 413	40	V	G	De1050(75)
Methane	Argon	Z (4A, Linde)		306	101	D	G	Ru27(80)
Methane	n-Butane	AC (Columbia G)	1157	313-343	14-13700	V	T	Pa363(68)
Methane	n-Butane	AC (BPL, Pittsburgh)	1040	298	0-5300		T	Gr490(66)
Methane	n-Butane	SG (Davison 3)	807-830	311	6900	D	G	Ma1097(66)
Methane	Carbon dioxide	AC (AR-3)		298		D	G	Ar219(74)
Methane	Carbon dioxide	AC (BPL, Pittsburgh)	1053	298	345	V	T	Wi14(83)
Methane	Carbon dioxide	AC (PCB, Calgon)		294-296	807-2263	V	T	Ri1679(87)
Methane	Carbon dioxide	Z (13X)		298	6500	D	G	Ro616(80)
Methane	Carbon dioxide	Z (5A)		298	6500	D	G	Ro616(80)
Methane	Carbon monoxide	AC (BPL, Pittsburgh)	1053	298	345	V	T	Wi14(83)
Methane	Carbon monoxide	AC (PCB, Calgon)		293-295	859-2697	V	T	Ri1679(87)
Methane	Carbon monoxide	CMS (Cecalite, Ceca)		298		V	T	Ve1417(81)
Methane	Carbon monoxide	Z (4A, Linde)		309	101	D	G	Ru27(80)
Methane	Carbon monoxide	Z (5A)		298		V	T	Ve1417(81)
Methane	Carbon monoxide	Z (NK20B, Ceca)		298		V	T	Ve1417(81)
Methane	Ethane	AC (40, Comp. Esp. C. A.)	700	293	1-107	V	T	Co5(81)

Adsorbate No. 1	Adsorbate No. 2	Adsorbent	A, m2/g	Temp., K	P, kPa	Method	Data	Reference
Methane	Ethane	AC (BPL, Pittsburgh)	988	213-301	140-2000	V	T	Re336(80)
Methane	Ethane	AC (Nuxit-AL)		293	100	V	T	Sz245(63)
Methane	Ethane	CMS (5A, Takeda)		303	13, 40	V	T	Ko394(85)
Methane	Ethane	CMS (Cecalite, Ceca)		298		V	T	Ve1417(81)
Methane	Ethane	SG (Davison 15)	804	278-308	0-9700	D	T	Ma197(68)
Methane	Ethane	SG (MSM)		298		D	G	Ar219(74)
Methane	Ethane	Z (5A)		298		V	T	Ve1417(81)
Methane	Ethane	Z (5A, Linde)		298-348	101	D	G	Ru27(80)
Methane	Ethane	Z (NK20B, Ceca)		298		V	T	Ve1417(81)
Methane	Ethylene	AC (40, Comp. Esp. C. A.)	700	293	1-107	V	T	Co5(81)
Methane	Ethylene	AC (BPL, Pittsburgh)	988	213-301	138-2100	V	T	Re336(80)
Methane	Ethylene	AC (Nuxit-AL)		293	100	V	T	Sz245(63)
Methane	Helium	AC (IG-I, Barneby-Cheney)	904	76	267-3680	V	T	Ki949(70)
Methane	n-Hexane	AC (BPL, Pittsburgh)	1040	298	0-7000		T	Gr490(66)
Methane	n-Hexane	SG (Davison 3)	807-830	311	6900	D	G	Ma1097(66)
Methane	Hydrogen	Z (5A, Linde)		76	500-8500	V	G	Ki58(66)
Methane	Nitrogen	Z (4A, Linde)		306	101	D	G	Ru27(80)
Methane	n-Pentane	SG (Davison 3)	807-830	311	6900	D	G	Ma1097(66)
Methane	Propane	AC (Columbia G)	1157	303	345-13000	V	T	Pa363(68)
Methane	Propane	AC (BPL, Pittsburgh)	1040	298	0-7000		T	Gr490(66)
Methane	Propane	SG		293	700-6900	D	G	Gi702(65)
Methane	Propane	SG (Davison 15)	804	273-313	700-6900	D	T	Ha546(67)
Methanol	Acetone	AC (G-2X, Takeda)		303	4	V	G	Ko398(85)
Methanol	Acetone	CMS (5A, Takeda)		303	4	V	G	Ko398(85)
Methanol	Acetone	Z (13X, Linde)		303	4	V	G	Ko398(85)
Methanol	Benzene	AC (G-2X, Takeda)		303	4	V	G	Ko398(85)
Methanol	Benzene	CMS (5A, Takeda)		303	4	V	G	Ko398(85)
Methanol	Benzene	Z (13X, Linde)		303	4	V	G	Ko398(85)
Methanol	n-Hexane	AC (G-2X, Takeda)		303	4	V	G	Ko398(85)
Methanol	n-Hexane	CMS (5A, Takeda)		303	4	V	G	Ko398(85)
Methanol	Water	AC (HGI-780)	724	303	4-7	V	T	Ok209(78)
Methanol	Water	AC (Shirasagi-S)	972	303	4-7	V	T	Ok209(78)

Adsorbate No. 1	Adsorbate No. 2	Adsorbent	A, m2/g	Temp., K	P, kPa	Method	Data	Reference
Nitrogen	Argon	Z (5A, Linde)		304	101	D	G	Ru27(80)
Nitrogen	Carbon monoxide	Z (10X, Linde)	672	172-273	101	V	T	No112(81)
Nitrogen	Carbon monoxide	Z (4A, Linde)		307	101	D	G	Ru27(80)
Nitrogen	Helium	AC (IG-J, Barneby-Cheney)	904	76	409-6500	V	T	Ki949(70)
Nitrogen	Hydrogen	Z (5A, Linde)		76	500-8500	V	G	Ki58(66)
Nitrogen	Oxygen	Z (10A, Linde)	676	144	101	V	T	Da515(69)
Nitrogen	Oxygen	Z (5A, Linde)		283-323	101	D	G	Va1(73)
Nitrogen	Oxygen	Z (5A, Linde)	572	144	101	V	T	Da515(69)
Nitrogen	Oxygen	Z (5A, Linde)		298, 304	101	V	G	Ru27(80)
Nitrogen	Oxygen	Z (5A, Silliporite NK20, Ceca)		299-320	101-404	V	T	Ve66(85)
Oxygen	Carbon monoxide	Z (10X, Linde)	672	172-273	101	V	T	No112(81)
Oxygen	Nitrogen	Z (10X, Linde)	672	172-273	101	V	T	No112(81)
Oxygen	Nitrogen	Z (5A, Laporte)		278-303	170, 440	V	G	So1517(83)
Oxygen	Ozone	SG (KSS)	365	156-273	0.1-16	F-N	G	Ko869(74)
Ozone	Oxygen	SG	360	156-293		D	G	At1490(72)
n-Pentane	Z (CaA)			383	40	V	G	De1050(75)
Phenol	Toluene	AC (B4)		303		G	G	Ho154(78)
Propane	n-Butane	AC (Nuxit-AL)		293	100	V	T	Sz245(63)
Propane	n-Butane	CMS (5A, Takeda)		303	13	V	T	Ko394(85)
Propane	n-Butane	Z (5A)		273	1-53	G	T	Ve1071(75)
Propane	n-Butane	Z (CaA)		293, 383	40	V	G	De1050(75)
Propane	Carbon dioxide	Z (5A, Linde)		297	101	V,G	T	Cr339(80)
Propane	Carbon dioxide	Z (H-mordenite, Norton)		303	1-61	V	T	Ta1263(86)
Propane	Ethane	AC (4LXC, Columbia)	1130	298	1	D	G	Le461(85)
Propane	Hydrogen sulfide	Z (H-mordenite, Norton)		303	2-18	V	T	Ta1263(86)
Propane	Hydrogen sulfide	Z (NaX)		298, 313	7	V,G	G	Be2480(67)
iso-Propylbenzene	m-Xylene	Z (KNaY)		443	101		G	Pa2250(87)
Propylene	Carbon dioxide	Z (5A, Linde)		297	98	V,G	T	Cr339(80)
Propylene	Propane	AC (Black pearls I, Cabot)	705	298	101	V	T	Le1153(50)
Propylene	Propane	AC (Nuxit-AL)		293	100	V	T	Sz245(63)
Propylene	Propane	GCB (Sterling FTG-D5, Cabot)	13	298	13-93	G	G	Fr439(72)
Propylene	Propane	SG (Davison)	751	298	101	V	T	Le1153(50)

Adsorbate No. 1	Adsorbate No. 2	Adsorbent	A, m2/g	Temp., K	P, kPa	Method	Data	Reference
Propylene	Propane	SG (Davison)		298	33-133	V	G	Le1319(50)
Toluene	Water	AC (Shirasagi-S)	972	303	3	V	T	Ok209(78)
m-Xylene	Toluene	Z (KNaY)		423	101		G	Pa2250(87)
p-Xylene	iso-Propylbenzene	Z (KNaY)		443	101		G	Pa2250(87)
p-Xylene	Toluene	Z (KNaY)		423	101		G	Pa2250(87)
p-Xylene	m-Xylene	Z (KNaY)		423, 443	101		G	Pa2250(87)

Adsorbates	Adsorbent	A, m2/g	Temp., K	P, kPa	Method	Data	Reference
Ethane	AC (40, Comp. Esp. C. A.)	700	293		V	T	Co5(81)
Ethylene							
Methane							
Ethane	AC (40, Comp. Esp. C. A.)	700	293		V	T	Co5(81)
Ethylene							
Propylene							
Helium	AC (IG-1, Barneby-Cheney)	904	76	202-1100	V	G	Ki949(70)
Methane							
Nitrogen							
Ethane	AC (BPL, Pittsburgh)	988	213, 301	124-2964	V	T	Re336(80)
Ethylene							
Methane							
Ethane	AC (Nuxit-AL)		293	101	V	T	Sz245(63)
Hydrogen							
Methane							
Ethane	AC (Nuxit-AL)		293	101	V	T	Sz245(63)
Ethylene							
Hydrogen							
Carbon dioxide	AC (PCB, Calgon)		295, 297	1482-2435	V	T	Ri1679(87)
Hydrogen							
Methane							
Hydrogen	AC (PCB, Calgon)		293, 296	2049, 2200	V	T	Ri1679(87)
Hydrogen sulfide							
Methane							
Carbon monoxide	AC (PCB, Calgon)		291-298	1057-2614	V	T	Ri1679(87)
Hydrogen							
Methane							

Adsorbates	Adsorbent	A, m2/g	Temp., K	P, kPa	Method	Data	Reference
Carbon dioxide	AC (PCB, Calgon)		294-297	1965-2772	V	T	Ri1679(87)
Carbon monoxide							
Hydrogen sulfide							
Methane							
Carbon dioxide	AC (PCB, Calgon)		293-295	1358, 1951	V	T	Ri1679(87)
Carbon monoxide							
Hydrogen							
Hydrogen sulfide							
Methane							
Ethylene	AC (Columbia G)		298	101	V	G	Le1319(50)
Propane							
Propylene							
Carbon dioxide	AC (BPL, Pittsburgh)	1053	298	345	V	T	Wi14(83)
Carbon monoxide							
Methane	AC (BPL, Pittsburgh)	1053	298	345	V	T	Wi14(83)
Carbon monoxide							
Hydrogen							
Methane							
Ethylene	SG (Davison)		298	101	V	G	Le1319(50)
Propane							
Propylene							
Carbon monoxide	Z (10X, Linde)	676	144	101	V	T	Do30(75)
Nitrogen							
Oxygen							

Adsorbates	Adsorbent	A, m2/g	Temp., K	P, kPa	Method	Data	Reference
Argon	Z (5A)		203-297	22-420	V	T	Mi194(87)
Nitrogen							
Oxygen							
Toluene	Z (KNaY)		423	101		T	Pa2250(87)
m-Xylene							
p-Xylene							
iso-Propylbenzene	Z (KNaY)		443	101	T	T	Pa2250(87)
m-Xylene							
p-Xylene							
Carbon dioxide	Z (H-mordenite, Norton)		303	0.6-14	V	T	Ta1263(86)
Hydrogen sulfide							
Propane							

APPENDIX F

COMPILATION OF REFERENCES
FOR LIQUID MIXTURES

The monograph of Kipling [Ki(65)] contains an index for adsorption of liquid mixtures covering the period up to 1963.

Explanation of Table

Adsorbate   Names of liquid adsorbates, listed alphabetically by adsorbate no. 1. Multicomponent mixtures (ternary and higher) listed alphabetically by adsorbent.

Adsorbent   Name of adsorbent.
      AC   Activated carbon
      AL   Alumina
      CB   Carbon black
      CMS   Carbon molecular sieve
      GCB   Graphitized carbon black
      SG   Silica gel
      SI   Silica
      Z   Zeolite

A, m^2/g   Surface area of adsorbent, m^2/g.

Temp., K   Temperature or range of temperatures studied, K.

Data   Method of presenting experimental data.
      G   Graphical
      T   Tabular

Reference   Abbreviation of reference. For example, Ev2605(86) refers to D. H. Everett and A. J. P. Fletcher, *J. Chem. Soc., Faraday Trans. 1*, **82**, 2605 (1986).

Adsorbate No. 1	Adsorbate No. 2	Adsorbent	A, m2/g	Temp., K	Data	Reference
Acetic acid	Benzene	AC		293	G	Bl4103(55)
Acetic acid	Benzene	AC		293	G	Ki4123(52)
Acetic acid	Water	AC			T	Fo25(74)
Acetic acid	Water	AC (Darco)	667	278-313	T	Wa345(79)
Acetone	Benzene	SG (Woelm)	631	303	G	Th259(77)
Acetone	Carbon tetrachloride	SG (Woelm)	631	303	G	Th259(77)
Acetone	Water	AC (Filtrasorb 300, Calgon)	1000	298	T	Ra445(72)
Acetone	Water	AC (Filtrasorb 300, Calgon)	1000	298	G	Ra761(72)
Acetone	Water	AC (HGI-780)	724	303	G	Ok209(78)
Acetone	Water	AC (Shirasagi S)	972	303	G	Ok209(78)
Adipic acid	Water	AC (Darco)	595	278, 313	T	Le395(87)
n-Amyl alcohol	Carbon tetrachloride	SG	417	298	G	Zh185(85)
Aniline	Benzene	SG (Merck, GFR)	497	293	G	Go392(85)
Anthracene	Carbon tetrachloride	CB (Spheron 6)	119	303	G	Wr1908(72)
Anthracene	Carbon tetrachloride	GCB (Graphon)	89	303	G	Wr1908(72)
Anthracene	Cyclohexane	CB (Spheron 6)	119	303	G	Wr1908(72)
Anthracene	Cyclohexane	GCB (Graphon)	89	303	G	Wr1908(72)
Anthracene	Cyclohexanone	CB (Spheron 6)	119	303	G	Wr1908(72)
Anthracene	Cyclohexanone	GCB (Graphon)	89	303	G	Wr1908(72)
Benzene	1-Butanol	SG		293	G	Ki4054(57)
Benzene	Carbon tetrachloride	CB (Spheron 6)	119	303	G	Wr1908(72)
Benzene	Carbon tetrachloride	CMS (5A, Takeda)		298	G	Ta268(77)
Benzene	Carbon tetrachloride	GCB (Graphon)	89	303	G	Wr1908(72)
Benzene	Carbon tetrachloride	SG (Woelm)	631	303	G	Th259(77)
Benzene	Chlorobenzene	SG	420	293	T	Pe1384(79)
Benzene	Cyclohexane	AC		293	G	Bl3819(54)
Benzene	Cyclohexane	AC (BPL, Pittsburgh)	1100	303	T	Mi453(73)
Benzene	Cyclohexane	AL (Boehmite)		293	T	Ki4828(56)
Benzene	Cyclohexane	AL (E-Merck)	65	293-313	G	Pa18(85)
Benzene	Cyclohexane	CB (Acheson Colloids Ltd.)	120	293	T	Bl2373(57)
Benzene	Cyclohexane	CB (Spheron 6)	119	303	G	Wr1908(72)
Benzene	Cyclohexane	CB (Spheron 6, Cabot)		293	T	Bl2373(57)
Benzene	Cyclohexane	CMS (5A, Takeda)		298	G	Ta268(77)

Adsorbate No. 1	Adsorbate No. 2	Adsorbent	A, m2/g	Temp., K	Data	Reference
Benzene	Cyclohexane	GCB (Graphon)	89	303	G	Wr1908(72)
Benzene	Cyclohexane	GCB (Graphon, Cabot)	86	298-328	G	As239(73)
Benzene	Cyclohexane	GCB (Graphon, Cabot)	89	303	T	Ma162(73)
Benzene	Cyclohexane	SG (Davison 37)	350	273-333	T	Si249(72)
Benzene	Cyclohexane	SG (PA 400, Davison)	740	303	T	Si2828(70)
Benzene	Cyclohexane	SG (PA 400, Davison)	660	303	G	My3412(72)
Benzene	Cyclohexane	SI (Cab-O-Sil, Cabot)	223	303	T	Ma162(73)
Benzene	Cyclohexane	SI (Hi-Sil 215, PPG)	143	303	T	Ma162(73)
Benzene	Cyclohexanone	CB (Spheron 6)	119	303	G	Wr1908(72)
Benzene	Cyclohexanone	GCB (Graphon)	89	303	G	Wr1908(72)
Benzene	1,2-Dichloroethane	AC (BPL, Pittsburgh)		303	T	Si2828(70)
Benzene	1,2-Dichloroethane	SG (PA 400, Davison)	740	303	T	Si2828(70)
Benzene	1,2-Dichloroethane	SG (PA 400, Davison)	660	303	G	My3412(72)
Benzene	Dioxane	AL (E-Merck)	65	293-313	G	Pa18(85)
Benzene	p-Dioxane	SG (Davison 40)	616	296	G	Ma134(74)
Benzene	Ethanol	AC			T	Fo25(74)
Benzene	Ethanol	AC (BPL, Pittsburgh)		303	T	Si2828(70)
Benzene	Ethanol	GCB (Graphon, Cabot)	86	283-333	T	Br883(75)
Benzene	Ethanol	GCB (Graphon, Cabot)	89	303	T	Ma162(73)
Benzene	Ethanol	SG		293-333	G	Ki4054(57)
Benzene	Ethanol	SI (Cab-O-Sil, Cabot)	223	303	T	Ma162(73)
Benzene	Ethanol	SI (Hi-Sil 215, PPG)	143	303	T	Ma162(73)
Benzene	Ethyl acetate	AC (BPL, Pittsburgh)	1100	303	T	Mi453(73)
Benzene	n-Heptane	GCB (Graphon, Cabot)	86	283-343	G	As239(73)
Benzene	n-Heptane	GCB (Graphon, Cabot)	81	298	G	Pa1780(64)
Benzene	n-Heptane	SG	360	298	T	Fo269(73)
Benzene	n-Heptane	SG (PA 400, Davison)	740	303	T	Si2828(70)
Benzene	n-Heptane	SG (PA 400, Davison)	660	303	G	My3412(72)
Benzene	n-Heptane	Z (Dealuminated Y)		298	G	De261(86)
Benzene	n-Heptane	Z (NaY)		298	G	De261(86)
Benzene	Nitrobenzene	Al	65	303	T	Su301(70)
Benzene	Nitrobenzene	SG	412	293	T	Su301(70)
Benzene	Nitromethane	AL	65	303	T	Su301(70)

Adsorbate No. 1	Adsorbate No. 2	Adsorbent	A, m2/g	Temp., K	Data	Reference
Benzene	Nitromethane	SG	412	293	T	Su301(70)
Benzene	Toluene	SG	420	293	T	Pe1384(79)
Benzene	Water	Z (4A, Nishio Kogyo Co.)		283, 333	G	Go451(84)
Benzoic acid	Carbon tetrachloride	CB (Spheron 6)		303	G	Wr1461(74)
Benzoic acid	Carbon tetrachloride	GCB (Graphon)		303	G	Wr1461(74)
Benzoic acid	Cyclohexane	AL (Alcoa C730)	22	303	G	Wr1461(74)
Benzoic acid	Cyclohexane	AL (Alon C)	90	303	G	Wr1461(74)
Benzoic acid	Cyclohexane	CB (Spheron 6)		303	G	Wr1461(74)
Benzoic acid	Cyclohexane	GCB (Graphon)		303	G	Wr1461(74)
Benzoic acid	Cyclohexane	SI (Cab-O-Sil)	190	303	G	Wr1461(74)
Benzoic acid	Cyclohexanone	CB (Spheron 6)		303	G	Wr1461(74)
Benzoic acid	Cyclohexanone	GCB (Graphon)		303	G	Wr1461(74)
Benzoic acid	Water	AC	1500	293	G	Fr721(81)
Benzoic acid	Water	AC (B10, Lurgi)	1450	293	G	Jo1097(78)
Benzoic acid	Water	AC (Nuxit LBS)	510	298	T	Sc289(79)
Benzoic acid	Water	AC (Nuxit-AL III)	860	298	T	Sc289(79)
Benzoic acid	Water	AC (Takeda X-7100)	1000	303	G	Mu34(82)
Benzyl alcohol	Water	Z (5A)		298	T	Ja202(65)
p-Bromophenol	Benzene	AL (Alpha)	132	308	G	Da417(74)
p-Bromophenol	Benzene	SG (Sorbsil M 60)	432	308	G	Da1117(73)
1-Butanol	Water	AC	978-1420	298	G	Ab201(83)
2-Butanol	Water	AC	978-1420	298	G	Ab201(83)
n-Butanol	Benzene	AL (Alpha)		293	T	Ki4828(56)
n-Butanol	Benzene	AL (Boehmite)		293-333	T	Ki4828(56)
n-Butanol	Benzene	AL (Gibbsite)		293	T	Ki4828(56)
n-Butanol	Benzene	GCB		298	G	Co185(73)
n-Butanol	Benzene	Z (Dealuminated Y)		298	G	De261(86)
n-Butanol	Benzene	Z (NaY)		298	G	De261(86)
n-Butanol	Cyclohexane	SG	417	298	G	Zh185(85)
n-Butanol	Water	AC (Nuxit-AL I)	620	298	T	Sc289(79)
n-Butanol	Water	AC (Nuxit-AL III)	840	298	T	Sc289(79)
n-Butyl amine	Benzene	AC		293	T	Bl2373(57)
tert-Butyl chloride	Carbon tetrachloride	AC		293	T	Bl2373(57)

Adsorbate No. 1	Adsorbate No. 2	Adsorbent	A, m2/g	Temp., K	Data	Reference
p-t-Butylphenol	Benzene	AL (Alpha)	132	308	G	Da417(74)
p-t-Butylphenol	Benzene	SG (Sorbsil M 60)	432	308	G	Da1117(73)
n-Butyric acid	Water	AC (Darco)	667	278-313	T	Wa345(79)
Carbon tetrachloride	2,3-Dimethylbutane	SG (Davison 40)	616	296	G	Ma134(74)
p-Chloranilin	Water	AC (Hydraffin 71, Lurgi)		293	G	Se79(88)
Chlorobenzene	Benzene	SG (Merck, GFR)	497	293	G	Go392(85)
Chlorobenzene	Bromobenzene	GCB (Graphon)	84	273-323	T	Ev14(81)
Chlorobenzene	Bromobenzene	SG	420	293	T	Pe1384(79)
Chlorobenzene	n-Heptane	SG (Merck, GFR)	497	293	G	Go392(85)
Chloroform	Acetone	AC		293	G	Bl3819(54)
Chloroform	Benzene	AL (Boehmite)		293	T	Ki4828(56)
Chloroform	Carbon tetrachloride	AC		293	T	Bl2373(57)
o-Chlorophenol	Water	AC (Filtrasorb F400, Calgon)		293	G	Pe26(81)
p-Chlorophenol	Benzene	AL (Alpha)	132	308	G	Da417(74)
p-Chlorophenol	Benzene	SG (Sorbsil M 60)	432	308	G	Da1117(73)
p-Chlorophenol	Water	AC	1500	293	G	Fr721(81)
p-Chlorophenol	Water	AC (B10, Lurgi)	1450	293	G	Jo1097(78)
p-Chlorophenol	Water	AC (Filtrasorb 300, Calgon)	1000	298	T	Ra445(72)
p-Chlorophenol	Water	AC (Filtrasorb 300, Calgon)	1000	298	G	Ra761(72)
p-Chlorophenol	Water	AC (Takeda X-7100)	1000	303	G	Mu34(82)
Coronene	Cyclohexane	AL (Kaiser Aluminun Co.)	200-350	298	G	Ch725(83)
p-Cresol	Water	AC (CAL, Calgon)		308	G	Ok286(80)
p-Cresol	Water	AC (Filtrasorb 300, Calgon)	1000	298	T	Ra445(72)
p-Cresol	Water	AC (Filtrasorb 300, Calgon)	1000	298	G	Ra761(72)
p-Cyanophenol	Benzene	AL (Alpha)	132	308	G	Da417(74)
p-Cyanophenol	Benzene	SG (Sorbsil M 60)	432	308	G	Da1117(73)
Cyclohexane	Antracene	AC (Chemviron, CAL)	1050	298	T	Sc289(79)
Cyclohexane	Benzene	SG (Davison 40)	616	296	G	Ma134(74)
Cyclohexane	1,2-Dichloroethane	SG (PA 400, Davison)	660	303	G	My3412(72)
Cyclohexane	Dioxane	AL (E-Merck)	65	293-313	T	Pa284(82)
Cyclohexane	Dioxane	SG (H, E-Merck)	412	293-313	T	Pa284(82)
Cyclohexane	Ethanol	AC (BPL, Pittsburgh)		303	T	Si2828(70)
Cyclohexane	Ethanol	GCB (Graphon, Cabot)	89	303	T	Ma162(73)

Adsorbate No. 1	Adsorbate No. 2	Adsorbent	A, m2/g	Temp., K	Data	Reference
Cyclohexane	Ethanol	SI (Cab-O-Sil, Cabot)	223	303	T	Ma162(73)
Cyclohexane	Ethanol	SI (Hi-Sil, PPG)	143	303	T	Ma162(73)
Cyclohexane	n-Heptane	GCB (Graphon, Cabot)	86	283-343	G	As123(75)
Cyclohexane	n-Heptane	SG (PA 400, Davison)	740	303	T	Si2828(70)
Cyclohexane	Nitrobenzene	AL	65	303	T	Su301(70)
Cyclohexane	Nitrobenzene	SG	412	293	T	Su301(70)
Cyclohexanol	Water	AC	978-1420	298	G	Ab201(83)
Cyclohexanone	Carbon tetrachloride	SG	417	298	G	Zh185(85)
Cyclopentanol	Water	AC	978-1420	298	G	Ab201(83)
n-Decane	n-Dodecane	Z (5A, Linde)		303-363	T	Su223(68)
n-Decane	n-Tetradecane	Z (5A, Linde)		303-363	T	Su223(68)
Dichloroethane	Cyclohexane	CMS (5A, Takeda)		298	G	Ta268(77)
1,2-Dichloroethane	Benzene	GCB (Graphon)	84	273-333	T	Ev14(81)
2,4-Dichlorophenol	Water	AC	1500	293	G	Fr721(81)
2,4-Dichlorophenol	Water	AC (B10, Lurgi)	1450	293	G	Jo1097(78)
2,2-Dimethyl-1-propanol	Water	AC	978-1420	298	G	Ab201(83)
Dioxane	Nitrobenzene	AL	65	303	T	Su301(70)
Dioxane	Nitrobenzene	SG	412	293	T	Su301(70)
Dioxane	Nitromethane	AL	65	303	T	Su301(70)
Dioxane	Nitromethane	SG	412	293	T	Su301(70)
p-Dioxane	Water	Z (5A)		298	T	Ja202(65)
Diphenyl	n-Heptane	SG		293	G	El889(72)
Diphenyl	n-Heptane	SI (Dehydroxylated)		293	G	El889(72)
n-Dodecane	n-Tetradecane	Z (5A, Linde)		303-363	T	Su223(68)
Dodecyl benzol sulfonic acid	Water	AC (B10, Lurgi)	1450	293	G	Jo1097(78)
Dodecylbenzene sulfonic acid	Water	AC	1500	293	G	Fr721(81)
Ethanol	Benzene	AC		293	G	Ki4123(52)
Ethanol	Benzene	AC		293-333	T	Ki4828(56)
Ethanol	Benzene	AL (Alpha)		293, 333	T	Ki4828(56)
Ethanol	Benzene	AL (Boehmite)		293-333	T	Ki4828(56)
Ethanol	Benzene	AL (Gibbsite)		293, 333	T	Ki4828(56)
Ethanol	Benzene	Z (Dealuminated Y)		298	G	De261(86)
Ethanol	Benzene	Z (NaY)		298	G	De261(86)

Adsorbate No. 1	Adsorbate No. 2	Adsorbent	A, m²/g	Temp., K	Data	Reference
Ethanol	n-Butanol	AC (CS-120A, Futamura)	1350	298	G	Oz1059(84)
Ethanol	Cyclohexane	SG	417	298	G	Zh185(85)
Ethanol	n-Propanol	AC (CS-120A, Futamura)	1350	298	G	Oz1059(84)
Ethanol	Water	AC (Fisher Scientific Co.)		303-333	G	Bu1209(85)
Ethanol	Water	GCB (Graphon, Cabot)	84	298-348	T	Ev2605(86)
Ethanol	Water	Z (4A, Nishio Kogyo Co.)		283, 333	G	Go451(84)
Ethanol	Water	Z (Silicalite, Linde)	390-450	303-333	G	Bu1209(85)
Ethyl acetate	Cyclohexane	AC (BPL, Pittsburgh)	1100	303	T	Mi453(73)
Ethyl dichloride	Benzene	AC		293	G	Ki4123(52)
Ethyl p-hydroxybenzoate	Benzene	SG (Sorbsil M 60)	432	308	G	Da1117(73)
Ethyl p-hydroxylbenzoate	Benzene	AL (Alpha)	132	308	G	Da417(74)
Ethylene dichloride	Benzene	AL (Alpha)		293	T	Ki4828(56)
Ethylene dichloride	Benzene	AL (Boehmite)		293-333	T	Ki4828(56)
Ethylene dichloride	Benzene	SG		293-333	G	Ki4054(57)
Ethylenediamine	Water	Z (4A)		298	T	Ja202(65)
Glucose	Water	AC (Fisher Scientific Co.)		303	G	Bu1209(85)
Glutaric acid	Water	AC (Darco)	595	278, 313	T	Le395(87)
n-Heptane	Benzene	Z (5A, Linde)		279-315	G	Gu14(80)
n-Heptane	Butylbenzene	GCB (Graphon, Cabot)	81	298	G	Pa1780(64)
n-Heptane	Cyclohexane	SG (PA 400, Davison)	660	303	G	My3412(72)
n-Heptane	n-Decane	Z (5A, Linde)		303-363	T	Su223(68)
n-Heptane	Decylbenzene	GCB (Graphon, Cabot)	81	298	G	Pa1780(64)
n-Heptane	1,2-Dichloroethane	SG (PA 400, Davison)	660	303	G	My3412(72)
n-Heptane	n-Dodecane	Z (5A, Linde)		303-363	T	Su223(68)
n-Heptane	Dodecylbenzene	GCB (Graphon, Cabot)	81	298	G	Pa1780(64)
n-Heptane	Ethanol	AC (CS-120A, Futamura)	1350	298	G	Oz1059(84)
n-Heptane	Ethanol	GCB (Graphon, Cabot)	86	282-333	T	Br883(75)
n-Heptane	Ethylbenzene	GCB (Graphon, Cabot)	81	298	G	Pa1780(64)
n-Heptane	Hexadecylbenzene	GCB (Graphon, Cabot)	81	298	G	Pa1780(64)
n-Heptane	Hexylbenzene	GCB (Graphon, Cabot)	81	298	G	Pa1780(64)
n-Heptane	Octadecylbenzene	GCB (Graphon, Cabot)	94	298	G	Pa1780(64)
n-Heptane	n-Octane	Z (5A)		279-315	T	Gu28(81)
n-Heptane	Octylbenzene	GCB (Graphon, Cabot)	81	298	G	Pa1780(64)

Adsorbate No. 1	Adsorbate No. 2	Adsorbent	A, m2/g	Temp., K	Data	Reference
n-Heptane	n-Tetradecane	Z (5A, Linde)		303-363	T	Su223(68)
n-Heptane	Tetradecylbenzene	GCB (Graphon, Cabot)	81	298	G	Pa1780(64)
n-Heptanoic acid	Water	AC (Darco)	667	278-313	T	Wa345(79)
n-Heptanol	Carbon tetrachloride	SG	417	298	G	Zh185(85)
n-Hexane	Benzene	Z (5A, Linde)		279-315	G	Gu14(80)
n-Hexane	1-Decene	Z (NaX)		293	G	He280(81)
n-Hexane	1-Dodecene	Z (NaX)		293	G	He280(81)
n-Hexane	n-Heptane	Z (5A)		279-315	T	Gu28(81)
n-Hexane	1-Hexadecene	Z (NaX)		293	G	He280(81)
n-Hexane	n-Hexanol	SG (Davison 62)	340	298	G	Wa153(78)
n-Hexane	1-Octene	Z (NaX)		293	G	He280(81)
n-Hexane	1-Tetradecene	Z (NaX)		293	G	He280(81)
n-Hexanoic acid	Water	AC (Darco)	667	278-313	T	Wa345(79)
1-Hexanol	Water	AC	978-1420	298	G	Ab201(83)
n-Hexanol	Cyclohexane	SG	417	298	G	Zh185(85)
1-Hexene	n-Hexane	Z (NaX)		293-313	G	He559(84)
Hydrogen sulfide	Water	AC (BPL, Pittsburgh)	1050-1150	298	G	Ra17(78)
p-Hydroxyacetophenone	Benzene	AL (Alpha)	132	308	G	Da417(74)
p-Hydroxyacetophenone	Benzene	SG (Sorbsil M 60)	432	308	G	Da1117(73)
p-Hydroxybenzaldehyde	Benzene	SG (Sorbsil M 60)	432	308	G	Da1117(73)
p-Hydroxylbenzaldehyde	Benzene	AL (Alpha)	132	308	G	Da417(74)
Indol	Water	AC (Hydraffin 71, Lurgi)		293	G	Se79(88)
Lauric acid	Benzene	GCB (Graphon)		293	G	Ki3382(63)
Lauric acid	Carbon tetrachloride	CB (Spheron 6)	91	293	G	Ki3382(63)
Malonic acid	Water	AC (Darco)	595	278, 313	T	Le133(86)
Methanol	Benzene	AL (Alpha)		293	T	Ki4828(56)
Methanol	Benzene	AL (Boehmite)		293-333	T	Ki4828(56)
Methanol	Benzene	AL (Gibbsite)		293	T	Ki4828(56)
Methanol	Benzene	GCB		298	G	Co176(73)
Methanol	Benzene	Z (Dealuminated Y)		298	G	De261(86)
Methanol	Benzene	Z (NaY)		298	G	De261(86)
Methanol	n-Butanol	AC (CS-120A, Futamura)	1350	298	G	Oz1059(84)
Methanol	n-Propanol	AC (CS-120A, Futamura)	1350	298	G	Oz1059(84)

Adsorbate No. 1	Adsorbate No. 2	Adsorbent	A, m2/g	Temp., K	Data	Reference
Methanol	Water	AC (HGI-780)	724	303	G	Ok209(78)
Methanol	Water	AC (Shirasagi S)	972	303	G	Ok209(78)
Methanol	Water	GCB (Graphon, Cabot)	84	298-338	T	Ev2605(86)
p-Methoxyphenol	Benzene	AL (Alpha)	132	308	G	Da417(74)
p-Methoxyphenol	Benzene	SG (Sorbsil M 60)	432	308	G	Da1117(73)
Methyl acetate	Benzene	AC		293	T	Bl2373(57)
Methyl acetate	Benzene	AL (Boehmite)		293-333	T	Ki4828(56)
Methyl acetate	Benzene	SG		293, 333	T	Ki4054(57)
Methyl acetate	Ehylene dichloride	AL (Boehmite)		293	T	Ki4828(56)
Methyl cyclohexane	Methylcyclopentane	AC (CS-120A, Futamura)	1350	298	G	Oz1059(84)
Methyl cyclohexane	n-Propanol	AC (CS-120A, Futamura)	1350	298	G	Oz1059(84)
Methyl cyclopentane	n-Butanol	AC (CS-120A, Futamura)	1350	298	G	Oz1059(84)
Methyl cyclopentane	n-Heptane	AC (CS-120A, Futamura)	1350	298	G	Oz1059(84)
Methyl cyclopentane	n-Hexane	AC (CS-120A, Futamura)	1350	298	G	Oz1059(84)
Methyl cyclopentane	iso-Propanol	AC (CS-120A, Futamura)	1350	298	G	Oz1059(84)
Methyl cyclopentane	n-Propanol	AC (CS-120A, Futamura)	1350	298	G	Oz1059(84)
Methyl ethyl ketone	Carbon tetrachloride	SG	417	298	G	Zh185(85)
p-Methyl phenol	Benzene	SG (Sorbsil M 60)	432	308	G	Da1117(73)
2-Methyl-1-butanol	Water	AC	978-1420	298	G	Ab201(83)
3-Methyl-1-butanol	Water	AC	978-1420	298	G	Ab201(83)
2-Methyl-1-propanol	Water	AC	978-1420	298	G	Ab201(83)
2-Methyl-2-butanol	Water	AC	978-1420	298	G	Ab201(83)
3-Methyl-2-butanol	Water	AC	978-1420	298	G	Ab201(83)
2-Methyl-2-propanol	Water	AC	978-1420	298	G	Ab201(83)
p-Methylphenol	Benzene	AL (Alpha)	132	308	G	Da417(74)
Naphthalene	Carbon tetrachloride	CB (Spheron 6)	119	303	G	Wr1908(72)
Naphthalene	Carbon tetrachloride	GCB (Graphon)	89	303	G	Wr1908(72)
Naphthalene	Cyclohexane	AL (Kaiser Aluminun Co.)	200-350	298	G	Ch725(83)
Naphthalene	Cyclohexane	CB (Spheron 6)	119	303	G	Wr1908(72)
Naphthalene	Cyclohexane	GCB (Graphon)	89	303	G	Wr1908(72)
Naphthalene	Cyclohexanone	CB (Spheron 6)	119	303	G	Wr1908(72)
Naphthalene	Cyclohexanone	GCB (Graphon)	89	303	G	Wr1908(72)
1-Naphthoic acid	Carbon tetrachloride	AL (Alcoa C730)	22	303	G	Wr1461(74)

Adsorbate No. 1	Adsorbate No. 2	Adsorbent	A, m2/g	Temp., K	Data	Reference
1-Naphthoic acid	Carbon tetrachloride	CB (Spheron 6)		303	G	Wr1461(74)
1-Naphthoic acid	Carbon tetrachloride	GCB (Graphon)		303	G	Wr1461(74)
1-Naphthoic acid	Carbon tetrachloride	SI (Cab-O-Sil)	190	303	G	Wr1461(74)
2-Naphthoic acid	Carbon tetrachloride	AL (Alcoa C730)	22	303	G	Wr1461(74)
2-Naphthoic acid	Carbon tetrachloride	CB (Spheron 6)		303	G	Wr1461(74)
2-Naphthoic acid	Carbon tetrachloride	GCB (Graphon)		303	G	Wr1461(74)
2-Naphthoic acid	Carbon tetrachloride	SI (Cab-O-Sil)		303	G	Wr1461(74)
Nitrobenzene	Benzene	SG (Merck, GFR)	497	293	G	Go392(85)
Nitrobenzene	n-Heptane	SG (Merck, GFR)	497	293	G	Go392(85)
p-Nitrophenol	Benzene	AL (Alpha)	132	308	G	Da417(74)
p-Nitrophenol	Benzene	SG (Sorbsil M 60)	432	308	G	Da1117(73)
p-Nitrophenol	Water	AC	1000, 1500	293	G	Fr721(81)
p-Nitrophenol	Water	AC		293	G	Fr1279(74)
p-Nitrophenol	Water	AC (B10, Lurgi)	1450	293	G	Jo1097(78)
p-Nitrophenol	Water	AC (CAL, Calgon)		308	G	Ok286(80)
p-Nitrophenol	Water	AC (Hydraffin 71, Lurgi)		293	G	Se79(88)
p-Nitrophenol	Water	AC (X-7100, Takeda)	1000	303	G	Mu34(82)
Octa-ethylporphyrin	Cyclohexane	AL (Kaiser Aluminun Co.)	200-350	298	G	Ch725(83)
n-Octane	Benzene	Z (5A, Linde)		279-315	G	Gu14(80)
Octanoic acid	Benzene	GCB (Graphon)		293	G	Ki3382(63)
Octanoic acid	Carbon tetrachloride	CB (Spheron 6)	91	293	G	Ki3382(63)
Octanoic acid	Carbon tetrachloride	GCB (Graphon)		293	G	Ki3382(63)
n-Octanol	Cyclohexane	SG	417	298	G	Zh185(85)
1-Octene	n-Octane	Z (NaX)		293	G	He227(84)
Oxalic acid	Water	AC (Darco)	595	278, 313	T	Le133(86)
Palmitic acid	Benzene	GCB (Graphon)		293	G	Ki3382(63)
Palmitic acid	Carbon tetrachloride	CB (Spheron 6)	91	293	G	Ki3382(63)
Palmitic acid	Carbon tetrachloride	GCB (Graphon)		293	G	Ki3382(63)
Palmitic acid	Cyclohexane	CB (Spheron 6)	91	293	G	Ki3382(63)
Palmitic acid	Cyclohexane	GCB (Graphon)		293	G	Ki3382(63)
n-Pentane	Benzene	Z (5A, Linde)		279-303	G	Gu14(80)
n-Pentane	n-Heptane	Z (5A)		279-303	T	Gu28(81)
n-Pentane	n-Hexane	Z (5A)		279-303	T	Gu28(81)

Adsorbate No. 1	Adsorbate No. 2	Adsorbent	A, m2/g	Temp., K	Data	Reference
n-Pentane	n-Octane	Z (5A)		279-303	T	Gu28(81)
1-Pentanol	Water	AC	978-1420	298	G	Ab201(83)
2-Pentanol	Water	AC	978-1420	298	G	Ab201(83)
3-Pentanol	Water	AC	978-1420	298	G	Ab201(83)
Phenol	Benzene	AL (Alpha)	132	308	G	Da417(74)
Phenol	Benzene	SG (Sorbsil M 60)	432	308	G	Da1117(73)
Phenol	Carbon tetrachloride	AL (Alpha)	132	308	G	Da417(74)
Phenol	Carbon tetrachloride	SG (Sorbsil M 60)	423	308	G	Da1117(73)
Phenol	Carbon tetrachloride	SI (Aerosil)		298	G	Ma2478(75)
Phenol	Cyclohexane	AL (Alpha)	132	308	G	Da417(74)
Phenol	Cyclohexane	SG (Sorbsil M 60)	423	308	G	Da1117(73)
Phenol	Dioxan	AL (Alpha)	132	308	G	Da417(74)
Phenol	n-Hexane	SG (Sorbsil M 60)	423	308	G	Da1117(73)
Phenol	Tetrahydrofuran	AL (Alpha)	132	308	G	Da417(74)
Phenol	Toluene	SG (Sorbsil M 60)	423	308	G	Da1117(73)
Phenol	Water	AC	1000, 1500	293	G	Fr721(81)
Phenol	Water	AC		293	G	Fr1279(74)
Phenol	Water	AC (B10, Lurgi)	1450	293	G	Jo1097(78)
Phenol	Water	AC (CAL)	1000	298	G	Mi169(70)
Phenol	Water	AC (CAL, Calgon)		308	G	Ok286(80)
Phenol	Water	AC (Filtrasorb 400)	595	287-291	G	Ga1231(75)
Phenol	Water	AC (Filtrasorb F400, Calgon)		293	G	Pe26(81)
Phenol	Water	AC (Hydraffin 71, Lurgi)		293	G	Se79(88)
Phenol	Water	CB (Aquezo, Co. Ltd. Tokyo)	886	298	G	As348(84)
Phenyl acetic acid	Water	AC	1500	293	G	Fr721(81)
Phenyl acetic acid	Water	AC (B10, Lurgi)	1450	293	G	Jo1097(78)
o-Phenylphenol	Water	AC (B10, Lurgi)	1450	293	G	Jo1097(78)
Pimelic acid	Water	AC (Darco)	595	278, 313	T	Le395(87)
Propanol	Benzene	AL (Boehmite)		293	G	Ki4828(56)
2-Propanol	Water	AC (Filtrasorb 300, Calgon)	1000	273-343	T	Ra445(72)
n-Propanol	Benzene	Z (Dealuminated Y)		298	G	De261(86)
n-Propanol	Benzene	Z (NaY)		298	G	De261(86)
Propionic acid	Water	AC (Darco)	667	278-313	T	Wa345(79)

Adsorbate No. 1	Adsorbate No. 2	Adsorbent	A, m2/g	Temp., K	Data	Reference
Propionitrile	Water	AC (Filtrasorb 300, Calgon)	1000	273-343	T	Ra445(72)
Propionitrile	Water	AC (Filtrasorb 300, Calgon)	1000	298	G	Ra761(72)
Pyridine	Ethanol	AC		293	G	Bl3819(54)
Pyridine	Water	AC		293	G	Bl3819(54)
Stearic acid	Benzene	CB (Spheron 6)	91	293	G	Ki3382(63)
Stearic acid	Carbon tetrachloride	CB (Spheron 6)	91	293	G	Ki3382(63)
Stearic acid	Cyclohexane	CB (Spheron 6)	91	293	G	Ki3382(63)
Stearic acid	Ethanol	CB (Spheron 6)	91	293	G	Ki3382(63)
Succinic acid	Water	AC (Darco)	595	278, 313	T	Le133(86)
Tetra-phenylporphyrin	Cyclohexane	AL (Kaiser Aluminun Co.)	200-350	298	G	Ch725(83)
1-Tetradecene	n-Dodecane	Z (BaX)		293	G	He565(84)
1-Tetradecene	n-Dodecane	Z (CaX)		293	G	He565(84)
1-Tetradecene	n-Dodecane	Z (KX)		293	G	He565(84)
1-Tetradecene	n-Dodecane	Z (NaCsX)		293	G	He565(84)
1-Tetradecene	n-Dodecane	Z (NaRbX)		293	G	He565(84)
1-Tetradecene	n-Dodecane	Z (NaX)		293	G	He565(84)
1-Tetradecene	n-Dodecane	Z (SrX)		293	G	He565(84)
Toluene	Bromobenzene	SG	420	293	T	Pe1384(79)
Toluene	Carbon Tetrachloride	CMS (5A, Takeda)		298	G	Ta268(77)
Toluene	Chlorobenzene	SG	420	293	T	Pe1384(79)
Toluene	Cyclohexane	CMS (5A, Takeda)		298	G	Ta268(77)
1,1,1-Trichloroethane	tert-Butyl chloride	AC		293	T	Bl2373(57)
1,1,1-Trichloroethane	Carbon tetrachloride	AC		293	T	Bl2373(57)
p-Xylene	Cyclohexane	CMS (5A, Takeda)		298	G	Ta268(77)

Adsorbates	Adsorbent	A, m2/g	Temp., K	Data	Reference
p-Nitrophenol	AC		293	G	Fr1279(74)
Phenol					
Water					
p-Chlorophenol	AC	1500	293	G	Fr721(81)
p-Nitrophenol					
Water					
Benzoic acid	AC	1500	293	G	Fr721(81)
p-Nitrophenol					
Water					
p-Nitrophenol	AC	1000, 1500	293	G	Fr721(81)
Phenol					
Water					
p-Chlorophenol	AC	1500	293	G	Fr721(81)
Phenyl acetic acid					
Water					
2,4-Dichlorophenol	AC	1500	293.15	G	Fr721(81)
Dodecylbenzene sulfonic acid					
Water					
Acetic acid	AC (Darco)	667	278, 298	T	Wa350(81)
Propionic acid					
Water					
Acetic acid	AC (Darco)	667	278, 298	T	Wa350(81)
n-Butanoic acid					
Water					
n-Butanoic acid	AC (Darco)	667	278, 298	T	Wa350(81)
Propionic acid					
Water					

Adsorbates	Adsorbent	A, m2/g	Temp., K	Data	Reference
p-Nitrophenol	AC (CAL, Calgon)		308	G	Ok286(80)
Phenol					
Water					
p-Cresol	AC (CAL, Calgon)		308	G	Ok286(80)
p-Nitrophenol					
Water					
Acetone	AC (Filtrasorb 300, Calgon)	1000	298	T	Ra761(72)
Propionitrile					
Water					
p-Chlorophenol	AC (Filtrasorb 300, Calgon)	1000	298	T	Ra761(72)
p-Cresol					
Water					
p-Nitrophenol	AC (B10, Lurgi)	1450	293	G	Jo1097(78)
Phenol					
Water					
Benzene	AC (BPL, Pittsburgh)	1100	303	T	Mi453(73)
Cyclohexane					
Ethyl acetate					
Indol	AC (Hydraffin 71, Lurgi)		293	G	Se79(88)
Phenol					
Water					
p-Chloranilin	AC (Hydraffin 71, Lurgi)		293	G	Se79(88)
Phenol					
Water					
p-Chloranilin	AC(Hydraffin 71, Lurgi)		293	G	Se79(88)
p-Nitrophenol					
Water					

Adsorbates	Adsorbent	A, m2/g	Temp., K	Data	Reference
Indol	AC (Hydraffin 71, Lurgi)		293	G	Se79(88)
p-Nitrophenol					
Phenol					
Water					
p-Nitrophenol	AC (Hydraffin 71, Lurgi)		293	G	Se79(88)
Phenol					
Water					
Benzene	SG (Merck GFR)	497	293	T	Go392(85)
Chlorobenzene					
n-Heptane					
Benzene	SG (Merck GFR)	497	293	T	Go392(85)
n-Heptane					
Nitrobenzene					
Aniline	SG (Merck GFR)	497	293	T	Go392(85)
Benzene					
n-Heptane					
n-Butanol	SG	417	298	G	Zh185(85)
Cyclohexane					
Ethanol					
n-Hexanol					
n-Octanol					
n-Amyl acetate	SG	417	298	G	Zh185(85)
Carbon tetrachloride					
Cyclohexanone					
n-Heptyl alcohol					
Methyl ethyl ketone					

Adsorbates	Adsorbent	A, m2/g	Temp., K	Data	Reference
n-Hexane	Z (5A)		279-303	T	Gu28(81)
n-Octane					
n-Pentane					
n-Heptane	Z (5A)		279-303	T	Gu28(81)
n-Octane					
n-Pentane					
n-Heptane	Z (5A)		279-303	T	Gu28(81)
n-Hexane					
n-Pentane					

APPENDIX G

BIBLIOGRAPHY

This appendix contains a bibliography of references in alphabetical order by the first author. Each reference is preceded by the abbreviation used in the text: first two letters of the last name of the first author, page number of article, and the last two digits of the year of publication. If the reference is a book, the page number of interest is indicated in the text.

Ab201(83)   Abe, I., K. Hayashi and T. Hirashima, *J. Colloid Interface Sci.*, **94**, 201 (1983).

Al658(81)   Al-Sahhaf, T. A., E. D. Sloan and A. L. Hines, *Ind. Eng. Chem. Process Des. Dev.* **20**, 658 (1981).

Al513(85)   Alzamora, L., *J. Colloid Interface Sci.* **106**, 513 (1985).

Al420(84)   Alzamora, L., *J. Colloid Interface Sci.* **98**, 420 (1984).

Am306(78)   Amelitcheva, T. M., A. G. Bezus, L. L. Bogomolova, A. V. Kiselev and N. K. Shoniya, *J. Chem. Soc., Trans. Faraday 1,* **74**, 306 (1978).

Ar219(74)   Arenkova, G. G., V. I. Lozgachev and M. L. Sazonov, *Russ. J. Phys. Chem.* **48**, 219 (1974).

As348(84)   Asakawa, T. and K. Ogino, *J. Colloid Interface Sci.* **102**, 348 (1984).

As123(75)   Ash, S. G., R. Bown and D. H. Everett, *J. Chem. Soc., Trans. Faraday 1,* **71**, 123 (1975).

As239(73)   Ash, S. G., R. Bown and D. H. Everett, *J. Chem. Thermodynamics* **5**, 239 (1973).

At1490(72)   Atyaksheva, L. F., G. I. Emel'yanova and N. I. Kobozev, *Russ. J. Phys. Chem.* **46**, 1490 (1972).

Av597(73)   Avery, R. G. and J. D. F. Ramsay, *J. Colloid Interface Sci.* **42**, 597 (1973).

Ba80(52)   Bartell, F. E. and J. E. Bower, *J. Colloid Sci.* **7**, 80 (1952).

Ba2147(73)   Barthomeuf, D. and B. H. Ha, *J. Chem. Soc., Trans. Faraday 1,* **69**, 2147 (1973).

Ba491(66)   Barto, J., J. L. Durham, V. F. Baston and W. H. Wade, *J. Colloid Interface Sci.* **22**, 491 (1966).

Ba141(60)   Basmadjian, D., *Can. J. Chem.* **38**, 141 (1960).

Ba649(68)   Bassett, D. R., E. A. Boucher and A. C. Zettlemoyer, *J. Colloid Interface Sci.* **27**, 649 (1968).

Be212(74)   Begun, L. B., V. M. Kisarov and A. I. Subbotin, *Sov. Chem. Ind.* **6**, 212 (1974).

Be381(72)   Bering, B. P., V. V. Serpinskii and S. I. Surinova, *Bull. Acad. Sci. USSR, Div. Chem. Sci.* **21**, 381 (1972).

Be2480(67)   Bering, B. P., V. V. Serpinskii and S. I. Surinova, *Bull. Acad. Sci. USSR, Div. Chem. Sci.* 2480 (1967).

Be753(65)   Bering, B. P., V. V. Serpinskii and S. I. Surinova, *Bull. Acad. Sci. USSR, Div. Chem. Sci.* 753 (1965).

Be1817(61)   Bering, B. P. and V. V. Serpinskii, *Bull. Acad. Sci. USSR, Div. Chem. Sci.* 1817 (1961).

Be1146(59)   Bering, B. P. and V. V. Serpinskii, *Bull. Acad. Sci. USSR, Div. Chem. Sci.* 1146 (1959).

Be223(72)   Bezus, A. G., A. V. Kiselev and P. Q. Du, *J. Colloid Interface Sci.* **40**, 223 (1972).

Bh109(72)   Bhambhani, M. R., P. A. Cutting, K. S. W. Sing and D. H. Turk, *J. Colloid Interface Sci.* **38**, 109 (1972).

Bl2373(57)   Blackburn, A., J. J. Kipling and D. A. Tester, *J. Chem. Soc.* 2373 (1957).

Bl3819(54)   Blackburn, A. and J. J. Kipling, *J. Chem. Soc.* 3819 (1954).

Bl4103(55)   Blackburn, A. and J. J. Kipling, *J. Chem. Soc.* 4103 (1955).

Bo286(87)   Boki, K., S. Tanada, O. Nobuhiro, S. Tsutsui, R.Yamasaki and M. Nakamura, *J. Colloid Interface Sci.* **120**, 286 (1987).

Bo1535(79)   Bomchil, G., N. Harris, M. Leslie, J. Tabony and J. W. White, *J. Chem. Soc., Trans. Faraday 1*, **75**, 1535 (1979).

Br883(75)   Brown, C. E., D. H. Everett and C. J. Morgan, *Trans. Faraday Soc.* **71**, 883 (1975).

Bu1209(85)   Bui, S., X. Verykios and R. Mutharasan, *Ind. Eng. Chem. Process Des. Dev.*, **24**, 1209 (1985)

Bü34(72)   Bülow, M., A. Grossmann and W. Schirmer, *Chem. Tech.* **24**, 34 (1972).

Ca339(66)   Camp, D. T. and L. N. Canjar, *AIChE Journal* **12**, 339 (1966).

Ch1679(83)   Chandwadkar, A. J. and S. B. Kulkarni, *J. Chem. Soc., Trans. Faraday 1*, **79**, 1679 (1983).

Ch725(83)   Chantong, A. and F. E. Massoth, *AIChE Journal* **29**, 725 (1983).

Ch382(87)   Choudhary, V. R. and K. R. Srinivasan, *Chem. Eng. Science* **42**, 382 (1987).

Ch1345(76)   Chuikina, V. K., A. V. Kiselev, L. V. Mineyeva and G. G. Muttik, *J. Chem. Soc., Trans. Faraday 1*, **72**, 1345 (1976).

Ci710(53)   Cines, M. R. and F. N. Ruehlen, *J. Phys. Chem.* **57**, 710 (1953).

Co405(67)   Cochrane, H., P. L. Walker, Jr., W. S. Diethorn and H. C. Friedman, *J. Colloid Interface Sci.* **24**, 405 (1967).

Co176(73)   Coltharp, M. T. and N. Hackerman, *J. Colloid Interface Sci.* **43**, 176 (1973).

Co185(73)   Coltharp, M. T. and N. Hackerman, *J. Colloid Interface Sci.* **43**, 185 (1973).

Co1351(86)   Cortés, J., M. Jensen and P. Araya, *J. Chem. Soc., Trans. Faraday 1*, **82**, 1351 (1986).

Co2127(84)   Cortés, J., L. Alzamora, G. Troncoso, A. L. Prieto and E. Valencia, *J. Chem. Soc., Trans. Faraday 1*, **80**, 2127 (1984).

Co5(81)   Costa, E., J. L. Sotelo, G. Calleja and C. Marron, *AIChE Journal* **27**, 5 (1981).

Co1969(79)   Coughlan, B. and W. A. McCann, *J. Chem. Soc., Trans. Faraday 1*, **75**, 1969 (1979).

Co1809(75)   Coughlan, B. and S. Kilmartin, *J. Chem. Soc., Trans. Faraday 1*, **71**, 1809 (1975).

Cr339(80)   Crosser, O. K. and B. V. Hong, *J. Chem. Eng. Data* **25**, 339 (1980).

Cu58(74)   Curthoys, G., V. Ya. Davydov, A. V. Kiselev, S. A. Kiselev and B. V. Kuznetsov, *J. Colloid Interface Sci.* **48**, 58 (1974).

Da2129(80)   Danner, R. P., M. P. Nicoletti and R. S. Al-Ameeri, *Chem. Eng. Science* **35**, 2129 (1980).

Da248(78)   Danner, R. P. and E. C. F. Choi, *Ind. Eng. Chem. Fundam.* **17**, 248 (1978).

Da515(69)   Danner, R. P. and L. A. Wenzel, *AIChE Journal* **15**, 515 (1969).

Da353(69)   Davis, B. W., *J. Colloid Interface Sci.* **31**, 353 (1969).

Da417(74)   Davis, K. M. C., *J. Chem. Soc., Trans. Faraday 1,* **70**, 417 (1974).
Da1117(73)   Davis, K. M. C., *J. Chem. Soc., Trans. Faraday 1,* **69**, 1117 (1973).
Da3791(66)   Day, J. J., *Diss. Abstr.* **26**, 3791 (1966).
De(44)   Deitz, V. R., *Bibliography of Solid Adsorbents*, National Bureau of Standards, Wash., D. C. (1944).
De191(42)   Deitz, V. R. and L. F. Gleysteen, *J. of Research of the National Bureau of Standards* **29**, 191 (1942).
Dé261(86)   Dékány, I., F. Szántó, L. G. Nagy and H. K. Beyer, *J. Colloid Interface Sci.* **112**, 261 (1986).
De1050(75)   Denisova, L. V. and A. M. Tolmachev, *Russ. J. Phys. Chem.* **49**, 1050 (1975).
De1685(85)   Derkaui, A., A. V. Kiselev and B. V. Kuznetsov, *J. Chem. Soc., Trans. Faraday 1,* **81**, 1685 (1985).
De1947(72)   Derrah, R. I., K. F. Loughlin and D. M. Ruthven, *J. Chem. Soc., Trans. Faraday 1,* **68**, 1947 (1972).
Do832(72)   Dollimore, D., G. R. Heal and D. R. Martin, *J. Chem. Soc., Trans. Faraday 1,* **68**, 832 (1972).
Do30(75)   Dorfman, L. R. and R. P. Danner, *AIChE Symp. Ser.* No. 152, **71**, 30 (1975)
Du628(76)   Dubinin, M. M., K. M. Nikolaev, N. S. Polyakov and Yu. N. Til'kunov, *Russ. J. Phys. Chem.* **50**, 628 (1976).
Dz2662(79)   Dzhigit, O. M., A. V. Kiselev and T. A. Rachmanova, *J. Chem. Soc., Trans. Faraday 1,* **75**, 2662 (1979).
Ea419(75)   Eagan, J. D. and R. B. Anderson, *J. Colloid Interface Sci.* **50**, 419 (1975).
Eb2404(63)   Eberly, P. E. Jr., *J. Phys. Chem.* **67**, 2404 (1963).
Eg22(73)   Egerton, T. A. and F. S. Stone, *J. Chem. Soc., Trans. Faraday 1,* **69**, 22 (1973).
El889(72)   Eltekov, Yu. A., V. V. Khopina and A. V. Kiselev, *J. Chem. Soc., Trans. Faraday 1,* **68**, 889 (1972).
Ev1803(64)   Everett, D. H, *Trans. Faraday Soc.,* **60**, 1803 (1964)
Ev2605(86)   Everett, D. H. and A. J. P. Fletcher, *J. Chem. Soc., Trans. Faraday 1,* **82**, 2605 (1986).
Ev14(81)   Everett, D. H. and R. T. Podoll, *J. Colloid Interface Sci.* **82**, 14 (1981).
Fe457(72)   Fernbacher, J. M. and L. A. Wenzel, *Ind. Eng. Chem. Fundam.* **11**, 457 (1972).
Fl205(74)   Flock, J. W. and D. N. Lyon, *J. Chem. Eng. Data* **19**, 205 (1974).
Fo25(74)   Foti, G., L. G. Nagy and G. Schay, *Acta Chim. Acad. Sci. Hung.* **80**, 25 (1974).
Fo269(73)   Foti, G., L. G. Nagy and G. Schay, *Acta Chim. Acad. Sci. Hung.* **76**, 269 (1973).
Fr439(72)   Friederich, R. O. and J. C. Mullins, *Ind. Eng. Chem. Fundam.* **11**, 439 (1972).
Fr721(81)   Fritz, W. and E. U. Schlünder, *Chem. Eng. Science* **36**, 721 (1981).
Fr1279(74)   Fritz, W. and E. U. Schlünder, *Chem. Eng. Science* **29**, 1279 (1974).

Fu269(68)   Fukunaga, P., K. D. Hwang, S. H. Davis, Jr. and J. Winnick, *Ind. Eng. Chem. Proc. Design and Develop.*, **7**, 269 (1968).

Ga1231(75)   Ganho, R., H. Gibert and H. Angelino, *Chem. Eng. Science* **30**, 1231 (1975).

Gi702(65)   Gilmer, H. B. and R. Kobayashi, *AIChE Journal* **11**, 702 (1965).

Gi797(64)   Gilmer, H. B. and R. Kobayashi, *AIChE Journal* **10**, 797 (1964).

Gl73(69)   Glessner, A. J. and A. L. Myers, *Chem. Eng. Prog., Symp. Ser.* **65**, 73 (1969).

Gm(77)   Gmehling, J. and U. Onken, *Vapor-Liquid Equilibrium Data Collection*, DECHEMA, Chemistry Data Series, Frankfurt, W. Germany (1977).

Go451(84)   Goto, S., H. Matsuda and Y. Nakamura, *J. Chem. Eng. Japan*, **17**, 451 (1984)

Go392(85)   Goworek, J., J. Oscik and R. Kusak, *J. Colloid Interface Sci.* **103**, 392 (1985).

Gr490(66)   Grant, R. J. and M. Manes, *Ind. Eng. Chem. Fundam.* **5**, 490 (1966).

Gr403(62)   Grant, R. J., M. Manes and S. B. Smith, *AIChE Journal* **8**, 403 (1962).

Gu28(81)   Gupta, R. K., D. Kunzru and D. N. Saraf, *Ind. Eng. Chem. Fundam.* **20**, 28 (1981).

Gu14(80)   Gupta, R. K., D. Kunzru and D. N. Saraf, *J. Chem. Eng. Data* **25**, 14 (1980).

Ha252(73)   Haber, J., J. Najbar, J. Pawelek and J. Pawlikowska-Czubak, *J. Colloid Interface Sci.* **45**, 252 (1973).

Ha301(86)   Hall, P. G. and R. T. Williams, *J. Colloid Interface Sci.* **113**, 301 (1986).

Ha2265(78)   Hall, P. G. and S. A. Müller, *J. Chem. Soc., Trans. Faraday 1,* **74**, 2265 (1978).

Ha165(78)   Hall, P. G. and S. A. Müller, *Surface Technology* **7**, 165 (1978).

Ha38(73)   Hanna, K. M., I. Odler, S. Brunauer, J. Hagymassy, Jr. and E. E. Bodor, *J. Colloid Interface Sci.* **45**, 38 (1973).

Ha546(67)   Haydel, J. J. and R. Kobayashi, *Ind. Eng. Chem. Fundam.* **6**, 546 (1967).

He559(84)   Herden, H., W. D. Einicke, M. Jusek, U. Messow and R. Schöllner, *J. Colloid Interface Sci.* **97**, 559 (1984).

He565(84)   Herden, H., W. D. Einicke, U. Messow, K. Quitzsch and R. Schöllner, *J. Colloid Interface Sci.* **97**, 565 (1984).

He227(84)   Herden, H., W. D. Einicke, U. Messow, E. Volkmann, R. Schöllner and J. Kärger, *J. Colloid Interface Sci.* **102**, 227 (1984).

He280(81)   Herden, H., W. D. Einicke and R. Schöllner, *J. Colloid Interface Sci.* **79**, 280 (1981).

Ho140(52)   Honig, J. M. and L. H. Reyerson, *J. Phys. Chem.*, **56**, 140 (1952)

Ho1169(82)   Hoppe, H. and E. Worch, *Z. Phys. Chemie (Leipzig)*, **263** 1169 (1982).

Ho154(78)   Hoppe, H., F. Winkler and E. Worch, *Z. Chem.* **18**, 154 (1978).

Ho470(83)   Horvath, G. and K. Kawazoe, *J. Chem. Eng. Japan* **16**, 470 (1983).

Hs490(74)   Hsing, H. H. and W. H. Wade, *J. Colloid Interface Sci.* **47**, 490 (1974).

Hy95(85)   Hyun, S. H. and R. P. Danner, *Ind. Eng. Chem. Fundam.* **24**, 95 (1985).

Hy196(82)   Hyun, S. H. and R. P. Danner, *J. Chem. Eng. Data* **27**, 196 (1982).

Iv108(76)   Ivchenko, B. I., V. S. Morozov, G. P. Gubanova and V. N. Pelipas, *Russ. J. Phys. Chem.* **50**, 108 (1976).

Ja202(65)   Jain, L. K., H. M. Gehrhardt and B. G. Kyle, *J. Chem. Eng. Data* **10**, 202 (1965).

Jo415(68)   Jones, B. R. and W. H. Wade, *J. Colloid Interface Sci.* **28**, 415 (1968).

Jo1097(78) Jossens, L., J. M. Prausnitz, W. Fritz, E. U. Schlünder and A. L. Myers, *Chem. Eng. Sci.* **33**, 1097 (1978).

Ka144(75)   Karger, J., M. Bülow and W. Schirmer, *Z. Phys. Chem. (Leipzig)* **256**, 144 (1975).

Ka211(62)   Kaser, J. D., L. O. Rutz and K. Kammermeyer, *J. Chem. Eng. Data* **7**, 211 (1962).

Ka928(87)   Kaul, B. K., *Ind Eng. Chem. Res.* **26**, 928 (1987).

Ka158(74)   Kawazoe, K., T. Kawai, Y. Eguchi and K. Itoga, *J. Chem. Eng. Japan* **7**, 158 (1974).

Ke743(70)   Kel'tsev, N. V. and Y. I. Shumyatskii, *Russ. J. Phys. Chem.* **44**, 743 (1970).

Ki949(70)   Kidnay, A. J. and M. J. Hiza, *AIChE Journal* **16**, 949 (1970).

Ki58(66)   Kidnay, A. J. and M. J. Hiza, *AIChE Journal* **12**, 58 (1966).

Ki(65)   Kipling, J. J., *Adsorption from Solutions of Non-Electrolytes*, Academic Press, London, England (1965).

Ki3382(63) Kipling, J. J. and E. H. M. Wright, *J. Chem. Soc.* 3382 (1963).

Ki4054(57) Kipling, J. J. and D. B. Peakall, *J. Chem. Soc.* 4054 (1957).

Ki4828(56) Kipling, J. J. and D. B. Peakall, *J. Chem. Soc.* 4828 (1956).

Ki4123(52) Kipling, J. J. and D. A. Tester, *J. Chem. Soc.* 4123 (1952).

Ko869(74)   Kobozev, N. I., G. I. Emel'yanova, E. V. Voloshanovskii and L. F. Atyaksheva, *Russ. J. Phys. Chem.* **48**, 869 (1974).

Ko394(85)   Konno, M., K. Shibata and S. Saito, *J. Chem. Eng. Japan* **18**, 394 (1985).

Ko398(85)   Konno, M., M. Terabayashi, Y. Takako and S. Saito, *J. Chem. Eng. Japan* **18**, 398 (1985).

Ko(88)   Kopatsis, A., A. Salinger and A. L. Myers, *AIChE Journal*, in press (1988)

Ko1049(74) Kowalczyk, H. and K. Karpinski, *Rocz. Chem.* **48**, 1049 (1974).

Ku330(85)   Kuo, S. L., E. O. Pedram and A. L. Hines, *J. Chem. Eng. Data* **30**, 330 (1985).

Ku588(84)   Kuro-Oka, M., T. Suzuki, T. Nitta and T. Katayama, *J. Chem. Eng. Japan* **17**, 588 (1984).

La104(69)   Lamond, T. G. and C. R. Price, *J. Colloid Interface Sci.* **31**, 104 (1969).

La1361(18) Langmuir, I., *J. Amer. Chem. Soc.* **40**, 1361 (1918).

La1025(71) Larionov, O. G. and A. L. Myers, *Chem. Eng. Science*, **26**, 1025 (1971)

La1927(70) Laukhuf, W. L. S., *Diss. Abstr. Int. B* **31**, 1927 (1970).

La48(69)	Laukhuf, W. L. S. and C. A. Plank, *J. Chem. Eng. Data.* **14**, 48 (1969)
Le30(64)	Lederman, P. B. and B. Williams, *AIChE Journal* **10**, 30 (1964).
Le395(87)	Lee, C. Y. C. and A. L. Hines, *J. Chem. Eng. Data* **32**, 395 (1987).
Le133(86)	Lee, C. Y. C., E. O. Pedram and A. L. Hines, *J. Chem. Eng. Data* **31**, 133 (1986).
Le1(84)	Lee, T. V., J. C. Huang, D. Rothstein and R. Madey, *Sep. Sci. Technol.* **19**, 1 (1984).
Le461(85)	Lee, T. V., R. Madey and J. C Huang, *Sep. Sci. Technol.* **20**, 461 (1985).
Le227(77)	Lemcoff, N. O. and K. S. W. Sing, *J. Colloid Interface Sci.* **61**, 227 (1977).
Le3163(82)	Leng, C. A. and A. T. Clark, *J. Chem. Soc., Trans. Faraday 1*, **78**, 3163 (1982).
Le1319(50)	Lewis, W. K., E. R. Gilliland, B. Chertow and W. P. Cadogan, *Ind. Eng. Chem.* **42**, 1319 (1950).
Le1326(50)	Lewis, W. K., E. R. Gilliland, B. Chertow and W. P. Cadogan, *Ind. Eng. Chem.* **42**, 1326 (1950).
Le1153(50)	Lewis, W. K., E. R. Gilliland, B. Chertow and W. H. Hoffman, *J. Am. Chem. Soc.* **72**, 1153 (1950).
Le1157(50)	Lewis, W. K., E. R. Gilliland, B. Chertow and W. Milliken, *J. Am. Chem. Soc.* **72**, 1157 (1950).
Le1160(50)	Lewis, W. K., E. R. Gilliland, B. Chertow and D. Bareis, *J. Am. Chem. Soc.* **72**, 1160 (1950).
Li420(84)	Lin, P. J., J. F. Parcher and K. J. Hyver, *J. Chem. Eng. Data* **29**, 420 (1984).
Lo11(55)	Lopez-Gonzalez, J. D., F. G. Carpenter and V. R. Deitz, *J. of Research of National Bureau of Standards* **55**, 11 (1955).
Lo331(72)	Loughlin, K. F. and D. M. Ruthven, *J. Colloid Interface Sci.* **39**, 331 (1972).
Ma134(74)	Maatman, R., M. Poel and P. Mahaffy, *J. Colloid Interface Sci.* **48**, 134 (1974).
Ma46(85)	Machin, W. D. and J. T. Stuckless, *J. Colloid Interface Sci.* **108**, 46 (1985).
Ma597(85)	Machin, W. D. and J. T. Stuckless, *J. Chem. Soc., Trans. Faraday 1*, **81**, 597 (1985).
Ma1591(82)	Machin, W. D., *J. Chem. Soc., Trans. Faraday 1*, **78**, 1591 (1982).
Ma129(69)	Mahajan, O. P. and P. L. Walker, Jr., *J. Colloid Interface Sci.* **29**, 129 (1969).
Ma2478(75)	Marshall, K. and C. H. Rochester, *J. Chem. Soc., Trans. Faraday 1*, **71**, 2478 (1975).
Ma286(72)	Maslan, F. and E. R. Aberth, *J. Chem. Eng. Data* **17**, 286 (1972).
Ma1097(66)	Mason, J. P. and C. E. Cooke, Jr., *AIChE Journal* **12**, 1097 (1966).
Ma197(68)	Masukawa, S. and R. Kobayashi, *J. Chem. Eng. Data* **13**, 197 (1968).
Ma162(73)	Matayo, D. R. and J. P. Wightman, *J. Colloid Interface Sci.* **44**, 162 (1973).

Mc577(67)   McClellan, A. L. and H. F. Harnsberger, *J. Colloid Interface Sci.* **23** 577 (1967).

Me259(67)   Meredith, J. M. and C. A. Plank, *J. Chem. Eng. Data* **12**, 259 (1967).

Mi260(75)   Mikhail, R. Sh. and T. El-Akkad, *J. Colloid Interface Sci.* **51**, 260 (1975).

Mi505(70)   Mikhail, R. Sh. and F. A. Shebl, *J. Colloid Interface Sci.* **32**, 505 (1970).

Mi45(68)   Mikhail, R. Sh., S. Brunauer and E. E Bodor, *J. Colloid Interface Sci.* **26**, 45 (1968).

Mi54(68)   Mikhail, R. Sh., S. Brunauer and E. E Bodor, *J. Colloid Interface Sci.* **26**, 54 (1968).

Mi169(70)   Miller, C. O. M. and C. W. Clump, *AIChE Journal* **16**, 169 (1970).

Mi194(87)   Miller, G. W., K. S. Knaebel and K. G. Ikels, *AIChE Journal* **33**, 194 (1987).

Mi453(73)   Minka, C. and A. L. Myers, *AIChE Journal* **19**, 453 (1973).

Mo2525(85)   Morishige, K., S. Kittaka and S. Ihara, *J. Chem. Soc., Trans. Faraday 1*, **81**, 2525 (1985).

Mo599(76)   Morozov, V. S., B. I. Ivchenko, G. P. Gubanova, L. V. Fartushnaya and V. N. Pelinas, *Russ. J. Phys. Chem.* **50**, 599 (1976).

Mo568(74)   Morozov, V. S., B. I. Ivchenko, G. P. Gubanova and L. V. Fartushnaya, *Russ. J. Phys. Chem.* **48**, 568 (1974).

Mo1662(72)   Morozov, V. S., B. I. Ivchenko, G. P. Gubanova and L. V. Fartushnaya, *Russ. J. Phys. Chem.* **46**, 1662 (1972).

Mu34(82)   Muraki, M., Y. Iwashima and T. Hayakawa, *J. Chem. Eng. Japan* **15**, 34 (1982).

My365(84)   Myers, A. L. in *Fundamentals of Adsorption*, ed. A. L. Myers and G. Belfort, Engineering Found., N.Y. (1984) pp. 365-381, ISBN 0-8169-0265-8.

My1(87)   Myers, A. L. in *Fundamentals of Adsorption*, ed. A. I. Liapis, Engineering Found., N.Y. (1987) pp. 1-25, ISBN 0-8169-0398-0.

My691(83)   Myers, A. L. *AIChE Journal* **29**, 691 (1983).

My35(86)   Myers, A. L., *J. Non-Equilib. Thermodyn.*, **11**, 35 (1986).

My97(82)   Myers, A. L., C. Minka and D. Y. Ou, *AIChE Journal* **28**, 97 (1982).

My666(73)   Myers, A. L. *AIChE Journal* **19**, 666 (1973).

My121(87)   Myers, A. L. and S. Sircar, *Langmuir*, **3**, 121 (1987).

My3415(72)   Myers, A. L. and S. Sircar, *J. Phys. Chem.* **76**, 3415 (1972).

My3412(72)   Myers, A. L. and S. Sircar, *J. Phys. Chem.* **76**, 3412 (1972).

My121(65)   Myers, A. L. and J. M. Prausnitz, *AIChE Journal* **11**, 121 (1965).

My392(86)   Myers, A. L. and D. P. Valenzuela, *J. Chem. Eng. Japan*, **19**, 392 (1986)

Na114(87)   Nakahara, T. and T. Wakai, *J. Chem. Eng. Data* **32**, 114 (1987).

Na202(84)   Nakahara, T., M. Hirata, G. Amagasa and T. Ogura, *J. Chem. Eng. Data* **29**, 202 (1984).

Na317(82)   Nakahara, T., M. Hirata and H. Mori, *J. Chem. Eng. Data.* **27**, 317 (1982).

Na161(81)   Nakahara, T., M. Hirata and S. Komatsu, *J. Chem. Eng. Data* **26**, 161
              (1981).
Na310(74)   Nakahara, T., M. Hirata and T. Omori, *J. Chem. Eng. Data.* **19**, 310
              (1974).
Ne308(65)   Nelder, J. A. and R. Mead, *Computer Journal*, **7**, 308 (1965).
Ne999(70)   Nemeth, E. J. and E. B. Stuart, *AIChE Journal* **16**, 999 (1970).
No112(81)   Nolan, J. T., T. W. McKeehan and R. P. Danner, *J. Chem. Eng. Data*
              26, 112 (1981).
OB1467(84)   O'Brien, J. A. and A. L. Myers, *J. Chem. Soc., Trans. Faraday 1*, **80**,
              1467 (1984)
OB1188(85)   O'Brien, J. A. and A. L. Myers, *Ind. Eng. Chem. Process Des. Dev.*,
              24, 1188 (1985).
OB451(87)   O'Brien, J. A. and A. L. Myers, in "Fundamentals of Adsorption", ed.
              A. I. Liapis, Engineering Found., N.Y. (1987) pp. 451-461, ISBN
              0-8169-0398-0.
OB(88)   O'Brien, J. A. and A. L. Myers, *Ind. Eng. Chem. Research,* in press
              (1988).
Ok262(81)   Okazaki, M., H. Tamon and R. Toei, *AIChE Journal* **27**, 262 (1981).
Ok286(80)   Okazaki, M., H. Kage and R. Toei, *J. Chem. Eng. Japan* **13**, 286
              (1980).
Ok209(78)   Okazaki, M., H. Tamon and R. Toei, *J. Chem. Eng. Japan* **11**, 209
              (1978).
Oz1059(84)   Ozawa, S., K. Kawahara, M. Yamabe, H. Unno and Y. Ogino, *J. Chem.
              Soc., Trans. Faraday 1*, **80**, 1059 (1984).
Pa2250(87)   Paludetto, R., G. Storti, G. Gamba, S. Carra and M. Morbidelli, *Ind.
              Eng. Chem. Res.* **26**, 2250 (1987).
Pa1057(75)   Parbuzin, V. S. and Y. N. Kuryakov, *Russ. J. Phys. Chem.* **49**, 1057
              (1975).
Pa1515(74)   Parbuzin, V. S., V. A. Yakoviev and G. M. Panchenkov, *Russ. J.
              Phys. Chem.* **48**, 1515 (1974).
Pa97(67)   Parbuzin, V. S., G. M. Panchenkov, B. Kh. Rakhmukov and G. V.
              Sandul, *Russ. J. Phys. Chem.* **41**, 97 (1967).
Pa1780(64)   Parfitt, G. D. and E. Willis, *J. Phys. Chem.* **68**, 1780 (1964).
Pa18(85)   Patel, M. and V. Ramakrishna, *J. Chem. Eng. Data* **30**, 18 (1985).
Pa284(82)   Patel, M., *J. Colloid Interface Sci.* **90**, 284 (1982).
Pa363(68)   Payne, H. K., G. A. Sturdevant and T. W. Leland, *Ind. Eng. Chem.
              Fundam.* **7**, 363 (1968).
Pe11(83)   Pedram, E. O. and A. L. Hines, *J. Chem. Eng. Data* **28**, 11 (1983).
Pe26(81)   Peel, R. G., A. Benedek and C. M. Crowe, *AIChE Journal* **27**, 26
              (1981).
Pe1384(79)   Peisen, L. and T. Gu, *Scientia Sinica* **22**, 1384 (1979).
Pe439(84)   Pendleton, P. and A. C. Zettlemoyer, *J. Colloid Interface Sci.* **98**, 439
              (1984).
Pe252(76)   Perfetti, G. A. and J. P. Wightman, *J. Colloid Interface Sci.* **55**, 252
              (1976).
Pe473(75)   Perfetti, G. A. and J. P. Wightman, *Carbon* **13**, 473 (1975).

Pe313(74)	Perfetti, G. A. and J. P. Wightman, *J. Colloid Interface Sci.* **49**, 313 (1974).
Pe570(62)	Peterson, D. L. and O. Redlich, *J. Chem. Eng. Data* **7**, 570 (1962).
Pr(86)	Prausnitz, J. M., R. N. Lichtenthaler and E. G. de Azevedo, *Molecular Thermodynamics of Fluid-Phase Equilibria,* Prentice-Hall, Englewood Cliffs, N.J. (1986).
Pr(86)	Press, W. H., B. P. Flannery, S. A. Teukolsky and W. T. Vetterling, *Numerical Recipes: The Art of Scientific Computing,* Cambridge U. Press, N.Y. (1986).
Pu459(75)	Putnam, F. A. and T. Fort, Jr., *J. Phys. Chem.* **79**, 459 (1975).
Ra761(72)	Radke, C. J. and J. M. Prausnitz, *AIChE Journal* **18**, 761 (1972).
Ra445(72)	Radke, C. J. and J. M. Prausnitz, *Ind. Eng. Chem. Fundam.* **11**, 445 (1972).
Ra17(78)	Ramachandran, P. A. and J. M. Smith, *Ind. Eng. Chem. Fundam.* **17**, 17 (1978).
Ra205(75)	Ramsay, J. D. F. and R. G. Avery, *J. Colloid Interface Sci.* **51**, 205 (1975).
Ra1315(50)	Ray, G. C. and E. O. Box Jr., *Ind. Eng. Chem.*, **42**, 1315 (1950).
Re336(80)	Reich, R., W. T. Ziegler and K. A. Rogers, *Ind. Eng. Chem. Process Des. Dev.* **19**, 336 (1980).
Re(77)	Reid, R. C., J. M. Prausnitz and B. E. Poling, *The Properties of Gases and Liquids*, Fourth Edition, McGraw-Hill, New York (1987).
Ri1759(76)	Richards, E., H. J. F. Stroud and N. G. Parsonage, *J. Chem. Soc., Trans. Faraday 1,* **72**, 1759 (1976).
Ri1679(87)	Ritter, J. A. and R. T. Yang, *Ind. Eng. Chem. Res.* **26**, 1679 (1987).
Ro541(74)	Roa, V. and J. Cortes, *J. Colloid Interface Sci.* **46**, 541 (1974).
Ro1495(73)	Rogers, K. A. *Diss. Abstr. Int. B* **34**, 1495 (1973).
Ro616(80)	Rolniak, P. D. and R. Kobayashi, *AIChE Journal* **26**, 616 (1980).
Ro319(55)	Ross, S. and W. Winkler, *J. Colloid Sci.* **10**, 319 (1955).
Ro330(55)	Ross, S. and W. Winkler, *J. Colloid Sci.* **10**, 330 (1955).
Ru1325(86)	Ruthven, D. M., N. S. Raghavan and M. M. Hassan, *Chem. Eng. Science* **41**, 1325 (1986).
Ru27(85)	Ruthven, D. M. and F. Wong, *Ind. Eng. Chem. Fundam.* **24**, 27 (1985).
Ru(84)	Ruthven, D. M. *Principles of Adsorption and Adsorption Processes,* Wiley, N.Y. (1984) ISBN 0-471-86606-7.
Ru654(81)	Ruthven, D. M. and L-K. Lee, *AIChE Journal* **27**, 654 (1981).
Ru27(80)	Ruthven, D. M. and R. Kumar, *Ind. Eng. Chem. Fundam.* **19**, 27 (1980).
Ru753(76)	Ruthven, D. M., *AIChE Journal* **22**, 753 (1976).
Ru1043(76)	Ruthven, D. M. and I. H. Doetsch, *J. Chem. Soc., Trans. Faraday 1,* **72**, 1043 (1976).
Ru397(75)	Ruthven, D. M. and R. I. Derrah, *J. Colloid Interface Sci.* **52**, 397 (1975).
Ru701(73)	Ruthven, D. M., K. F. Loughlin and K. A. Holborow, *Chem. Eng. Sci.* **28**, 701 (1973).

Ru696(72)   Ruthven, D. M. and K. F. Loughlin, *J. Chem. Soc., Trans. Faraday 1*, **68**, 696 (1972).

Ru1145(71)   Ruthven, D. M. and K. F. Loughlin, *Chem. Eng. Sci.* **26**, 1145 (1971).

Sa156(83)   Sakoda, A. and M. Suzuki, *J. Chem. Eng. Japan* **16**, 156 (1983).

Sa731(67)   Satterfield, C. N. and A. J. Frabetti, Jr., *AIChE Journal* **13**, 731 (1967).

Sa3121(82)   Savvakis, C., K. Tsimillis and J. H. Petropoulos, *J. Chem. Soc., Trans. Faraday 1*, **78**, 3121 (1982).

Sc289(79)   Schay, G., L. G. Nagy and G. Foti, *Acta Chim. Acad. Sci. Hung.* **100**, 289 (1979).

Se79(88)   Seidel, A. and D. Gelbin, *Chem. Eng. Sci.* **43**, 79 (1988).

Sh100(68)   Shen, J. and J. M. Smith, *Ind. Eng. Chem. Fundam.* **7**, 100 (1968).

Sh106(68)   Shen, J. and J. M. Smith, *Ind. Eng. Chem. Fundam.* **7**, 106 (1968).

Sh1(85)   Shiralkar, V. P. and S. B. Kulkarni, *J. Colloid Interface Sci.* **108**, 1 (1985).

Si919(86)   Sircar, S. and R. Kumar, *Sep. Sci. Technol.* **21**, 919 (1986).

Si1101(84)   Sircar, S., *J. Chem. Soc., Trans. Faraday 1*, **80**, 1101 (1984).

Si55(84)   Sircar, S. and A. L. Myers, *AIChE Symposium Series* **80**, 55 (1984).

Si159(73)   Sircar, S. and A. L. Myers, *AIChE Journal* **19**, 159 (1973).

Si186(71)   Sircar, S. and A. L. Myers, *AIChE Journal* **17**, 186 (1971).

Si2828(70)   Sircar, S. and A. L. Myers, *J. Phys. Chem.* **74**, 2828 (1970).

Si489(73)   Sircar, S. and A. L. Myers, *Chem. Eng. Sci.* **28**, 489 (1973).

Si69(85)   Sircar, S. and A. L. Myers, *Ads. Sci. and Technol.*, **2**, 69 (1985).

Si535(86)   Sircar, S. and A. L. Myers, *Sep. Sci. and Tech.*, **21**, 535 (1986).

Si249(72)   Sircar, S., J. Novosad and A. L Myers, *Ind. Eng. Chem. Fundam.* **11**, 249 (1972).

Sk476(71)   Skalny, J., E. E. Bodor and S. Brunauer, *J. Colloid Interface Sci.* **37**, 476 (1971).

Sl347(75)   Sloan, E. D. Jr. and J. C. Mullins, *Ind. Eng. Chem. Fundam.* **14**, 347 (1975).

So1517(83)   Sorial, G. A., W. H. Granville and W. O. Daly, *Chem. Eng. Science* **38**, 1517 (1983).

Sp78(69)   Springer, C., C. J. Major and K. Kammermeyer, *J. Chem. Eng. Data* **14**, 78 (1969).

St1(70)   Stacy, T. D., E. W. Hough and W. D. McCain Jr., *J. Chem. Eng. Data* **15**, 1 (1970).

St74(68)   Stacy, T. D., E. W. Hough and W. D. McCain Jr., *J. Chem. Eng. Data* **13**, 74 (1968).

St112(85)   Stakebake, J. L. and J. Fritz, *J. Colloid Interface Sci.* **105**, 112 (1985).

St942(76)   Stroud, H. J. F., E. Richards, P. Limcharoen and N. G. Parsonage, *J. Chem. Soc., Trans. Faraday 1*, **72**, 942 (1976).

Su223(68)   Sundstrom, D. W. and F. G. Krautz, *J. Chem. Eng. Data* **13**, 223 (1968).

Su301(70)   Suri, S. K. and V. Ramakrishna, *Acta Chim. Acad. Sci. Hung.* **63**, 301 (1970).

Su279(82)   Suzuki, M. and A. Sakoda, *J. Chem. Eng. Japan* **15**, 279 (1982).

Sz37(63)	Szepezy, L. and V. Illes, *Acta Chim. Hung.* **35**, 37 (1963).
Sz53(63)	Szepezy, L. and V. Illes, *Acta Chim. Hung.* **35**, 53 (1963).
Sz245(63)	Szepezy, L. and V. Illes, *Acta Chim. Hung.* **35**, 245 (1963).
Ta671(74)	Takaishi, T., A. Yusa, Y. Ogino and S. Ozawa, *J. Chem. Soc., Trans. Faraday 1,* **70**, 671 (1974).
Ta268(77)	Takeuchi, Y. and E. Furuya, *J. Chem. Eng. Japan* **10**, 268 (1977).
Ta1263(86)	Talu, O. and I. Zwiebel, *AIChE Journal* **32**, 1263 (1986).
Ta136(81)	Tamon, H., S. Kyotani, H. Wada, M. Okazaki and R. Toei, *J. Chem. Eng. Japan* **14**, 136 (1981).
Ta347(69)	Taylor, A. H. and R. A. Ross, *J. Chem. Eng. Data* **14**, 347 (1969).
Th259(77)	Thomas, G. and C. H. Eon, *J. Colloid Interface Sci.,* **62**, 259 (1977)
Th240(71)	Thomas, W. J. and J. L. Lombardi, *Trans. Inst. Chem. Eng.* **49**, 240 (1971).
Th279(68)	Thompson, J. K. and H. A. Resing, *J. Colloid Interface Sci.* **26**, 279 (1968).
To311(71)	Toth, J., *Acta Chim. Acad. Sci. Hung.* **69**, 311 (1971).
To1(62)	Toth, J., *Acta Chim. Acad. Sci. Hung.* **30**, 1 (1962).
Va1(73)	Van der Vlist, E. and J. Van der Meijden, *J. Chromatogr.* **79**, 1 (1973).
Va153(84)	Valenzuela, D. P. and A. L. Myers, *Sep. Pur. Methods* **13**, 153 (1984).
Va1371(81)	Vansant, E. F. and R. Voets, *J. Chem. Soc., Trans. Faraday 1,* **77**, 1371 (1981).
Va2547(75)	Vashchenko, L. A. and V. V. Serpinskii, *Bull. Acad. Sci. USSR, Div. Chem. Sci.* 2547 (1975).
Va526(81)	Vavlitis, A. P., D. M. Ruthven and K. F. Loughlin, *J. Colloid Interface Sci.* **84**, 526 (1981).
Ve66(85)	Vereist, H. and G. V. Baron, *J. Chem. Eng. Data* **30**, 66 (1985).
Ve1417(81)	Veyssiere, M. C., C. Clavaud, P. Cartraud and A. Cointot, *J J. Chem. Soc., Trans. Faraday 1,* **77**, 1417 (1981).
Ve1071(75)	Veyssiere, M. C. and A. Cointot, *Bull. Soc. Chim. Fr.* **5-6**, 1071 (1975).
Wa399(81)	Wakasugi, Y., S. Ozawa and Y. Ogino, *J. Colloid Interface Sci.* **79**, 399 (1981).
Wa153(78)	Wang, H. L., J. L. Duda and C. J. Radke, *J. Colloid Interface Sci.* **66**, 153 (1978).
Wa350(81)	Wang, S. W., A. L. Hines and E. Pedram, *Ind. Eng Chem. Fundam.* **20**, 350 (1981).
Wa345(79)	Wang, S. W., A. L. Hines and D. S. Farrier, *J. Chem. Eng. Data* **24**, 345 (1979).
Wi14(83)	Wilson, R. J. and R. P. Danner, *J. Chem. Eng. Data* **28**, 14 (1983).
Wo474(82)	Worch, E., *Chem. Techn.* **34**, 474 (1982).
Wr1461(74)	Wright, E. H. M. and N. C. Pratt, *J. Chem. Soc., Trans. Faraday 1,* **70**, 1461 (1974).
Wr1908(72)	Wright, E. H. M. and A. V. Powell, *J. Chem. Soc., Trans. Faraday 1,* **68**, 1908 (1972).

Yo308(71)   Youngquist, G. R., J. L. Allen and J. Eisenberg, *Ind. Eng. Chem. Prod. Res. Develop.* **10**, 308 (1971).

Yo(62)   Young D. M. and A. D. Crowell, *Physical Adsorption of Gases,* Butterworths, London (1962).

Zh185(85)   Zhao, Z. and T. Gu, *J. Chem. Soc., Trans. Faraday 1,* **81**, 185 (1985).

Zh1172(70)   Zhukova, Z. A. and N. V. Kel'tsev, *Russ. J. Phys. Chem.* **44**, 1172 (1970).

SUBJECT INDEX